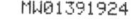

BEGINNING ALGEBRA

Kathy R. Autrey
Kathleen M. Chadick
Northwestern State University

Third Edition

KENDALL/HUNT PUBLISHING COMPANY
4050 Westmark Drive   Dubuque, Iowa 52002

Copyright © 2000, 2003, 2005 by Kendall/Hunt Publishing Company

ISBN 0-7575-2118-5

All rights reserved. No part of this publication may be reproduced, stored in a retrieval system, or transmitted, in any form or by any means, electronic, mechanical, photocopying, recording, or otherwise, without the prior written permission of the copyright owner.

Printed in the United States of America

10 9 8 7 6 5 4 3 2 1

Contents

0 THE FUNDAMENTAL OPERATIONS OF REAL NUMBERS 1

0.1 Symbols 1

0.2 Sets of Numbers 6

0.3 Basic Properties of Real Numbers 12

0.4 Addition and Subtraction of Signed Numbers 17

0.5 Multiplication and Division of Signed Numbers 22

0.6 Operations on Fractions 26

0.7 Decimal Place Value and Rounding Off 35

0.8 Addition and Subraction of Decimals 41

0.9 Multipication of Decimals 48

0.10 Division of Decimals 53

0.11 Fraction-Decimal Conversions 60

0.12 Areas of Plane Figures 65

0.13 Perimeters of Plane Figures 69

0.14 Volumes and Surface Areas 72

1 EXPONENTS AND POLYNOMIALS 83

1.1 Multiplication with Exponents 83

1.2 Division with Exponents 88

1.3 Common Roots and Radicals 95

1.4 Properties of Radicals 101

1.5 Addition and Subtraction of Radicals 105

1.6 Multiplication of Radicals 108

1.7 Division of Radicals 110

1.8 Applications of Exponents and Radicals 113

2 Polynomials and Factoring 119

- 2.1 Polynomials 119
- 2.2 Polynomial Notation 122
- 2.3 Addition and Subtraction of Polynomials 124
- 2.4 Multiplication of Polynomials 128
- 2.5 The FOIL Method and Special Products 132
- 2.6 Division of Polynomials 136
- 2.7 Factoring Integers 142
- 2.8 Common Factors and Factoring by Grouping 144
- 2.9 Factoring the Difference of Two Squares and Perfect-Square Trinomials 150
- 2.10 Factoring Trinomials of the Form $x^2 + bx + c$ 154
- 2.11 Factoring Trinomials of the Form $ax^2 + bx + c, a \neq 1$ 158

3 Linear Equations and Inequalities in One Variable 167

- 3.1 Linear Equations in One Variable 167
- 3.2 Literal Equations and Formulas 174
- 3.3 Applied Problems 180
- 3.4 Linear Inequalities 191

4 Quadratic Equations 199

- 4.1 Solving Quadratic Equations 199
- 4.2 Solving Quadratic Equations by Using the Quadratic Formula 204
- 4.3 Applications of Quadratic Equations 208
- 4.4 Complex Solutions to Quadratic Equations (Optional) 212

5 Graphing and Systems of Linear Equations   217

    5.1   The Cartesian Coordinate System   217

    5.2   Graphs of Linear Equations   224

    5.3   Intercepts and Slope of a Line   231

    5.4   Graphing Linear Inequalities   238

Solutions to Selected Problems   247

Index   263

Chapter 0

THE FUNDAMENTAL OPERATIONS OF REAL NUMBERS

0.1 Symbols
0.2 Sets of Numbers
0.3 Basic Properties of Real Numbers
0.4 Addition and Subtraction of Signed Numbers
0.5 Multiplication and Division of Signed Numbers
0.6 Operations on Fractions
0.7 Decimal Place Value and Rounding Off
0.8 Addition and Subtraction of Decimals
0.9 Multiplication of Decimals
0.10 Division of Decimals
0.11 Fraction-Decimal Conversions
0.12 Areas of Plane Figures
0.13 Perimeters of Plane Figures
0.14 Volumes and Surface Areas

The background you have in arithmetic will help you to learn algebra. For this reason, we begin this book by reviewing the basic principles you learned in arithmetic. This chapter discusses symbols, the operations and properties of real numbers, and addition, subtraction, multiplication, and division of signed numbers.

0.1 Symbols

Algebra is a generalized form of arithmetic. In algebra we perform the same basic operations as in arithmetic, except that letters are used to represent specific numbers. Many of the symbols used in arithmetic are also used in algebra.

For example, if we let a and b represent any two numbers, the four basic operations are written as follows:

Operation	Symbols	Words
Addition	$a + b$	The **sum** of a and b
Subtraction	$a - b$	The **difference** of a and b
Multiplication	$a \cdot b$, $a \times b$, $(a)(b)$, $(a)b$, $a(b)$, ab	The **product** of a and b
Division	$a \div b$, $\dfrac{a}{b}$, a/b	The **quotient** of a and b

From *Beginning Algebra, 4th edition* by M.A. Munem and W. Tschirhart. Copyright © 2000 by Kendall/Hunt Publishing Company. Reprinted by permission.

Another important use of symbols is to represent numbers. In arithmetic the most common symbols used to represent numbers are called **numerals.** For example, the number ten is represented by the numeral 10. Using symbols, the word statement "six plus four" can be written as the mathematical statement $6 + 4$.

Additional symbols are used to compare mathematical quantities.

Comparison	Symbols	Words
Equality	$a = b$	a is **equal to** b
Inequality	$a \neq b$	a is **not equal to** b
	$a < b$	a is **less than** b
	$a > b$	a is **greater than** b
	$a \leq b$	a is **less than** or **equal to** b
	$a \geq b$	a is **greater than** or **equal to** b

Example 1 Write each statement in mathematical symbols.
(a) Seven equals four plus three.
(b) Seven times three equals twenty-one.
(c) Nine is not equal to five.
(d) Thirty divided by five equals six.
(e) Five is less than six plus four.
(f) Twenty is greater than thirty less twelve.

Solution
(a) $7 = 4 + 3$
(b) $7 \times 3 = 21$ or $7 \cdot 3 = 21$
(c) $9 \neq 5$
(d) $30 \div 5 = 6$
(e) $5 < 6 + 4$
(f) $20 > 30 - 12$

Example 2 Indicate whether each statement is true or false.
(a) $13 \leq 17$
(b) $24 \geq 31$
(c) $4 \geq 4$

Solution
(a) Since $13 < 17$, the statement $13 \leq 17$ is true.
(b) Both $24 > 31$ and $24 = 31$ are false; therefore, $24 \geq 31$ is false.
(c) Since $4 = 4$, the statement $4 \geq 4$ is true.

In the statement $7 = 4 + 3$, the number 7 is called the *sum* of 4 and 3. In the statement $2 = 8 - 6$, 2 is called the *difference* of 8 and 6.

In the statement $7 \times 3 = 21$, the number 21 is called the *product* of 7 and 3, and 7 and 3 are said to be **factors** of 21. In general, when we multiply two or more numbers, the result of the multiplication is called the **product** of the numbers.

In the statement $30 \div 5 = 6$, the number 6 is called the *quotient* of 30 and 5. In general, when we divide one number by another number, the result of the division is called the **quotient** of the two numbers.

It is important to keep in mind that there are several ways to write the product or quotient of two numbers, as illustrated in the chart on page 1.

Order of Operations

An expression containing numerals and operation symbols is called a **numerical expression.** For instance, $3 + 12 \div 4 + 2 \times 6$ is a numerical expression. To perform the computations in

$$5 \times 2 + 3,$$

it is not obvious which operation is to be performed first, addition or multiplication. We could add 2 and 3, to get 5, and then multiply by 5, to get 25. This would give us

$$5 \times 2 + 3 = 5 \times 5 = 25.$$

Or, we could multiply 5 and 2, to get 10, and then add 3, to get 13.

$$5 \times 2 + 3 = 10 + 3 = 13.$$

To decide which result is correct, we must agree on the following convention for the **order of operations.**

(i) First, perform all multiplications and divisions working from left to right.
(ii) Then, perform all additions and subtractions working from left to right.

The result of performing the operations in a numerical expression is a number, called the **value** of the expression. Thus, the value of $5 \times 2 + 3$ is

$$5 \times 2 + 3 = 10 + 3 = 13.$$

In Examples 3–5, find the value of each expression.

Example 3 $5 + 7 \times 4$

Solution   First, we perform the multiplication, and then we add:

$$\begin{aligned} 5 + 7 \times 4 &= 5 + 28 &&\text{(We multiplied.)} \\ &= 33. &&\text{(We added.)} \end{aligned}$$

Example 4 $5 \times 11 - 8 + 2$

Solution

$$\begin{aligned} 5 \times 11 - 8 + 2 &= 55 - 8 + 2 &&\text{(We multiplied.)} \\ &= 47 + 2 &&\text{(We subtracted.)} \\ &= 49. &&\text{(We added.)} \end{aligned}$$

Example 5 $7 + 6 \times 5 \div 15 - 3$

Solution

$$\begin{aligned} 7 + 6 \times 5 \div 15 - 3 &= 7 + 30 \div 15 - 3 &&\text{(We performed the multiplication.)} \\ &= 7 + 2 - 3 &&\text{(We performed the division.)} \\ &= 9 - 3 &&\text{(We performed the addition.)} \\ &= 6. &&\text{(We subtracted.)} \end{aligned}$$

Grouping Symbols

When we write mathematical expressions, we often use parentheses ( ) to group numbers and operations together. Parentheses are among the most commonly used **grouping symbols.** (**Braces** { } and **brackets** [ ] are also frequently used as grouping symbols; we provide examples of their use on page 16.)

For example, suppose that in the expression

$$5 \times 2 + 3$$

we want to show that the sum of 2 and 3 is to be multiplied by 5. We can rewrite the expression as $5 \times (2 + 3)$. So its value is

$$5 \times (2 + 3) = 5 \times 5 = 25.$$

The parentheses around 2+3 indicate that this sum is to be multiplied by 5.

We often indicate multiplication by the absence of an operation symbol. For example, the operation $5 \times (2 + 3)$ may be written as $5(2 + 3)$.
In general:

> In any expression containing parentheses, the value of the expression can be obtained by first performing the operations inside the parentheses, and then following the order of operations.

Example 6 Find the value of $3 \times (5 + 2) - 4$.

Solution
$$\begin{aligned}
3 \times (5 + 2) - 4 &= 3 \times 7 - 4 \quad &\text{(We added the numbers inside the parentheses.)} \\
&= 21 - 4 \quad &\text{(We multiplied.)} \\
&= 17. \quad &\text{(We subtracted.)}
\end{aligned}$$

Another grouping symbol is the **fraction bar,** which indicates division. For example, $6 \div 3$ can also be written as the **fraction** $\frac{6}{3}$. The **numerator** of the fraction is the number above the fraction bar, and the **denominator** of the fraction is the number below the bar.

In the case of an expression such as

$$\frac{7 + 5}{4 + 2}$$

the value of the fraction is obtained by dividing the value of the numerator by the value of the denominator. Thus,

$$\frac{7 + 5}{4 + 2} = \frac{12}{6}$$
$$= 2.$$

The expression

$$\frac{7 + 5}{4 + 2}$$

can also be expressed as

$$(7 + 5) \div (4 + 2).$$

The parentheses are used here to indicate which values should be determined first.

Example 7 Find the value of the expression

$$\frac{5(7-3)+4}{3 \times 16 \div 8}$$

Solution We find the values of the numerator and the denominator separately. The value of the numerator is

$$5(7-3)+4 = 5(4)+4 \quad \text{(We subtracted the numbers inside the parentheses.)}$$
$$= 20 + 4 \quad \text{(We multiplied.)}$$
$$= 24. \quad \text{(We added.)}$$

The value of the denominator is

$$3 \times 16 \div 8 = 48 \div 8 \quad \text{(We multiplied first.)}$$
$$= 6 \quad \text{(We divided.)}$$

Therefore,

$$\frac{5(7-3)+4}{3 \times 16 \div 8} = \frac{24}{6} = 4.$$

PROBLEM SET 0.1

In problems 1–14, write each statement in mathematical symbols.

1. Ten equals three plus seven.
2. Thirty equals six times five.
3. Twelve is not equal to sixteen.
4. Fifteen equals ninety divided by six.
5. Eleven minus two equals nine.
6. Twenty plus four is not equal to twenty-two.
7. Sixteen divided by two equals eight.
8. Five times the sum of three and four equals thirty-five.
9. Three is less than ten.
10. Forty is greater than fifteen.
11. Twenty-seven is greater than the sum of six and nine.
12. Eight is less than the product of nine and two.
13. One plus two is less than seven minus two.
14. The product of four and nine is greater than the quotient of twenty-four and six.

In problems 15–24, indicate whether each statement is true or false.

15. $7 + 3 < 11$
16. $7 \neq 8 - 1$
17. $-9 < 0$
18. $4 \leq 4$
19. $-26 \geq 0$
20. $0 \leq 6$
21. $13 \geq 13$
22. $4 + 2 \geq 3 \cdot 2$
23. $7 \neq 4 + 2$
24. $3 \cdot 6 \leq 3 + 4 + 11$

In problems 25–50, use the order of operations to find the value of each expression.

25. $4 \times 7 + 8$
26. $5 \times 8 - 4$
27. $2 + 7 \times 5 - 2$
28. $17 - 3 \times 5 + 7$
29. $7 \times 2 + 5 \times 4$
30. $9 \times 6 + 11 - 2$
31. $4 + 4 \times 4 - 4 \div 4$
32. $5 + (5 \times 5) - (5 \div 5) + (5 \times 5 \div 5)$
33. $4 \times 8 \times 3 \div 12 + 10 \times 2 - 8 \times 3$
34. $17 \times 3 - 3 \times 2 \times 5 \div 15 + 7 \times 2$
35. $24 \div 2 \times 3 \quad 2$
36. $5 \times 2 + 4 + 3 \times 16 + 4$
37. $8(9 - 2) + 6$
38. $5(8 + 5) - 2$
39. $8 + (16 + 4) \div 5$
40. $(6 + 4) \div 5 + 3$
41. $(6 - 4) \div 2 + 5$
42. $8 - (4 \div 2) + 7$
43. $\dfrac{8 + 4}{3 + 3}$
44. $\dfrac{18 + 3}{4 + 3}$
45. $\dfrac{24 + 6 \div 3}{7 + 6 \times 3}$
46. $\dfrac{8 + 12 \div 3 - 2}{10 \div 2}$
47. $\dfrac{3(3 + 2) - 3(3) + 2}{5(2) - 2(2 - 1)}$
48. $\dfrac{(6 + 24 \div 4) \div (8 \times 4 - 20)}{7 - 2 \times 2}$

49. $\dfrac{5 \times 10 \div 2 \times 3 + 6 \div 2 \times 100}{2 + 2 + 1}$

50. $\dfrac{25 + 18 \div 2 \times 3 + 8 \times 5}{2 \times 2}$

51. There were 20 passengers on a bus. At the first stop 4 more got on. At the second stop half of the passengers on the bus got off. At the third stop twice as many people got on as had gotten on at the first stop. Write an expression (using parentheses) to describe the situation and determine how many people were on the bus after the third stop.

52. A gambler begins with $80. After two hours he triples his money. An hour later he loses $140. Write an expression to describe this situation and find its value.

0.2 Sets of Numbers

In this section, we group or classify numbers in some useful ways. In order to accomplish this, we begin by introducing the idea of a set.

A **set** may be thought of as a well-defined collection of objects. Any one of the objects in a set is called an **element** or a **member** of the set. We use capital letters, such as $A, B, C,$ to represent sets. We also represent sets by putting braces { } around the members of each set. Thus, if we write {1, 2, 3, 4, 5}, we mean the set whose elements are the numbers 1, 2, 3, 4, and 5. Two sets are **equal** if they contain precisely the same elements.

Example 1 Write the set A consisting of all the even numbers between 1 and 9.

Solution $A = \{2, 4, 6, 8\}$

If a set consists of a large number of elements, and the elements fall into a pattern, we often use "..." to represent some of the elements of the set. For example,

$$D = \{1, 2, 3, \ldots, 13\}$$

is the set D of all whole numbers between 0 and 14

Any set, regardless of how large, that has a *definite* number of elements, is called a **finite set.**

Example 2 Write the set B consisting of all the odd numbers between and including 1 and 999.

Solution $B = \{1, 3, 5, 7, \ldots, 999\}$

If a set has an *unlimited* number of elements, we use three dots "..." to indicate the missing elements. For example, we can write

$$A = \{5, 10, 15, 20, 25, \ldots\},$$

where "..." means "and so on." In this example, the symbol "..." represents an unending set of numbers that have the same pattern as the numbers listed. Such sets are called **infinite sets.**

Example 3 Write the set C consisting of all even whole numbers greater than zero.

Solution $C = \{2, 4, 6, 8, \ldots\}$

The Number Line and Real Numbers

A set of numbers can be represented as a collection of points on a **number line,** also called a **coordinate axis.** We can construct a number line by drawing a line with arrowheads on each end to show that the line extends endlessly in both directions.

Next, we choose a point on the line to associate with the number 0. This point is called the **origin.** Now we choose another point to the right of the origin to associate with the number 1 (Figure 1).

Figure 1

The distance between the two points labeled 0 and 1 is called the **unit length** of the number line. By measuring one unit length to the right of the point 1, we find a point to associate with the number 2. Repeating this process, we can find points to associate with the numbers 3, 4, 5, . . . (Figure 2).

Figure 2

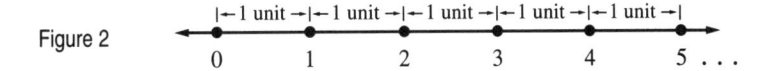

Think, for a moment, of the number line as divided into two sides, one to the left of the zero point, or origin, and the other to its right. The side to the right of the origin, containing the point associated with the number 1, is called the **positive side** of the number line, and the direction from the origin to the point labeled 1 is called the **positive direction** on the number line. The point associated with a number on a number line is called the **graph** of that number, and the number is called the **coordinate** of that point. For example, the coordinates of points A, B, C, and D on the number line below are 0, 2, 4, and 5, respectively (Figure 3).

Figure 3

The points to the left or the origin on the number line represent **negative numbers.** To graph the number -1 (which is read "negative one"), we start at the origin and move one unit to the left (Figure 4). To graph -2 (read "negative two"), we move another unit to the left, and so on. Thus, the coordinates of points E, F, G, and H on the number line are -4, -3, -2, and -1.

Figure 4

Note that a negative number is a coordinate of a point on the line to the left of zero, and it is written as a numeral with a negative sign in front of it. The side of the line that contains the points with negative coordinates is called the **negative side** of the number line, and the direction from 0 to -1 is called the **negative direction.**

Note that we have introduced a new meaning for the "minus" sign—we now use this symbol to indicate a negative number. We also use this symbol to show a subtraction. Thus, we use the mathematical symbols 8 − 5 to mean "subtract 5 from 8"; and we use the symbol −5 to represent "negative 5"—a point that lies five units to the left of the origin on the number line.

We now list some important infinite sets or numbers

1. The **natural numbers** (also called **counting numbers** or **positive integers**) are the numbers 1, 2, 3, 4, 5, and so on. This set is represented by the symbol ℕ. Thus,

$$\mathbb{N} = \{1, 2, 3, 4, 5, \ldots\}.$$

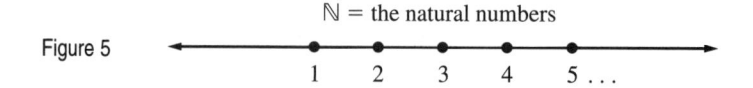

Figure 5

2. The **whole numbers** (also called **nonnegative integers**) are the numbers 0, 1, 2, 3, 4, and so on. This set is represented by W. Thus,

$$W = \{0, 1, 2, 3, 4, \ldots\}.$$

Figure 6

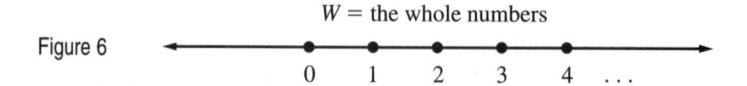

3. The **integers** are represented by the symbol **I**, where

$$\mathbf{I} = \{\ldots, -4, -3, -2, -1, 0, 1, 2, 3, 4, \ldots\}.$$

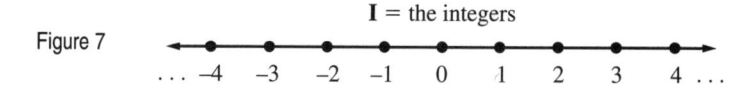

Figure 7

4. The **positive real numbers** correspond to points to the right of the origin 0 (Figure 8a), and the **negative real numbers** correspond to points to the left of the origin (Figure 8b). The set of **real numbers** is represented by the symbol ℝ (Figure 8c).

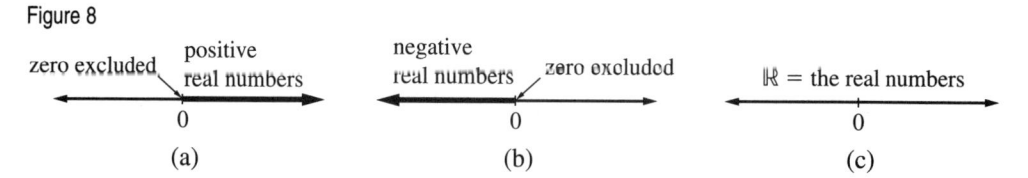

Figure 8

5. The **rational numbers** are the real numbers that can be written in the form

$$\frac{a}{b},$$ where a and b are integers and $b \neq 0$.

Since b may equal 1, every integer is a rational number. Other examples of rational numbers are

$$\frac{7}{2}, \quad \frac{3}{5}, \quad \text{and} \quad -\frac{22}{7}.$$

The set of all rational numbers is represented by ℚ.

6. The **irrational numbers** are the real numbers that are not rational, that is, the numbers that cannot be written in the form a/b, where a and b are integers. Examples of irrational numbers are

$$\sqrt{2}, \quad \sqrt{3}, \quad \text{and} \quad \pi.$$

The set of irrational numbers is represented by L.

Forms of Rational Numbers

A rational number that is not an integer is usually written as a **fraction** or as a **decimal**.

It is often useful to change rational numbers into decimal form. A fraction whose denominator is 10, 100, 1,000, or any power of ten, can easily be written as a decimal. For example,

$$\frac{3}{10} = 0.3, \quad \frac{7}{100} = 0.07, \quad \text{and} \quad \frac{1,432}{1,000} = 1,432.$$

A fraction with a denominator that is *not* a power of ten can also be expressed as a decimal. To express a rational number as a decimal, we divide the numerator of the fraction by the denominator. For example,

$$\frac{1}{2} = 1 \div 2 = 0.5, \quad \frac{4}{5} = 4 \div 5 = 0.8, \quad \text{and} \quad \frac{5}{16} = 5 \div 16 = 0.3125.$$

The decimal form of a rational number may be **terminating** or **non-terminating** but **repeating**. For example,

$$\frac{2}{5} = 2 \div 5 = 0.4 \quad \text{and} \quad -\frac{3}{4} = -3 \div 4 = -0.75$$

are terminating decimals, whereas

$$\frac{2}{3} = 0.6666\ldots, \quad \frac{7}{9} = 0.7777\ldots, \quad \frac{1}{7} = 0.1428571428571\ldots$$

are nonterminating repeating decimals.

A repeating decimal such as $0.6666\ldots$ is often written as $0.\overline{6}$, where the overbar indicates the block of digits that repeat. For instance,

$0.666\ldots = 0.\overline{6}, \quad 0.777\ldots = 0.\overline{7}, \quad \text{and} \quad 0.142857142857\ldots = 0.\overline{142857}.$

Irrational numbers such as $\sqrt{2}, \sqrt{3}$, and π, have decimal representations that are **nonterminating** and **nonrepeating**. A (partial) decimal representation for each of these numbers is as follows:

$$\sqrt{2} = 1.4142136\ldots, \quad \sqrt{3} = 1.7320508\ldots, \quad \pi = 3.1415926\ldots.$$

Example 4 Express each rational number as a decimal.

(a) $\dfrac{3}{4}$ (b) $-\dfrac{5}{8}$ (c) $\dfrac{3}{16}$ (d) $-\dfrac{17}{99}$

Solution Since

$$\frac{a}{b} = a \div b,$$

we can express each rational number as a decimal by dividing its numerator by its denominator.

(a) If you divide 3 by 4, you will obtain

$$\frac{3}{4} = 3 \div 4 = 0.75.$$

(b) $-\dfrac{5}{8} = -0.625$

(c) $\dfrac{3}{16} = 0.1875$

(d) $-\dfrac{17}{99} = -0.171717\ldots = -0.\overline{17}$

Example 5 Express each decimal in the form of a/b, where a and b are integers.

(a) 0.7   (b) -0.61 (c) -0.003 (d) 1.075

Solution (a) $0.7 = \dfrac{7}{10}$ (b) $-0.61 = -\dfrac{61}{100}$

(c) $-0.003 = -\dfrac{3}{1,000}$ (d) $1.075 = \dfrac{1,075}{1,000} = \dfrac{43}{40}$

In Section 0.11, you will see how to rewrite a repeating decimal in the form a/b.

A **percent** is a form of a rational number that implies a denominator of 100. The word percent means "per 100," so that 6 percent, denoted by 6%, means $\frac{6}{100}$, 12% means $\frac{12}{100}$, and 157% means $\frac{157}{100}$. Since percents are used in applications that appear later in the text, we review here the rules for changing percents to decimals or to fractions.

To change a percent to a decimal, omit the % sign and multiply by 0.01.

Example 6 Change each percent to a decimal.

(a) 8%   (b) 13%   (c) 19.7%   (d) 200%

Solution (a) $8\% = 8 \times 0.01 = 0.08$ (b) $13\% = 13 \times 0.01 = 0.13$
(c) $19.7\% = 19.7 \times 0.01 = 0.197$ (d) $200\% = 200 \times 0.01 = 2.00$

To change a percent to a fraction, omit the % sign and multiply by $\frac{1}{100}$.

Example 7 Change each percent to a fraction.

(a) 11%   (b) 17%   (c) $9\frac{3}{4}\%$ (d) 12.3%   (e) 325%

Solution (a) $11\% = 11 \times \dfrac{1}{100} = \dfrac{11}{100}$

(b) $17\% = 17 \times \dfrac{1}{100} = \dfrac{17}{100}$

(c) $9\dfrac{3}{4}\% = 9\dfrac{3}{4} \times \dfrac{1}{100} = \dfrac{39}{4} \times \dfrac{1}{100} = \dfrac{39}{400}$

(d) $12.3\% = 12.3 \times \dfrac{1}{100} = \dfrac{12.3}{100} = \dfrac{123}{1,000}$

(e) $325\% = 325 \times \dfrac{1}{100} = \dfrac{325}{100} = \dfrac{13}{4}$

In applications involving percents, we are frequently required to find a certain percent of some number. To do so, we change the percent to a decimal or fraction and then multiply.

Example 8 Find the value of each expression. (Note that "of" denotes multiplication.)
(a) 10% of 150 (b) 23% of 289 (c) $6\tfrac{1}{4}\%$ of 1,000

Solution (a) 10% of $150 = \dfrac{10}{100} \times 150 = \dfrac{1}{10} \times 150 = 15$

(b) 23% of $289 = 0.23 \times 289 = 66.47$

(c) $6\tfrac{1}{4}\%$ of $1,000 = 6.25\%$ of $1,000 = 0.0625 \times 1,000 = 62.5$

Example 9 A company deposits 5% of each employee's annual salary into a special account. If an employee's annual salary is $42,500, what is the annual amount of the company's deposit for that employee?

Solution The company's annual deposit is 5% of $42,500; that is

$$5\% \text{ of } 42,500 = 0.05 \times 42,500 = 2,125.$$

Therefore, the amount deposited for the year into the special account is $2,125.

PROBLEM SET 0.2

In problems 1–10, write the following sets.

1. The set of integers between -3 and 5.
2. The set of integers between 0 and 100.
3. The set of even integers between 4 and 14.
4. The set of rational numbers that have a denominator of 3 and that have a value between 0 and 2.
5. The set of negative integers between -50 and 0.
6. The set of multiples of 5 greater than 5.
7. The set of multiples of 3 between 4 and 20.
8. The set of rational numbers that have a denominator of 4 and that have a value between -2 and 2.
9. The set of integers that are greater than 1 and that are exactly divisible by 4.
10. The set of integers that are greater than 10 and that are exactly divisible by 7.

In problems 11–14, graph each set on a number line.

11. $\{-3, -2, -1, 0, 1, 2, 3\}$
12. $\{-4, -3, -2, -1, 0\}$
13. $\{-5, -4, -1, 0, 3, 4\}$
14. $\{-6, -3, 0, 3, 6\}$

In problems 15–28, express each rational number in decimal notation.

15. $-\dfrac{7}{2}$
16. $-\dfrac{67}{100}$
17. $-\dfrac{4}{5}$
18. $-\dfrac{5}{4}$
19. $\dfrac{6}{5}$
20. $\dfrac{5}{8}$
21. $\dfrac{7}{8}$
22. $-\dfrac{5}{6}$
23. $\dfrac{7}{20}$
24. $-\dfrac{11}{12}$
25. $-\dfrac{17}{25}$
26. $\dfrac{7}{35}$
27. $\dfrac{8}{15}$
28. $-\dfrac{18}{12}$

In problems 29–38, express each decimal as a rational number in the form of a/b, where a and b are integers.

29. 0.43
30. 0.54
31. 2.64
32. −0.125
33. −0.008
34. −40.08
35. −36.63
36. −0.582
37. 0.694
38. −0.209

In problems 39–50, change each percent to a decimal.

39. 16%
40. 22%
41. 14.7%
42. 0.9%
43. 6.5%
44. 13.8%
45. 29%
46. 51%
47. 30.3%
48. 120%
49. 180%
50. 3,500%

In problems 51–62, change each percent to a fraction.

51. 19%
52. 21%
53. $23\frac{3}{4}$%
54. $37\frac{5}{8}$%
55. $7\frac{2}{5}$%
56. 18.3%
57. 19.3%
58. 37.7%
59. 24.9%
60. 17.41%
61. 125%
62. 175%

In problems 63–72, find the indicated value.

63. 12% of 200
64. 16% of 350
65. 18% of 1,000
66. 22% of 789
67. 14.5% of 500
68. 21.7% of 896
69. $8\frac{1}{2}$% of 420
70. $10\frac{3}{8}$% of 10,000
71. $13\frac{3}{4}$% of 670
72. $14\frac{5}{8}$% of 1,200

73. A company estimates that its profit for a year is 20% of its sales for that year. If sales for the year total $1,715,000, what is the company's annual profit?

74. A mathematics professor scales her examinations so that 7/8 of the students receive grades of C or higher. What percentage of the class receive C or higher?

75. Sociologists estimate that of all the couples who get married in Camelot this year, 30% will be divorced within 15 years. If 24,000 couples marry, how many will probably be divorced within 15 years?

76. A petroleum distributor sells gasohol containing 12.3% alcohol. If his storage tank contains 300,000 gallons of gasohol, how many gallons of alcohol does it contain?

77. A chemist finds that a beaker of sea water contains 0.34% of magnesium chloride by weight. What is the weight of magnesium chloride in a beaker that contains 2,067.65 grams of sea water?

78. A metallurgist estimates that the amount of pure copper in bornite is 63.7%. If a sample of bornite weighs 14.75 pounds, what is the weight of the copper it contains?

0.3 Basic Properties of Real Numbers

In this section, we introduce the properties of real numbers. These properties serve as a foundation for the algebraic steps introduced in later chapters. We state these properties with letter symbols representing real numbers. Using the letters a and b to represent any real numbers, we write the *sum* and the *product* of a and b as

$$a + b \quad \text{and} \quad a \cdot b.$$

When we write a product with letters, we normally indicate the operation with a multiplication dot · rather than with the multiplication symbol ×. This is because the symbol of operation × might be confused with the letter x.

The basic properties of real numbers can be expressed in terms of the operations of addition and multiplication. Let a, b, and c denote real numbers:

1 The Commutative Properties

(i) For addition:
$$a + b = b + a$$

(ii) For multiplication:
$$a \cdot b = b \cdot a$$

In other words, the order in which you add or multiply two numbers does not affect the result. For example,

$$3 + 5 = 5 + 3,$$

since

$$3 + 5 = 8 \quad \text{and} \quad 5 + 3 = 8.$$

Also,
$$4 \cdot 6 = 6 \cdot 4,$$
since
$$4 \cdot 6 = 24 \quad \text{and} \quad 6 \cdot 4 = 24.$$

2 The Associative Properties

(i) For addition:	(ii) For multiplication:
$a + (b + c) = (a + b) + c$	$a \cdot (b \cdot c) = (a \cdot b) \cdot c$

The associative properties can be thought of as grouping properties. For example,
$$2 + (3 + 4) = (2 + 3) + 4,$$
since
$$2 + (3 + 4) = 2 + 7 = 9 \quad \text{and} \quad (2 + 3) + 4 = 5 + 4 = 9.$$
Also,
$$4 \cdot (3 \cdot 5) = (4 \cdot 3) \cdot 5$$
since
$$4 \cdot (3 \cdot 5) = 4 \cdot 15 = 60 \quad \text{and} \quad (4 \cdot 3) \cdot 5 = 12 \cdot 5 = 60.$$

3 The Distributive Properties

(i) $a \cdot (b + c) = a \cdot b + a \cdot c$	(ii) $(b + c) \cdot a = b \cdot a + c \cdot a$

The distributive properties involve both addition and multiplication. In this case, we say that multiplication distributes over addition. For example,
$$6 \cdot (3 + 2) = 6 \cdot 3 + 6 \cdot 2,$$
since
$$6 \cdot (3 + 2) = 6 \cdot 5 = 30 \quad \text{and} \quad 6 \cdot 3 + 6 \cdot 2 = 18 + 12 = 30.$$
Also,
$$(4 + 7) \cdot 3 = 4 \cdot 3 + 7 \cdot 3$$
since
$$(4 + 7) \cdot 3 = 11 \cdot 3 = 33 \quad \text{and} \quad 4 \cdot 3 + 7 \cdot 3 = 12 + 21 = 33.$$

4 The Identity Properties

(i) For addition:	(ii) For multiplication:
$a + 0 = 0 + a = a$	$a \cdot 1 = 1 \cdot a = a$

That is, 0 is the **additive identity** element and 1 is the **multiplicative identity** element. For example,
$$6 + 0 = 0 + 6 = 6 \quad \text{and} \quad 3 \cdot 1 = 1 \cdot 3 = 3.$$

5 The Inverse Properties

(i) For addition:
For each real number a there is a real number $-a$ called the **additive inverse** of a (also called the **negative** of a or the **opposite** of a), such that
$$a + (-a) = (-a) + a = 0.$$

(ii) For multiplication:
For each real number $a \neq 0$ there is a real number $1/a$ called the **multiplicative inverse** of a (or the **reciprocal** of a), such that
$$a \cdot \frac{1}{a} = \frac{1}{a} \cdot a = 1.$$

For example,
$$7 + (-7) = (-7) + 7 = 0,$$
so that -7 is the additive inverse of 7, and 7 is the additive inverse of -7. Also,
$$5 \cdot \frac{1}{5} = \frac{1}{5} \cdot 5 = 1,$$
so that $\frac{1}{5}$ is the multiplicative inverse of 5, and 5 is the multiplicative inverse of $\frac{1}{5}$.

Example 1 State the property that is illustrated by each statement.

(a) $5 + 7 = 7 + 5$
(b) $9 + (8 + 2) = (9 + 8) + 2$
(c) $2 \cdot 5 = 5 \cdot 2$
(d) $14 \cdot (2 + 3) = 14 \cdot 2 + 14 \cdot 3$
(e) $3 + (-3) = 0$
(f) $8 \cdot \frac{1}{8} = 1$
(g) $(4 + 0) + 3 = 4 + 3$
(h) $7 \cdot 1 = 7$

Solution
(a) The commutative property for addition
(b) The associative property for addition
(c) The commutative property for multiplication
(d) The distributive property
(e) The inverse property for addition
(f) The inverse property for multiplication
(g) The identity property for addition
(h) The identity property for multiplication

Notice that we have different uses for the symbol "$-$". They are:
(i) To indicate subtraction.
(ii) To indicate a negative number.
(iii) To indicate the negative of a number (the additive inverse.)

In particular, the symbol $-a$ does not necessarily mean a negative number. In fact, the number can be positive, zero, or negative, depending on the value of a. For example,

if $a = 5$, then $-a = -5$;
if $a = 0$, then $-a = 0$;
if $a = -8$, then $-a = 8$.

Notice that $-(-a) = a$. For example, $-(-3) = 3$.

Example 2 Find the additive inverse and the multiplicative inverse of each number.

(a) 15   (b) $\frac{1}{7}$ (c) $\frac{2}{3}$

Solution

	Number	Additive Inverse	Multiplicative Inverse
(a)	15	-15	$\frac{1}{15}$
(b)	$\frac{1}{7}$	$-\frac{1}{7}$	7
(c)	$\frac{2}{3}$	$-\frac{2}{3}$	$\frac{3}{2}$

The following properties can be **derived** from the properties already discussed.

6 The Cancellation Properties

(i) For addition:
$$\text{If } a + c = b + c, \text{ then } a = b.$$
(ii) For multiplication:
$$\text{If } ac = bc, \text{ with } c \neq 0, \text{ then } a = b.$$

For example,
$$\text{if } a + 3 = b + 3, \text{ then } a = b,$$
and
$$\text{if } 4a = 4b, \text{ then } a = b.$$

7 The Zero-Factor Properties

(i) $a \cdot 0 = 0 \cdot a = 0$.
(ii) If $a \cdot b = 0$, then $a = 0$ or $b = 0$ (or both).

For example,
$$(-8) \cdot 0 = 0 \cdot (-8) = 0,$$
and if
$$5 \cdot x = 0, \quad \text{then} \quad x = 0.$$

Symbols of Inclusion

We saw in Section 0.1 that the value of an expression involving addition, subtraction, multiplication, or division may depend on the order in which the operations are performed. Thus, by using the rules for order of operations, we find the value of the expression $3(5 + 2) - 4$ as follows:

$$3(5 + 2) - 4 = 3(7) - 4$$
$$= 21 - 4$$
$$= 17.$$

Another way of saying "find the value of" is "simplify." For instance, in the previous example we simplified $3(5 + 2) - 4$ as 17.

Sometimes an expression is so complicated that additional grouping symbols are needed. When this occurs brackets [ ] and braces { } can be used.

> To **simplify** expressions containing different grouping symbols, we simplify the innermost groups first and work toward the outside.

Example 3 Simplify $3[40 - (5 + 3)]$.

Solution $3[40 - (5 + 3)] = 3[40 - 8]$ (We added the numbers in the innermost group first.)

$= 3[32]$ (We subtracted the numbers inside the brackets.)

$= 96.$

Example 3 Simplify $2\{3 + 4[2 + 3(5 + 1)]\}$.

Solution $2\{3 + 4[2 + 3(5 + 1)]\} = 2\{3 + 4[2 + 3(6)]\}$ We added the numbers in the innermost group first.)

$= 2\{3 + 4[2 + 18]\}$ (We multiplied the numbers in the innermost group.)

$= 2\{3 + 4[20]\}$ (We added the numbers in the innermost group.)

$= 2\{3 + 80\}$ (We multiplied.)
$= 2\{83\}$ (We added.)
$= 166.$

In a fraction such as

$$\frac{25 - 5}{7 - 3},$$

the fraction-bar symbol is used as a grouping symbol and as a division sign. Thus,

$$\frac{25 - 5}{7 - 3} = \frac{20}{4} = 5.$$

PROBLEM SET 0.3

In problems 1−10, verify the given statement by performing the operations.

1. $7 + 8 = 8 + 7$
2. $10 \cdot 8 = 8 \cdot 10$
3. $5 \cdot 9 = 9 \cdot 5$
4. $(6 + 9) \cdot 3 = 6 \cdot 3 + 9 \cdot 3$
5. $6 \cdot (2 + 3) = 6 \cdot 2 + 6 \cdot 3$
6. $7 + 11 = 11 + 7$
7. $5 + (7 + 2) = (5 + 7) + 2$
8. $6 \cdot (8 \cdot 3) = (6 \cdot 8) \cdot 3$
9. $4 \cdot (8 \cdot 7) = (4 \cdot 8) \cdot 7$
10. $15 + (17 + 24) = (15 + 17) + 24$

In problems 11−30, state the property that is illustrated by each statement.

11. $10 + 9 = 9 + 10$
12. $-15 \cdot 1 = -15$
13. $7 + (-7) = 0$
14. $(8 + 9) \cdot 6 = 8 \cdot 6 + 9 \cdot 6$
15. $3 \cdot (8 + 4) = 3 \cdot 8 + 3 \cdot 4$
16. $18 + (17 + 2) = (18 + 17) + 2$
17. $4 \cdot (9 \cdot 3) = (4 \cdot 9) \cdot 3$
18. $25 \cdot 30 = 30 \cdot 25$
19. $-8 \cdot \dfrac{1}{-8} = 1$
20. $0 + 13 = 13$
21. $(-3)(6) = (6)(-3)$
22. $-\dfrac{1}{2} + \dfrac{1}{2} = 0$
23. $(-7 + 5) \cdot 4 = -7 \cdot 4 + 5 \cdot 4$
24. $-6 \cdot (3 + 4) = (-6) \cdot 3 + (-6) \cdot 4$
25. $11 \cdot 1 = 11$
26. $\dfrac{1}{4} \cdot 4 = 1$
27. $-2 + (5 + 13) = (-2 + 5) + 13$
28. $(-14) + (-8) = (-8) + (-14)$
29. $17 + 0 = 17$
30. $-4 \cdot (7 \cdot 3) = (-4 \cdot 7) \cdot 3$

In problems 31−40, find the additive inverse and the multiplicative inverse of each number. (If an inverse does not exist, write "none.")

31. 3
32. 11
33. 6
34. 15
35. $\dfrac{1}{5}$
36. $\dfrac{1}{10}$
37. 0
38. 0.3
39. $\dfrac{3}{4}$
40. $\dfrac{7}{6}$

In problems 41−58, simplify each expression.

41. $4[20 - (7 - 2)]$
42. $6[13 + (10 - 3)]$
43. $2[(5 + 1) \cdot (8 \div 4)]$
44. $3[10 \div 5 \cdot 4 \cdot 4 \div 2]$
45. $5[(6 - 2) \cdot 3 + 1]$
46. $7[8 + 2(5 + 3)]$
47. $3 + 2[(5 - 3) + 2(1 + 3)]$
48. $29 - 7[3 + (15 - 6) - 2(5 - 1)]$
49. $7\{2 + [8 + 3(5 - 2)] + 4\}$
50. $2\{20 - [2(4 - 1) - 3(2 - 1)]\}$
51. $4\{3[10 + 2(7 - 4) + 5] + 2\}$
52. $16\{(16 - 6) + 2[24 - (20 - 3)] + 5\}$
53. $26 - 2[4 + 3(11 - 8) - 5(13 - 11)]$
54. $(50 - 13) + 7 + 2[4(9 - 5) - 2(8 - 5)]$
55. $[(31 - 23) - 2] + \{8 + [3 + 2(19 - 14) - (20 - 17)]\}$
56. $\{5 + 2(7 - 1) + 3[10 - 3(6 - 4)]\} + [9 - 2(8 - 3) + 5]$
57. $\dfrac{2(10 - 3)}{(2 \cdot 10) - 3}$
58. $\dfrac{5(6 - 3)}{15 \div (10 - 7)}$

0.4 Addition and Subtraction of Signed Numbers

As you learned in Section 0.2, each number on the number line represents a certain distance and direction from the origin. Points to the left of 0 represent negative numbers, and points to the right of 0 represent positive numbers. If a number is preceded by a negative sign (−), we know that the number is negative. If a number is not zero and is preceded by a plus sign (+) or no sign at all, we know that it is positive. Positive and negative numbers together are called **signed numbers.**

The *absolute value* of a number indicates how far the number is from the origin, regardless of whether it is to the left or to the right of the origin. Thus, if x is a real number, then the **absolute value** of x is denoted by $|x|$ and is given by

$$|x| = \text{the distance between } x \text{ and } 0.$$

For example, the number 6 is six units from 0 on the number line, so the absolute value of 6, written $|6|$, is 6. That is,

$$|6| = 6.$$

The absolute value of -4, written $|-4|$, is 4 because the number -4 is four units from the origin. That is,

$$|-4| = 4$$

since the absolute value indicates only the *distance* from zero and not the direction. The *absolute value of a number can never be negative.*

Example 1 Find the value of each expression.

(a) $|8|$ (b) $|-7|$ (c) $-|9|$ (d) $|-3| + |-7|$

Solution (a) $|8| = 8$, since 8 is a positive number that lies eight units from the origin.
(b) $|-7| = 7$, since -7 is a negative number that lies seven units from the origin.
(c) $-|9| = -(|9|) = (-9) = -9$.
(d) $|-3| = 3$; also, $|-7| = 7$, so $|-3| + |-7| = 3 + 7 = 10$.

Now we shall see how to add, subtract, multiply, and divide signed numbers.

Addition of Signed Numbers

The addition of signed numbers can be illustrated on a number line. First, we consider the addition of numbers with the same sign. For example, to add 2 and 3 on the number line (Figure 1), we start at the origin and move two units to the right. The number 2 is represented by an arrow from 0 to 2. From our position at 2 we then move three units to the right, so that the number 3 is represented by an arrow between 2 and 5. The sum of these two directed moves is $2 + 3 = 5$ and is represented by an arrow between 0 and 5 (Figure 1a). Figure 1b shows that $3 + 2 = 5$, a result already familiar to us from the commutative property.

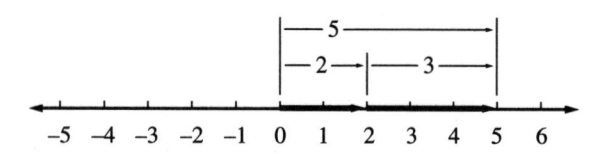

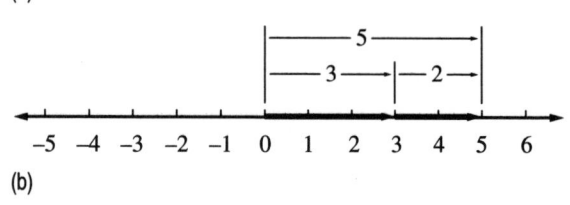

Figure 1 (a)

(b)

0.4 ADDITION AND SUBTRACTION OF SIGNED NUMBERS

Now consider the sum $(-3) + (-3)$. We find this sum by first moving two units to the left of the origin on the number line to -2. Then we move three units to the left of -2, to -5. The sum is $(-2) + (-3) = -5$ (Figure 2).

Figure 2

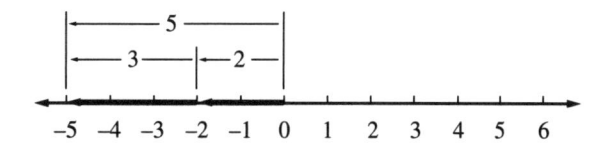

Note that in both cases just illustrated, the sums can be found by using the following rule:

> To **add** numbers with the same sign, add their absolute values and keep their common sign.

For example,
$$3 + 4 = |3| + |4| = 7$$
$$(-4) + (-6) = -(|-4| + |-6|) = -(4 + 6) = -10.$$

The addition of numbers that have different signs can also be illustrated on a number line. For example, the sum $(-3) + 8$ can be interpreted as a movement of eight units to the right of the number -3 (Figure 3a). The sum $(-3) + 8$ can also be interpreted as a movement of three units to the left of 8 (Figure 3b). The same sum is obtained in both cases, that is, $(-3) + 8 = 5$.

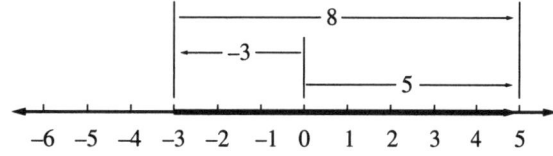

Figure 3 (a)

(b)

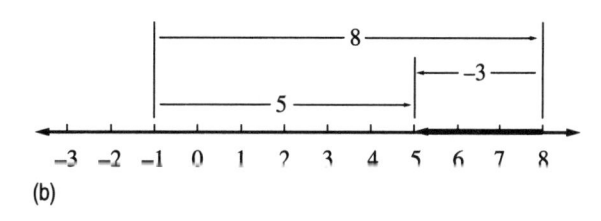

For another example, consider the sum $(-7) + 5$. Using the number line (Figure 4), we see that $(-7) + 5 = -2$.

Figure 4

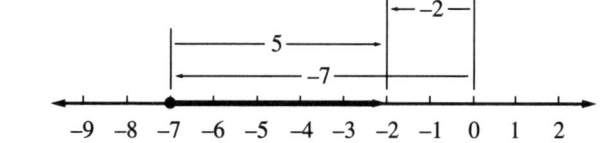

Note that in both examples just illustrated the sums can be found by using the following rule:

> To **add** two numbers with different signs, find the difference of their absolute values by subtracting the smaller absolute value from the larger. Retain the sign of the number with the larger absolute value.

Example 2 Perform each addition.
(a) $8 + 6$
(b) $(-4) + (-7)$
(c) $11 + (-11)$
(d) $(-4) + 21$
(e) $8 + (-23)$
(f) $3 + (-6) + 8$

Solution
(a) $8 + 6 = 14$
(b) $(-4) + (-7) = -(4 + 7) = -11$
(c) $11 + (-11) = 0$ (by the additive inverse property)
(d) $(-4) + 21 = +(21 - 4) = 17$
(e) $8 + (-23) = -(23 - 8) = -15$
(f) $3 + (-6) + 8 = 3 + 8 + (-6)$ (by the commutative property)
$= 11 + (-6) = +(11 - 6) = 5$

Example 3 If the temperature now is $-4°$ Fahrenheit (F) and goes up $12°$, what is the new temperature?

Solution
$$(-4) + 12 = +(12 - 4) = 8$$

The new temperature is $8°$F.

Example 4 Maria has $31 in her checking account. She makes a deposit of $50 and writes a check for $17. What is her new balance?

Solution Using signed numbers, we can represent the original balance as $+31$, the deposit as $+50$, and the amount of the check as -17, so that

$$(+31) + (+50) + (-17) = (31 + 50) + (-17)$$
$$= 81 + (-17)$$
$$= 64.$$

Therefore, Maria's new balance is $64.

Subtraction of Signed Numbers

To subtract b from a, an operation denoted by $a - b$, we use the following definition:

Definition **Subtraction**

> If a and b are real numbers, the difference $a - b$ is defined to be $a + (-b)$, where $-b$ is the additive inverse of b.

In other words, to subtract a number b, change its sign and add. This definition, together with the rules for adding signed numbers, provides a method for subtracting signed numbers. For example, the difference $(-7) - 3$ can be found as follows:

$$(-7) - 3 = (-7) + (-3)$$
$$= -(7 + 3)$$
$$= -10.$$

We can also illustrate this subtraction of signed numbers on a number line by interpreting the subtraction of 3 from -7 as a movement of three units to the left of the point whose coordinate is -7 (Figure 5).

Figure 5

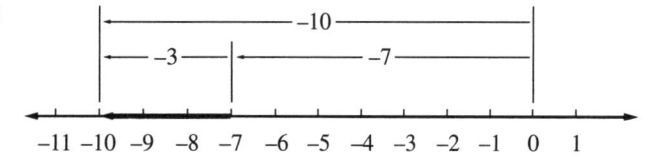

Now we restate the rule for subtracting signed numbers:

> To **subtract** one signed number from another, change the sign of the number to be subtracted, and then add by following the rules for adding signed numbers.

Example 5   Perform each subtraction.
(a) $5 - 3$
(b) $(-11) - 8$
(c) $12 - (-4)$
(d) $(-16) - (-19)$
(e) $4 - 8$

Solution
(a) $5 - 3 = 5 + (-3) = 2$
(b) $(-11) - 8 = (-11) + (-8) = -(11 + 8) = -19$
(c) $12 - (-4) = 12 + 4 = 16$
(d) $(-16) - (-19) = -16 + 19 = (19 - 16) = 3$
(e) $4 - 8 = 4 + (-8) = -8 - 4) = -4$

PROBLEM SET 0.4

In problems 1–10, find the value of each expression.

1. $|9|$
2. $|17|$
3. $|-11|$
4. $|-25|$
5. $-|5|$
6. $-|-83|$
7. $-|-8|$
8. $-|3|$
9. $|-4| + |4|$
10. $|-10| - |10|$

In problems 11–30, perform each addition.

11. $8 + 7$
12. $15 + 17$
13. $17 + 21$
14. $76 + 38$
15. $(-3) + (-8)$
16. $(-27) + (-18)$
17. $(-12) + (-23)$
18. $(-93) + (-46)$
19. $18 + (-7)$
20. $(-29) + 50$
21. $(-22) + 14$
22. $33 + (-42)$
23. $(-11) + (-9)$
24. $84 + (-69)$
25. $25 + (-13)$
26. $(-48) + (-21)$
27. $(-32) + 17$
28. $15 + (-33)$
29. $(-24) + (-19)$
30. $(-37) + (-62)$

In problems 31–50, perform each subtraction.

31. $18 - 13$
32. $65 - 39$
33. $16 - 9$
34. $20 - 14$
35. $7 - 15$
36. $13 - 22$
37. $11 - 27$
38. $37 - 43$
39. $18 - (-9)$
40. $27 - (-8)$

41. $31 - (-5)$
42. $38 - (-9)$
43. $(-25) - 4$
44. $(-49) - 17$
45. $(-65) - 15$
46. $(-18) - 12$
47. $(-17) - (-43)$
48. $(-22) - (-8)$
49. $(-25) - (-45)$
50. $(-11) - (-29)$

In problems 51–60, perform the indicated operations.

51. $(-17) + (-18) - 12$
52. $(-12) - (-6) - 10$
53. $(-22) - (-17) + 5$
54. $(-17) + 8 - (-9)$
55. $19 - (-22) - 8$
56. $13 + (-24) - (-17)$
57. $(-28) - 19 + (-13)$
58. $(-29) + 20 + (-21)$
59. $23 + (-17) - (-9)$
60. $33 - (-17) - 15$

In problems 61–64, use signed numbers to solve the word problems.

61. Before going to bed, Carlos noticed that the temperature was 18°F. When he awoke, he saw that the thermometer read -2°F. How many degrees did the temperature fall during the night?
62. If the temperature is 10° below 0°C and then goes up 25°, what is the new temperature?
63. David has a savings account with a balance of $270. If he makes a deposit of $100 and then two withdrawals of $40 and $65, what is his new balance?
64. In a series of three plays, a football team loses 4 yards, gains 7 yards, and then loses 5 yards. What is the team's position now in relation to the original line of scrimmage?
65. If we suppose that y is a negative number, what kind of a number is (a) $-y$? (b) $-(-y)$?
66. Show by an example whether the following statement is true or false:

If $|x| = |y|$, then $x = y$.

0.5 Multiplication and Division of Signed Numbers

We now develop rules for multiplying and dividing signed numbers.

Multiplication of Signed Numbers

The multiplication of two positive integers can be described as repeated addition. For instance, the product $4 \cdot 3$ can be interpreted as

$$4 + 4 + 4 \quad \text{or} \quad 3 + 3 + 3 + 3.$$

In both cases, the result is

$$4 \cdot 3 = 12.$$

Similarly,

$$4 \cdot 2 = 2 + 2 + 2 + 2 = 8$$
$$4 \cdot 1 = 1 + 1 + 1 + 1 = 4.$$

Using these examples, we see that *the product of two positive integers is positive.*

This allows us to state the following general rule:

The product of any two positive numbers is a positive number; that is, if a and b are positive numbers, then

$$a \cdot b = +(a \cdot b).$$

Recall that if a or b is zero, then

$$a \cdot 0 = 0 \quad \text{and} \quad 0 \cdot b = 0.$$

We can use a similar approach to find the product of a positive number and a negative number. For example, we can rewrite $4 \cdot (-3)$ as follows:

$$4 \cdot (-3) = (-3) + (-3) + (-3) + (-3) = -12.$$

Since multiplication is commutative, $(-3 \cdot 4)$ must also equal -12.

These examples suggest the following:

The product of a negative number and a positive number is a negative number. That is, if a and b are positive numbers, then

$$a(-b) = -(ab)$$

and

$$(-a)b = -(ab).$$

To examine the product of two negative numbers, consider the following pattern:

$$(-4) \cdot 3 = -12$$
$$(-4) \cdot 2 = -8$$
$$(-4) \cdot 1 = -4$$
$$(-4) \cdot 0 = 0.$$

Notice that as the second number decreases by 1, the product increases by 4. Thus, we would expect the following:

$$(-4) \cdot (-1) = 4$$

and

$$(-4) \cdot (-2) = 8.$$

The results of the last two examples suggest the following rule:

The product of two negative numbers is a positive number. That is, if a and b are positive numbers, then

$$(-a) \cdot (-b) = +(ab).$$

In Examples 1 and 2, find each product.

Example 1 (a) $(-9) \cdot 6$ (b) $5 \cdot (-8)$ (c) $6 \cdot (-15)$

Solution (a) $(-9) \cdot 6 = -(9 \cdot 6) = -54$ (b) $5 \cdot (-8) = -(5 \cdot 8) = -40$
(c) $6 \cdot (-15) = -(6 \cdot 15) = -90$

Example 2 (a) $9 \cdot 3$ (b) $(-4) \cdot (-2)$ (c) $(-5) \cdot (-3)$

Solution (a) $9 \cdot 3 = 27$ (b) $(-4) \cdot (-2) = 4 \cdot 2 = 8$
(c) $(-5) \cdot (-3) = 5 \cdot 3 = 15$

When multiplying more than two signed numbers, the numbers can be paired to determine the sign of the product. For example, consider the following product:

$$(-3) \cdot 2 \cdot (-5) \cdot 6$$

can be written as

$$[(-3) \cdot (-5)] \cdot (2 \cdot 6) = 15 \cdot 12$$
$$= 180.$$

Because there is an even number of negative numbers, the product is positive. In the case of the product

$$(-4) \cdot (-7) \cdot (-2) \cdot 3,$$

we have

$$[(-4) \cdot (-7)] \cdot [(-2) \cdot 3] = 28 \cdot (-6)$$
$$= -168.$$

Here, because there is an odd number of negative numbers, the product is negative. We can now state a more general rule for finding the product of signed numbers:

> The product of signed numbers is positive if there is an even number of negative numbers, and the product is negative if there is an odd number of negative numbers.

Thus, to multiply signed numbers, we multiply their absolute values and then use the preceding rule to determine the sign of the product.

Example 3 Determine each product.

(a) $3 \cdot (-2) \cdot 6 \cdot (-7) \cdot (-8) \cdot 2 \cdot (-4)$

(b) $(-1) \cdot 7 \cdot (-3) \cdot (-8) \cdot (-4) \cdot (-7)$

Solution (a) By counting we find that there are four negative factors. Since four is an even number, the product is positive:

$$3 \cdot (-2) \cdot 6 \cdot (-7) \cdot (-8) \cdot 2 \cdot (-4) = 16{,}128.$$

(b) Since there are five negative factors, an odd number, the product is negative:

$$(-1) \cdot 7 \cdot (-3) \cdot (-8) \cdot (-4) \cdot (-7) = -4{,}704.$$

Division of Signed Numbers

The division of signed numbers is the inverse operation of multiplication, that is,

$$15 \div 5 = 3 \quad \text{since} \quad 3 \cdot 5 = 15,$$

and

$$(-28) \div (-7) = 4 \quad \text{since} \quad 4 \cdot (-7) = -28.$$

More formally, we have the following definition:

Definition Division

> $a \div b$ is the number c (if there is one) for which $c \cdot b = a$. That is.
>
> $$a \div b = c \quad \text{if and only if} \quad c \cdot b = a.$$

For instance,

$$12 \div 4 = 3 \quad \text{because} \quad 3 \cdot 4 = 12$$
$$12 \div (-4) = -3 \quad \text{because} \quad (-3) \cdot (-4) = 12$$
$$(-12) \div 4 = -3 \quad \text{because} \quad (-3) \cdot 4 = -12$$
$$(-12) \div (-4) = 3 \quad \text{because} \quad 3 \cdot (-4) = -12.$$

These examples suggest the following rules:

> (i) If two numbers have the same signs, their quotient is positive.
>
> (ii) If two numbers have different signs, their quotient is negative.

Example 4 Perform each division.
(a) $45 \div (-9)$ (b) $(-42) \div 3$ (c) $(-65) \div (-13)$

Solution (a) $45 \div (-9) = -(45 \div 9) = -5$
(b) $(-42) \div 3 = -(42 \div 3) = -14$
(c) $(-65) \div (-13) = +(65 \div 13) = 5$

Example 5 Find each quotient.
(a) $\dfrac{21}{-7}$ (b) $\dfrac{-72}{24}$ (c) $\dfrac{-91}{-13}$

Solution (a) $\dfrac{21}{-7} = -\left(\dfrac{21}{7}\right) = -3$ (b) $\dfrac{-72}{24} = -\left(\dfrac{72}{24}\right) = -3$
(c) $\dfrac{-91}{-13} = \dfrac{91}{13} = 7$

Keep in mind, when you divide one number by another, that **division by zero is not allowed.** For instance,

$$5 \div 0 = \dfrac{5}{0} \quad \text{is undefined.}$$

Let's see why we must exclude zero from the denominator (or divisor). Suppose that there is a number c, such that $5 \div 0 = c$. We must guarantee that $c \cdot 0$ will equal 5. We ask, what number c multiplied by 0 will give 5? There is *no such number*. No matter what number we substitute for c, when we multiply c by 0, we cannot get 5 because $c \cdot 0 = 0$. Therefore, we say, in general,

$$\dfrac{a}{0} \quad \text{is undefined.}$$
$$0 \div 0 = \dfrac{0}{0} \quad \text{is } not \text{ allowed.}$$

On the other hand, the *numerator* of a fraction may be 0. If $b \neq 0$, then

$$0 \div b = \dfrac{0}{b} = 0 \cdot \dfrac{1}{b} = 0.$$

For example,

$$0 \div 3 = 0,$$

whereas

$$4 \div 0 \quad \text{is undefined,}$$

and

$$-8 \div 0 \quad \text{is also undefined.}$$

PROBLEM SET 0.5

In problems 1–20, find each product.

1. $3 \cdot 6$
2. $7 \cdot 12$
3. $(-4) \cdot (-5)$
4. $(-10) \cdot (-9)$
5. $2 \cdot (-8)$
6. $(-14) \cdot 6$
7. $(-11) \cdot 3$
8. $8 \cdot (-12)$
9. $(-15) \cdot (-5)$
10. $(-12) \cdot (-11)$
11. $9 \cdot (-7)$
12. $(-17) \cdot 3$
13. $(-25) \cdot (-4)$
14. $(-3) \cdot (-28)$
15. $(-1) \cdot (-1) \cdot (-1)$
16. $3 \cdot (-2) \cdot (-5)$
17. $(-3) \cdot 5 \cdot (-1)$
18. $(-1) \cdot (-2) \cdot (-3)$
19. $(-2) \cdot (-5) \cdot (-1) \cdot (-3)$
20. $(-2) \cdot (-3) \cdot (-4) \cdot (-5) \cdot (-1)$

In problems 21–44, find the indicated quotients if possible.

21. $20 \div 4$
22. $32 \div 8$
23. $(-18) \div 9$
24. $(-57) \div 19$
25. $(-28) \div 4$
26. $(-90) \div 15$
27. $36 \div (-12)$
28. $72 \div (-8)$
29. $51 \div (-17)$
30. $125 \div (-25)$
31. $(-32) \div (-8)$
32. $(-24) \div (-6)$
33. $(-54) \div (-18)$
34. $(-250) \div (-50)$
35. $\dfrac{-49}{7}$
36. $\dfrac{-39}{13}$
37. $\dfrac{216}{-6}$
38. $\dfrac{81}{-27}$
39. $\dfrac{-42}{-14}$
40. $\dfrac{-625}{-25}$
41. $\dfrac{0}{-5}$
42. $\dfrac{0}{4}$
43. $\dfrac{-15}{0}$
44. $\dfrac{-10}{0}$

In problems 45–60, use the order of operations to simplify each expression.

45. $(-3) \cdot (-4) + 8$
46. $(5) \cdot (-8) + (-10)$
47. $15 - (-4) \cdot (3)$
48. $(-12) \div 3 - (-2)$
49. $(-18) \div 3 + 4 \cdot (-2)$
50. $5 \cdot (-6) + (-12) \div (-4)$
51. $-3(5-7) + (-4)(3-6)$
52. $6(10-14) - (-5)(7-11)$
53. $-2[-3 - (-8)] - 5(-8 + 3)$
54. $-7(12 - 15) + (-2)[(-9) - (-3)]$
55. $(-11) \cdot (-6) \div (-3) + 15 \div (-5) - 2$
56. $(-48) \div (-3) \cdot 10 - 24 \div (-8) + 7$
57. $(-15) \div (-3) + 2 \cdot 4 - 8 \div (-4)$
58. $5 \cdot (-3) + 6 \div 2 \cdot 4 + 1$
59. $9 \cdot (-6) - 8 \cdot 2 \div 4 \cdot 5$
60. $7 \cdot 4 - 2 \cdot (-1) \div 2 \cdot 5 + 3$

0.6 Operations on Fractions

Recall from Section 0.2 that a rational number is any real number that can be written in the form

$$\frac{a}{b},$$ where a and b are integers and $b \neq 0$.

In this section, we review the operations on rational numbers that are written as fractions.

> Two fractions are said to be **equivalent** if one fraction can be obtained from the other by dividing out common factors from the numerator and denominator, or by multiplying the numerator and denominator by the same nonzero number.

For example,
$$\frac{12}{20} = \frac{3}{5},$$
since
$$\frac{12}{20} = \frac{12 \div 4}{20 \div 4} = \frac{3}{5}$$
or
$$\frac{3}{5} = \frac{3 \times 4}{5 \times 4} = \frac{12}{20}.$$

Note that we use an equals sign to denote equivalent fractions.

> A fraction is said to be **simplified** or **reduced to lowest terms** if its numerator and denominator have no common factors other than 1 or -1. Thus, to **reduce** a fraction to lowest terms, we first factor both the numerator and denominator, and then divide both by their common factor.

For example,
$$\frac{14}{21} = \frac{2 \cdot \cancel{7}}{3 \cdot \cancel{7}} = \frac{2}{3}.$$

Here, dividing the numerator and denominator by the common factor 7 is indicated by drawing slanted lines through the two common factors.

Example 9 Reduce each fraction to lowest terms.

(a) $\frac{25}{30}$ (b) $\frac{-27}{45}$ (c) $\frac{121}{-77}$

Solution (a) $\frac{25}{30} = \frac{5 \cdot \cancel{5}}{6 \cdot \cancel{5}} = \frac{5}{6}$ (b) $\frac{-27}{45} = \frac{(-3) \cdot \cancel{9}}{5 \cdot \cancel{9}} = \frac{-3}{5}$

(c) $\frac{121}{-77} = \frac{11 \cdot \cancel{11}}{(-7) \cdot \cancel{11}} = \frac{11}{-7}$

> Every fraction has three signs associated with it: the sign of the numerator, the sign of the denominator, and the sign of the entire fraction. Two of these signs may be changed without changing the value of the fraction.

For example,
$$\frac{-2}{3} = \frac{2}{-3} = -\frac{2}{3} = -\frac{-2}{-3}$$

and
$$\frac{4}{5} = \frac{-4}{-5} = -\frac{-4}{5} = -\frac{4}{-5}$$

28 CHAPTER 0 THE FUNDAMENTAL OPERATIONS OF REAL NUMBERS

Example 2 Reduce each fraction to lowest terms. Express the answer in the form $\frac{a}{b}$ or $-\frac{a}{b}$, where a and b are positive numbers.

(a) $\dfrac{-6}{9}$ (b) $\dfrac{-15}{-25}$ (c) $-\dfrac{27}{-12}$

Solution (a) $\dfrac{-6}{9} = -\dfrac{6}{9} = -\dfrac{2 \cdot \cancel{3}}{3 \cdot \cancel{3}} = -\dfrac{2}{3}$ (b) $\dfrac{-15}{-25} = \dfrac{15}{25} = \dfrac{3 \cdot \cancel{5}}{5 \cdot \cancel{5}} = \dfrac{3}{5}$

(c) $-\dfrac{27}{-12} = \dfrac{27}{12} = \dfrac{9 \cdot \cancel{3}}{4 \cdot \cancel{3}} = \dfrac{9}{4}$

Addition and Subtraction of Fractions

If two fractions have the same denominators, they are called **like fractions**. For example, $\frac{3}{5}$ and $\frac{2}{5}$ are like fractions. In order to add or subtract like fractions, we use the following rules:

$$\frac{a}{b} + \frac{c}{b} = \frac{a+c}{b}, \qquad b \neq 0$$

$$\frac{a}{b} - \frac{c}{b} = \frac{a-c}{b}, \qquad b \neq 0$$

Example 3 Perform the indicated operations and simplify the result. Express each answer in the reduced form $\frac{a}{b}$ or $-\frac{a}{b}$.

(a) $\dfrac{5}{24} + \dfrac{7}{24}$ (b) $\dfrac{13}{15} - \dfrac{7}{15}$ (c) $\left(-\dfrac{3}{8}\right) + \left(-\dfrac{1}{8}\right)$ (d) $\dfrac{4}{9} - \left(-\dfrac{7}{9}\right)$

Solution (a) $\dfrac{5}{24} + \dfrac{7}{24} = \dfrac{5+7}{24} = \dfrac{12}{24} = \dfrac{\cancel{12} \cdot 1}{\cancel{12} \cdot 2} = \dfrac{1}{2}$

(b) $\dfrac{13}{15} - \dfrac{7}{15} = \dfrac{13-7}{15} = \dfrac{6}{15} = \dfrac{2 \cdot \cancel{3}}{5 \cdot \cancel{3}} = \dfrac{2}{5}$

(c) $\left(-\dfrac{3}{8}\right) + \left(-\dfrac{1}{8}\right) = \dfrac{-3}{8} + \dfrac{-1}{8}$

$= \dfrac{(-3)+(-1)}{8} = \dfrac{-4}{8} = -\dfrac{4}{8} = -\dfrac{1 \cdot \cancel{4}}{2 \cdot \cancel{4}} = -\dfrac{1}{2}$

(d) $\dfrac{4}{9} - \left(-\dfrac{7}{9}\right) = \dfrac{4}{9} - \left(\dfrac{-7}{9}\right)$

$= \dfrac{4-(-7)}{9} = \dfrac{4+7}{9} = \dfrac{11}{9}$

If two fractions have different denominators, they are called **unlike fractions**.

To add or subtract unlike fractions, we first change the unlike fractions to equivalent like fractions, and then apply the rules for adding or subtracting like fractions.

For example,

$$\frac{2}{7} + \frac{3}{5} = \frac{2 \cdot 5}{7 \cdot 5} + \frac{3 \cdot 7}{5 \cdot 7} = \frac{10}{35} + \frac{21}{35} = \frac{10 + 21}{35} = \frac{31}{35}$$

and

$$\frac{9}{11} - \frac{2}{3} = \frac{9 \cdot 3}{11 \cdot 3} - \frac{2 \cdot 11}{3 \cdot 11} = \frac{27}{33} - \frac{22}{33} = \frac{27 - 22}{33} = \frac{5}{33}.$$

In the preceding examples, the denominator of the equivalent like fractions was obtained by simply multiplying the two given denominators. There are times when this approach leads to unnecessary complications. For example,

$$\frac{3}{8} + \frac{5}{12} = \frac{3 \cdot 12}{8 \cdot 12} + \frac{5 \cdot 8}{12 \cdot 8} = \frac{36}{96} + \frac{40}{96} = \frac{76}{96} = \frac{\not{4} \cdot 19}{\not{4} \cdot 24} = \frac{19}{24}.$$

This process is always correct, and it is necessary if the denominators have no common factors. When the denominators do have common factors, however, there is a simpler method. The addition just shown, for example, can be performed more simply as follows:

$$\frac{3}{8} + \frac{5}{12} = \frac{3 \cdot 3}{8 \cdot 3} + \frac{5 \cdot 2}{12 \cdot 2} = \frac{9}{24} + \frac{10}{24} = \frac{19}{24}.$$

Here, we found the smallest positive integer that contains both of the given denominators as factors. This number is called the **least common denominator (LCD)**. Thus, the LCD of $\frac{3}{8}$ and $\frac{5}{12}$ is 24, since 24 is the smallest positive integer that contains 8 and 12 as factors.

When adding or subtracting unlike fractions whose denominators are relatively small, we can usually determine the LCD by inspection. However, for those with relatively large denominators, the LCD may not be so obvious; therefore, a method to determine the LCD is useful. There are several schemes available to do this task. The one we illustrate here is also the scheme used for algebraic fractions. To determine the LCD of $\frac{3}{8}$ and $\frac{5}{12}$ in the preceding addition problem, we begin by factoring each denominator as follows:

$$8 = 2 \cdot 2 \cdot 2 \quad \text{and} \quad 12 = 2 \cdot 2 \cdot 3.$$

Next, we list each factor the greatest number of times it appears in either denominator. The greatest number of times the factor 2 appears in either denominator above is three, and the factor 3 appears only once. The product of these listed factors is the least common denominator. So the LCD is

$$2 \cdot 2 \cdot 2 \cdot 3 = 24.$$

Example 4 Determine the LCD for each set of fractions and change each fraction to an equivalent fraction with this denominator.

(a) $\frac{3}{25}, \frac{7}{30}$ (b) $\frac{19}{48}, \frac{11}{84}$

Solution (a) First we factor each denominator completely: $25 = 5 \cdot 5$ and $30 = 2 \cdot 3 \cdot 5$. Now, using the factored denominators, we have: LCD $= 2 \cdot 3 \cdot 5 \cdot 5 = 150$. Since $150 \div 25 = 6$, we have:

$$\frac{3}{25} = \frac{3 \cdot 6}{25 \cdot 6} = \frac{18}{150}.$$

Also, since $150 \div 30 = 5$, we have:

$$\frac{7}{30} = \frac{7 \cdot 5}{30 \cdot 5} = \frac{35}{150}.$$

(b) Factoring each denominator, we have:
$$48 = 2 \cdot 2 \cdot 2 \cdot 2 \cdot 3 \quad \text{and} \quad 84 = 2 \cdot 2 \cdot 3 \cdot 7$$
so that
$$\text{LCD} = 2 \cdot 2 \cdot 2 \cdot 2 \cdot 3 \cdot 7 = 336.$$
Since $336 \div 48 = 7$, we have
$$\frac{19}{48} = \frac{19 \cdot 7}{48 \cdot 7} = \frac{133}{336}.$$
Also, since $336 \div 84 = 4$, we have:
$$\frac{11}{48} = \frac{11 \cdot 4}{84 \cdot 4} = \frac{44}{336}.$$

Example 5 Perform the indicated operations using the LCD approach and simplify the answers.

(a) $\dfrac{5}{28} + \dfrac{4}{35}$ (b) $\dfrac{14}{165} - \dfrac{4}{99}$

(c) $\dfrac{7}{22} + \dfrac{5}{12} - \dfrac{1}{30}$ (d) $\dfrac{5}{12} + \left(-\dfrac{7}{18}\right) - \dfrac{2}{15}$

Solution

(a) $\dfrac{5}{28} + \dfrac{4}{35}$

$28 = 2 \cdot 2 \cdot 7$ and $35 = 5 \cdot 7$, so that LCD $= 2 \cdot 2 \cdot 5 \cdot 7 = 140$. Since $140 \div 28 = 5$ and $140 \div 35 = 4$, we have:

$$\frac{5}{28} + \frac{4}{35} = \frac{5 \cdot 5}{28 \cdot 5} + \frac{4 \cdot 4}{35 \cdot 4} = \frac{25}{140} + \frac{16}{140} = \frac{41}{140}.$$

(b) $\dfrac{14}{165} - \dfrac{4}{99}$

$165 = 3 \cdot 5 \cdot 11$ and $99 = 3 \cdot 3 \cdot 11$. Hence,
$$\text{LCD} = 3 \cdot 3 \cdot 5 \cdot 11 = 495.$$
Since $495 \div 165 = 3$ and $495 \div 99 = 5$, we have:

$$\frac{14}{165} - \frac{4}{99} = \frac{14 \cdot 3}{165 \cdot 3} - \frac{4 \cdot 5}{99 \cdot 5}$$
$$= \frac{42}{495} - \frac{20}{495}$$
$$= \frac{22}{495} = \frac{2 \cdot \cancel{11}}{45 \cdot \cancel{11}} = \frac{2}{45}.$$

(c) $\dfrac{7}{22} + \dfrac{5}{12} - \dfrac{1}{30}$

$22 = 2 \cdot 11$, $12 = 2 \cdot 2 \cdot 3$, $30 = 2 \cdot 3 \cdot 5$, so that
$$\text{LCD} = 2 \cdot 2 \cdot 3 \cdot 5 \cdot 11 = 660.$$
Since
$$660 \div 22 = 30, \quad 660 \div 12 = 55, \quad \text{and} \quad 660 \div 30 = 22,$$

we have:

$$\frac{7}{22} + \frac{5}{12} - \frac{1}{30} = \frac{7 \cdot 30}{22 \cdot 30} + \frac{5 \cdot 55}{12 \cdot 55} - \frac{1 \cdot 22}{30 \cdot 22}$$
$$= \frac{210}{660} + \frac{275}{660} - \frac{22}{660}$$
$$= \frac{210 + 275 - 22}{660}$$
$$= \frac{463}{660}.$$

(d) $\dfrac{5}{12} + \left(-\dfrac{7}{18}\right) - \dfrac{2}{15}$

Since

$$12 = 2 \cdot 2 \cdot 3, \quad 18 = 2 \cdot 3 \cdot 3, \quad \text{and} \quad 15 = 3 \cdot 5,$$

the LCD $= 2 \cdot 2 \cdot 3 \cdot 3 \cdot 5 = 180$. Also,

$$180 \div 12 = 15, \quad 180 \div 18 = 10, \quad \text{and} \quad 180 \div 15 = 12,$$

so that

$$\frac{5}{12} + \left(-\frac{7}{18}\right) - \frac{2}{15} = \frac{5}{12} + \frac{-7}{18} - \frac{2}{15}$$
$$= \frac{5 \cdot 15}{12 \cdot 15} + \frac{(-7) \cdot 10}{18 \cdot 10} - \frac{2 \cdot 12}{15 \cdot 12}$$
$$= \frac{75}{180} + \frac{-70}{180} - \frac{24}{180}$$
$$= \frac{75 + (-70) - 24}{180}$$
$$= \frac{-19}{180}$$
$$= -\frac{19}{180}.$$

Multiplication and Division of Fractions

The **product** of two fractions is a fraction whose numerator is the product of the two numerators and whose denominator is the product of the two denominators. That is, if a/b and c/d are fractions, then

$$\frac{a}{c} \cdot \frac{c}{d} = \frac{a \cdot c}{b \cdot d}.$$

Thus, to **multiply** two fractions, we multiply the numerators of the two fractions to obtain the numerator of the product, and we multiply the denominators of the two fractions to obtain the denominator of the product. As a final step, the product is simplified, if possible.

Example 6 Find the following products and simplify.

(a) $\dfrac{4}{15} \cdot \dfrac{5}{12}$ (b) $\dfrac{7}{13} \cdot \left(\dfrac{-26}{21}\right)$

Solution

(a) $\dfrac{4}{15} \cdot \dfrac{5}{12} = \dfrac{4 \cdot 5}{15 \cdot 12} = \dfrac{20}{180} = \dfrac{1 \cdot \cancel{20}}{9 \cdot \cancel{20}} = \dfrac{1}{9}$

(b) $\dfrac{7}{13} \cdot \left(\dfrac{-26}{21}\right) = \dfrac{7 \cdot (-26)}{13 \cdot 21} = \dfrac{-182}{273} = \dfrac{(-2) \cdot \cancel{91}}{3 \cdot \cancel{91}} = -\dfrac{2}{3}$

In practice, it is often easier to divide out factors common to both the numerator and denominator before completing the multiplication. For example, the multiplications in Example 6 could be performed as follows:

$$\dfrac{4}{5} \cdot \dfrac{5}{12} = \dfrac{\overset{1}{\cancel{4}} \cdot \overset{1}{\cancel{5}}}{\underset{3}{\cancel{15}} \cdot \underset{3}{\cancel{21}}} = \dfrac{1 \cdot 1}{3 \cdot 3} = \dfrac{1}{9}$$

and

$$\dfrac{7}{13} \cdot \left(\dfrac{-26}{21}\right) = \dfrac{\overset{1}{\cancel{7}} \cdot \overset{-2}{\cancel{(-26)}}}{\underset{1}{\cancel{13}} \cdot \underset{3}{\cancel{21}}} = \dfrac{1 \cdot (-2)}{1 \cdot 3} = \dfrac{-2}{3} = -\dfrac{2}{3}.$$

In order to develop a rule for dividing fractions, recall that $A \div B = A/B = A \cdot 1/B$ is true for all real numbers A and B, where $B \neq 0$. Thus,

$$\dfrac{a}{b} \div \dfrac{c}{d} = \dfrac{a}{b} \cdot \dfrac{1}{c/d}.$$

Since,

$$\dfrac{1}{c/d} = \dfrac{1 \cdot d}{(c/d) \cdot d} = \dfrac{d}{c},$$

we see that the reciprocal of c/d os obtained by inverting c/d. Thus, we have

$$\boxed{\dfrac{a}{b} \div \dfrac{c}{d} = \dfrac{a}{b} \cdot \dfrac{d}{c} = \dfrac{a \cdot d}{b \cdot c}.}$$

Hence, to **divide** one fraction by another, invert the second fraction and multiply.

Example 7 Find each quotient and simplify.

(a) $\dfrac{2}{3} \div \dfrac{8}{9}$ (b) $\left(-\dfrac{6}{7}\right) \div \dfrac{15}{28}$

Solution

(a) $\dfrac{2}{3} \div \dfrac{8}{9} = \dfrac{2}{3} \cdot \dfrac{9}{8} = \dfrac{\overset{1}{\cancel{2}} \cdot \overset{3}{\cancel{9}}}{\underset{1}{\cancel{3}} \cdot \underset{4}{\cancel{8}}} = \dfrac{1 \cdot 3}{1 \cdot 4} = \dfrac{3}{4}$

(b) $\left(-\dfrac{6}{7}\right) \div \dfrac{15}{28} = \left(-\dfrac{6}{7}\right) \cdot \dfrac{28}{15}$

$= -\left(\dfrac{6}{7} \cdot \dfrac{28}{15}\right) = -\dfrac{\overset{2}{\cancel{6}} \cdot \overset{4}{\cancel{28}}}{\underset{1}{\cancel{7}} \cdot \underset{5}{\cancel{15}}} = -\dfrac{2 \cdot 4}{1 \cdot 5} = -\dfrac{8}{5}$

The order of operations discussed in Section 0.1 can now be applied to fractions.

In Examples 8–10, perform the indicated operations and simplify.

Example 8 $\quad \dfrac{2}{3} \cdot \dfrac{9}{8} + \dfrac{4}{5}$

Solution Following the rules for the order of operations, we perform the multiplication first, and then the addition. Thus,

$$\dfrac{2}{3} \cdot \dfrac{9}{8} + \dfrac{4}{5} = \dfrac{\overset{1}{2} \cdot \overset{3}{9}}{\underset{1}{3} \cdot \underset{4}{8}} + \dfrac{4}{5}$$

$$= \dfrac{3}{4} + \dfrac{4}{5}$$

$$= \dfrac{3 \cdot 5}{4 \cdot 5} + \dfrac{4 \cdot 4}{5 \cdot 4} = \dfrac{15}{20} + \dfrac{16}{20}$$

$$= \dfrac{15 + 16}{20} = \dfrac{31}{20}.$$

Example 9 $\quad \dfrac{3}{4} \div \dfrac{1}{2} \cdot \dfrac{5}{9}$

Solution Following the rules for the order of operations, we perform the division and the multiplication from left to right.

$$\dfrac{3}{4} \div \dfrac{1}{2} \cdot \dfrac{5}{9} = \dfrac{3}{4} \cdot \dfrac{2}{1} \cdot \dfrac{5}{9} = \dfrac{3 \cdot \overset{1}{2}}{\underset{2}{4} \cdot 1} \cdot \dfrac{5}{9}$$

$$= \dfrac{3}{2} \cdot \dfrac{5}{9}$$

$$= \dfrac{\overset{1}{3} \cdot 5}{2 \cdot \underset{3}{9}}$$

$$= \dfrac{5}{6}.$$

Example 10 $\quad \dfrac{3}{5} \cdot \dfrac{2}{3} + \dfrac{5}{8} \div \dfrac{3}{4}$

Solution

$$\dfrac{3}{5} \cdot \dfrac{2}{3} + \dfrac{5}{8} \div \dfrac{3}{4} = \dfrac{3}{5} \cdot \dfrac{2}{3} + \dfrac{5}{8} \cdot \dfrac{4}{3}$$

$$= \dfrac{3 \cdot 2}{5 \cdot 3} + \dfrac{5 \cdot 4}{8 \cdot 3}$$

$$= \dfrac{2}{5} + \dfrac{5}{6}$$

$$= \dfrac{2 \cdot 6}{5 \cdot 6} + \dfrac{5 \cdot 5}{6 \cdot 5}$$

$$= \dfrac{12}{30} + \dfrac{25}{30}$$

$$= \dfrac{37}{30}.$$

CHAPTER 0 THE FUNDAMENTAL OPERATIONS OF REAL NUMBERS

PROBLEM SET 0.6

In problems 1–12, reduce each fraction to lowest terms.

1. $\dfrac{15}{20}$
2. $\dfrac{9}{12}$
3. $\dfrac{-12}{16}$
4. $\dfrac{-18}{42}$
5. $\dfrac{-24}{-36}$
6. $\dfrac{32}{-40}$
7. $\dfrac{13}{39}$
8. $-\dfrac{132}{99}$
9. $-\dfrac{14}{56}$
10. $\dfrac{-91}{-28}$
11. $\dfrac{120}{45}$
12. $\dfrac{63}{144}$

In problems 13–26, perform the additions or subtractions of like fractions and simplify the results.

13. $\dfrac{2}{7} + \dfrac{3}{7}$
14. $\dfrac{5}{6} + \dfrac{3}{6}$
15. $\dfrac{9}{11} - \dfrac{3}{11}$
16. $\dfrac{3}{18} - \dfrac{1}{18}$
17. $\left(-\dfrac{4}{5}\right) + \left(-\dfrac{3}{5}\right)$
18. $\left(-\dfrac{2}{3}\right) - \left(-\dfrac{7}{3}\right)$
19. $\dfrac{12}{13} - \left(-\dfrac{6}{13}\right)$
20. $\left(-\dfrac{14}{15}\right) + \dfrac{4}{15}$
21. $\left(-\dfrac{17}{24}\right) - \dfrac{1}{24}$
22. $\left(-\dfrac{11}{30}\right) - \left(-\dfrac{6}{30}\right)$
23. $\dfrac{3}{5} + \dfrac{1}{5} - \dfrac{2}{5}$
24. $\left(-\dfrac{3}{8}\right) + \dfrac{5}{8} - \dfrac{1}{8}$
25. $\left(-\dfrac{3}{10}\right) - \dfrac{1}{10} + \dfrac{7}{10}$
26. $\dfrac{7}{12} + \left(-\dfrac{5}{12}\right) - \left(-\dfrac{1}{12}\right)$

In problems 27–50, perform the additions or subtractions of unlike fractions and simplify the results.

27. $\dfrac{3}{4} + \dfrac{2}{5}$
28. $\dfrac{4}{7} + \dfrac{3}{5}$
29. $\dfrac{5}{7} - \dfrac{3}{5}$
30. $\dfrac{8}{11} - \dfrac{3}{4}$
31. $\dfrac{11}{12} + \dfrac{5}{8}$
32. $\dfrac{13}{24} + \dfrac{17}{40}$
33. $\dfrac{7}{16} - \dfrac{3}{20}$
34. $\dfrac{11}{26} - \dfrac{5}{39}$
35. $\dfrac{17}{30} + \dfrac{8}{21}$
36. $\dfrac{7}{72} + \dfrac{2}{27}$
37. $\dfrac{25}{54} - \dfrac{13}{40}$
38. $\dfrac{14}{45} - \dfrac{13}{30}$
39. $\dfrac{3}{26} + \dfrac{5}{39}$
40. $\dfrac{3}{34} + \dfrac{1}{51}$
41. $\dfrac{5}{42} - \dfrac{6}{35}$
42. $\dfrac{17}{57} - \dfrac{15}{38}$
43. $\dfrac{2}{3} + \dfrac{7}{8} - \dfrac{5}{6}$
44. $\dfrac{4}{5} + \dfrac{5}{6} - \dfrac{3}{10}$
45. $\dfrac{4}{5} - \dfrac{7}{12} + \dfrac{11}{30}$
46. $\dfrac{3}{7} - \dfrac{1}{5} - \dfrac{1}{4}$
47. $\dfrac{3}{5} - \dfrac{2}{3} - \left(-\dfrac{3}{10}\right)$
48. $\dfrac{4}{11} + \left(-\dfrac{3}{8}\right) - \dfrac{5}{22}$
49. $\dfrac{7}{24} + \left(-\dfrac{5}{12}\right) - \dfrac{11}{16}$
50. $\dfrac{13}{30} - \left(-\dfrac{8}{27}\right) - \dfrac{2}{45}$

In problems 51–80, perform the indicated operations and simplify the results.

51. $\dfrac{3}{7} \cdot \dfrac{14}{15}$
52. $\dfrac{4}{9} \cdot \dfrac{12}{5}$
53. $\dfrac{5}{6} \cdot \left(\dfrac{-18}{25}\right)$
54. $\dfrac{3}{8} \cdot \left(-\dfrac{16}{21}\right)$
55. $\dfrac{3}{5} \div \dfrac{9}{20}$
56. $\dfrac{15}{16} \div \dfrac{35}{24}$
57. $\left(\dfrac{-16}{21}\right) \div \dfrac{8}{7}$
58. $\dfrac{8}{17} \div \left(\dfrac{-3}{34}\right)$
59. $\dfrac{100}{121} \cdot \dfrac{22}{45}$
60. $\dfrac{72}{75} \cdot \left(-\dfrac{25}{36}\right)$
61. $\left(-\dfrac{8}{9}\right) \div \left(-\dfrac{12}{5}\right)$
62. $\left(\dfrac{-26}{35}\right) \cdot \left(\dfrac{14}{-39}\right)$
63. $\dfrac{72}{45} \cdot \left(\dfrac{2}{-27}\right)$
64. $\dfrac{108}{121} \div \dfrac{81}{77}$
65. $\left(\dfrac{-14}{27}\right) \cdot \left(\dfrac{18}{-21}\right)$
66. $\left(\dfrac{-42}{25}\right) \div \left(\dfrac{-21}{35}\right)$
67. $\dfrac{25}{64} \div \left(\dfrac{-15}{16}\right)$
68. $\left(-\dfrac{58}{13}\right) \cdot \dfrac{39}{29}$
69. $\left(-\dfrac{76}{121}\right) \div \left(\dfrac{57}{-132}\right)$
70. $\dfrac{121}{144} \div \dfrac{55}{72}$
71. $\dfrac{3}{4} \cdot \dfrac{8}{9} + \dfrac{5}{6}$
72. $\left(-\dfrac{4}{5}\right) \div \dfrac{3}{10} - \dfrac{7}{15}$
73. $\dfrac{2}{3} \div \dfrac{4}{9} \cdot \dfrac{10}{12}$
74. $\dfrac{4}{7} \cdot \dfrac{21}{2} \div \dfrac{5}{12}$
75. $\dfrac{1}{2} \cdot \dfrac{3}{4} + \dfrac{2}{3} \cdot \dfrac{1}{4}$
76. $\dfrac{3}{5} \div \dfrac{6}{7} - \dfrac{2}{3} \cdot \dfrac{5}{4}$
77. $\dfrac{11}{12} - \dfrac{3}{8} \div \dfrac{6}{7}$
78. $\dfrac{5}{9} \cdot \left(-\dfrac{3}{4}\right) \div \dfrac{2}{3} + \dfrac{3}{16}$
79. $\dfrac{4}{5} \div \dfrac{2}{15} - \dfrac{3}{8} \cdot \dfrac{4}{9}$
80. $\dfrac{3}{7} - \left(-\dfrac{4}{5}\right) \div \dfrac{2}{3} \div \dfrac{3}{5}$

81. In an algebra class, $\frac{1}{9}$ of the students missed no classes during a semester, $\frac{2}{7}$ or the students missed only one class, and $\frac{1}{5}$ of the students missed only two classes. What fraction of the students missed fewer than three classes?

82. A biologist has $\frac{5}{6}$ liter of a certain solution. If $\frac{3}{4}$ liter of water is added to dilute the solution, how much liquid is contained in the bottle?

83. A hardware store sells wrenches with a set of 16 sockets that range in diameter from $\frac{3}{16}$ inch to $\frac{7}{8}$ inch. What is the difference in diameter between the largest and the smallest socket?

84. A spray painter sprays $\frac{19}{20}$ of a fluid ounce of paint per second. How much paint is sprayed in $\frac{7}{9}$ of a second?

85. A homeowner wishes to install modular lighting in her living room along one wall that is $\frac{21}{2}$ feet long. If 5 lamps are to be installed, one at each end of the wall and all an equal distance apart, find the distance between each lamp.

0.7 Decimal Place Value and Rounding Off

For thousands of years, fractions were the only way to express a part of a whole. Then, about A.D. 1200 Arab and Jewish traders brought a new numbering system to Europe. This system used **decimals.**

This system is just like the usual base-10 number system except that it includes fractions. Each column has its own value. The **decimal point** is used as a reference point to separate the whole numbers (on the left) from the fractions (on the right).

Figure 1 shows the column values. To the left of the decimal point are the usual whole-number place values: 1, 10, 100, 1000, and so on. To the right of the decimal point are the **fractional place values:** $\frac{1}{10}, \frac{1}{100}, \frac{1}{1,000}, \frac{1}{10,000}$, and so on.

Ten-thousands	Thousands	Hunderds	Tens	Ones	Decimal Point	Tenths	Hundredths	Thousandths	Ten-thousandths	Hundred-thousandths	Millionths
10000	1000	100	10	1	.	$\frac{1}{10}$	$\frac{1}{100}$	$\frac{1}{1000}$	$\frac{1}{10000}$	$\frac{1}{100,000}$	$\frac{1}{1,000,000}$
	2	7	8	3	.	6	5	1	9		

Figure 1 Decimal System

As with whole numbers, each digit has a place value. For example, in the number shown the 6 has a value of $\frac{6}{10}$, the 5 has a value of $\frac{5}{100}$, and so on.

This number, 2783.6519, can be written out in an expanded form.

$$
\begin{aligned}
2783.6519 = \quad & 2 \times 1000 & = & \ 2000 \\
+ & 7 \times 100 & = & \ 700 \\
+ & 8 \times 10 & = & \ 80 \\
+ & 3 \times 1 & = & \ 3 \\
+ & 6 \times \frac{1}{10} & = & \ \frac{6000}{10,000} \\
+ & 5 \times \frac{1}{100} & = & \ \frac{500}{10,000} \\
+ & 1 \times \frac{1}{1000} & = & \ \frac{10}{10,000} \\
+ & 9 \times \frac{1}{10,000} & = & \ \frac{9}{10,000} \\
& & & 2783\frac{6519}{10,000}
\end{aligned}
$$

From *Basic Mathematics* (Revised Printing) by Howard Silver. © 1997 by Kendall/Hunt Publishing Company. Used with permission.

To determine the value of a digit in a decimal number:
1. Determine the column values starting at the decimal.
2. Multiply the digit by the column value.

Example 1   In 34.8721, what is the value of the 2?

Solution   We first determine the column values.

10	1	$\frac{1}{10}$	$\frac{1}{100}$	$\frac{1}{1000}$	$\frac{1}{10,000}$
3	4	8	7	2	1

← Determine column values

value of $2 = 2 \times \frac{1}{1000} = \frac{2}{1000}$ ← Multiply digit by column value

Example 2   In 2.9082, what is the value of the 9?

Solution   We first determine the column values.

1	$\frac{1}{10}$	$\frac{1}{100}$	$\frac{1}{1000}$	$\frac{1}{10,000}$
2	9	0	8	2

← Determine column values

value of $9 = 9 \times \frac{1}{10} = \frac{9}{10}$ ← Multiply digit by column value

Example 3   In 30.852, what is the value of the 5?

Solution   We first determine the column values.

10	1	$\frac{1}{10}$	$\frac{1}{100}$	$\frac{1}{1000}$
3	0	8	5	2

← Determine column values

value of $5 = 5 \times \frac{1}{100} = \frac{5}{100}$ ← Multiply digit by column value

Practice Problems

See Examples 1 to 3

1. In 23.759, what is the value of the 7?
2. In 48.9021, what is the value of the 2?
3. In 1.41456, what is the value of the 5?
4. In 467.92, what is the value of the 9?
5. In 9.08321, what is the value of the 2?

6. In 2950.36, what is the value of the 6?
7. In 1.234567, what is the value of the 7?
8. In 360.74832, what is the value of the 8?
9. In 4209.39275, what is the value of the 7?
10. In 10.23965, what is the value of the 2?

Practice Problem Answers

1. $\dfrac{7}{10}$ 2. $\dfrac{2}{1000}$ 3. $\dfrac{5}{10{,}000}$ 4. $\dfrac{9}{10}$ 5. $\dfrac{2}{10{,}000}$

6. $\dfrac{6}{100}$ 7. $\dfrac{7}{1{,}000{,}000}$ 8. $\dfrac{8}{1000}$ 9. $\dfrac{7}{10{,}000}$ 10. $\dfrac{2}{10}$

Sometimes, we have a decimal that has more digits than we need. When we simplify, we **round off**. For instance, 3.14159 is between two simpler numbers:

$$3.1 \rightarrow 3.10000$$
$$3.14159 \rightarrow 3.14159$$
$$3.2 \rightarrow 3.20000$$

Since our number is closer to 3.10000, so we round 3.14159 to 3.1.

> To round off decimal places:
> 1. Determine what place is to be rounded.
> 2. Go one extra place to the right.
> 3a. If this digit is 4 or less (0, 1, 2, 3 or 4), drop this digit and all others to the right. (Round down.)
> 3b. If this digit is 5 or more (5, 6, 7, 8 or 9), make the rounded place one more, and drop all digits to the right. (Round up.)

Example 4   Round 85.07276 to two decimal places.

Solution   85.07276   ← *Determine place to be rounded*
                ↑
    85.07⌑2⌑76   ← *Go one place right*
              ↑
    85.07   ← *Round down since digit is 2*

Example 5   Round 17.01392 to the nearest thousandth.

Solution   The thousandth's place is the third decimal place.

    17.01392   ← *Determine place to be rounded*
         ↑
    17.013⌑9⌑2   ← *Go one place right*
             ↑
    17.014   ← *Round up since digit is 9*

Example 6   Round 1.95046 to the nearest tenth.

Solution   The tenth's place is the first decimal place.

1.95046   ← *Determine place to be rounded*
↑
1.9⬚5⬚046   ← *Go one place right*
↑
2.0   ← *Round up since digit is 5*

When we round 1.9 up, we get 2.0.

Practice Problems

See Examples 4 to 6

1. Round 2.753 to two decimal places.
2. Round 43.7582 to one decimal place.
3. Round 47.90327 to three decimal places.
4. Round 3.27451 to two decimal places.
5. Round 4.18653 to one decimal place.
6. Round 567.89012 to four decimal places.
7. Round 567.83425 to three decimal places.
8. Round 45.938 to the nearest tenth.
9. Round 123.4567 to the nearest thousandth.
10. Round 38.29048 to the nearest hundredth.
11. Round 8438.86734 to the nearest ten.
12. Round 26.36211 to the nearest one.
13. Round 1574.5901 to the nearest hundred.
14. Round 406,947.329747 to the nearest thousand.

Practice Problem Answers

1. 2.75	**2.** 43.8	**3.** 47.903	**4.** 3.27	**5.** 4.2
6. 567.8901	**7.** 567.834	**8.** 45.9	**9.** 123.457	**10.** 38.29
11. 8440	**12.** 26	**13.** 1600	**14.** 407,000	

When working with decimals, it is often useful to have all the decimals in a particular problem have the same number of decimal places. Fortunately, this is easy to do by simply attaching zeros on the right.

Any decimal number can have extra zeros inserted on the right.

To write a whole number as a decimal:
1. Put a decimal point to the right of the number.
2. Insert as many zeros to the right as needed.

Example 7 (a) 2.6 = 2.60 = 2.600 = 2.6000 = ...
(b) 28.43 = 28.430 = 28.4300 = ...
(c) 36 = 36.0 = 36.00 = 36.000 = ...
(d) 500 = 500.0 = 500.00 = 500.000 = ...

We can use this to help us compare the sizes of decimals. For instance, consider 0.093 and 0.11. Let us rewrite them with the same number of decimal places. This is like finding the LCD. Recall that < means less than, as in 10 < 15.

$$0.093 \qquad 0.11$$
$$\downarrow \qquad \downarrow$$
$$0.093 \qquad 0.110 \qquad \leftarrow \textit{Write both decimals with three places}$$
$$\downarrow \qquad \downarrow$$
$$\frac{93}{1000} \quad < \quad \frac{110}{1000} \qquad \leftarrow \textit{Think of each as a fraction and compare}$$
$$\uparrow$$
$$\boxed{\text{larger}}$$

So, 0.093 < 0.11.

We do not have to write the decimals as fractions. By writing them with the same number of places, we can compare the digits (as if the decimal point were not there).

To compare two (or more) decimals:
1. If necessary, fill in zeros so that all decimals have the same number of places.
2. Temporarily ignore the decimal point.
3. Compare as you would compare whole numbers.

Example 8 Which decimal is larger, 0.5 or 0.444?

Solution We write both of these with three places and then compare.

$$0.5 \qquad 0.444 \qquad \leftarrow \textit{Write both decimals with three places}$$
$$\downarrow \qquad \downarrow$$
$$0.500 \quad > \quad 0.444 \qquad \leftarrow \textit{Think of each as a fraction and compare}$$
$$\downarrow$$
$$\boxed{\text{larger}}$$

So, 0.5 > 0.444.

Example 9 Write the following decimals in order (smallest to largest): 0.25, 0.3, 0.088, 0.2242.

Solution We write all these decimals with the same number of places, four.

$$0.25 \quad 0.3 \quad 0.088 \quad 0.2242 \quad \leftarrow \quad \textit{Write all with four places}$$
$$\downarrow \quad \downarrow \quad \downarrow \quad \downarrow$$
$$0.2500 \quad 0.3000 \quad 0.0880 \quad 0.2242 \quad \leftarrow \quad \textit{Compare like whole numbers;}$$
$$\uparrow \quad \uparrow \quad \uparrow \quad \uparrow \quad \quad ① = \textit{smallest}$$
$$③ \quad ④ \quad ① \quad ② \quad \quad ④ = \textit{largest}$$

$0.0880 < 0.2242 < 0.2500 < 0.3000$
(smallest)   (Largest)

Like fractions, decimals generally appear on the number line between the whole number markings. For instance, consider $A = 1.4$ below.

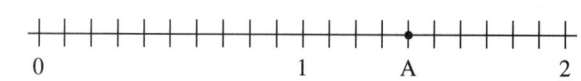

Practice Problems

For the following sets of decimals, determine the larger.

See Examples 7 to 9

1. 0.7, 0.66
2. 0.03, 0.004
3. 1.2, 1.09
4. 2.02, 2.007
5. 0.001, 0.0012
6. 0.041, 0.0288

Put the following sets of decimals into order (smallest to largest).

7. 0.5, 0.529, 0.4831
8. 2.1, 2.09, 2.111
9. 0.8, 0.88, 0.888
10. 0.1, 0.01, 0.001, 0.0001

Practice Problem Answers

1. 0.7
2. 0.03
3. 1.2
4. 2.02
5. 0.0012
6. 0.041
7. 0.4831, 0.5, 0.529
8. 2.09, 2.1, 2.111
9. 0.8, 0.88, 0.888
10. 0.0001, 0.001, 0.01, 0.1

PROBLEM SET 0.7

Determine the values of the indicated digits.
See Examples 1 to 3

1. In 34.928, what is the value of the 2?
2. In 7.49201, what is the value of the 2?
3. In 39.1852, what is the value of the 1?
4. In 869.03815, what is the value of the 5?

Round the following numbers to the indicated accuracy.
See Examples 4 to 6

5. Round 4.9372 to a one-place decimal.
6. Round 76.03957 to a two-place decimal.
7. Round 1.38592 to a three-place decimal.
8. Round 47.93752 to the nearest tenth.
9. Round 5.029456 to the nearest thousandth.
10. Round 579.0846523 to the nearest hundredth.
11. Round 4673.8 to the nearest hundred.
12. Round 23,694.73 to the nearest ten thousand.

For the following sets of decimals, determine the largest.
See Examples 8 and 9

13. 0.67, 0.666
14. 3.15, 3.14159

15. 0.02, 0.2, 0.0002
16. 0.9, 0.99, 0.999, 0.9999.

Business application

17. A company's gross revenue for a year was $12,732,934.92. Found this number to the nearest hundred thousand dollars.

Technical application

18. A pipe has an inside diameter of 3.067 inches. Round this to the nearest tenth of an inch.

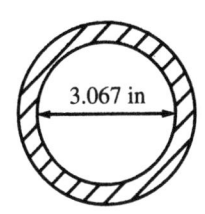

Health application

19. A nurse calculates that a patient needs 0.5555 milliliter of atropine sulfate. Round this to the nearest hundredth of a milliliter.
20. When writing checks, most amounts are two-place decimals such as $438.19. This would be written "four hundred thirty-eight and $\frac{19}{100}$." Fill in the following imaginary checks.

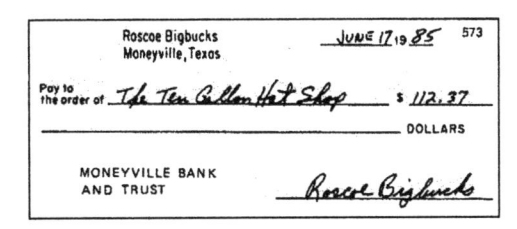

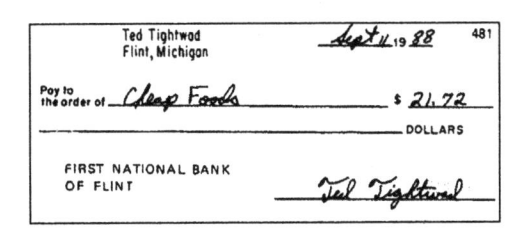

0.8 Addition and Subtraction of Decimals

When we **add and subtract decimals,** we line up the decimal points. It is usually helpful if all the numbers contain the same number of decimal places.

> To add (or subtract) decimals:
> 1. Line up the decimal points in a column.
> 2. If necessary, fill in zeros so that all decimals have the same number of places.
> 3. Add (or subtract) the numbers as if they were whole numbers.
> 4. The decimal point for the answer is just below the other decimal points.

Example 1 Bob is a big spender and is now up to his ears in bills. Today he is going to write checks for $125, $23.69, $12.38, and $89.47. What is the total?

Solution This is a whole-and-parts problem. We have all the parts and want the whole. The word *total* almost always means add. Let us add these figures, filling in zeros so that they all have two decimal places.

$125.00 ← *Line up decimal points, filling in zeros*
  23.69
  12.38 ← *Add as if they were whole numbers*
 +89.47
─────────
$270.54 ← *Put decimal point below others*

Example 2 Add 1.6001 + 21 + 9.04 + 8.731 on a hand calculator.

Solution We simply enter these decimals just as they appear and use the + key for addition.

PRESS	DISPLAY	COMMENT
1.6001	1.6001	1.6001
⊞ 21	21.	plus 21
⊞ 9.04	9.04	plus 9.04
⊞ 8.731	8.731	plus 8.731
⊟	40.3711	Answer

Example 3 Find the perimeter of the triangle shown.

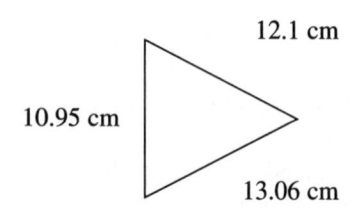

12.1 cm

10.95 cm

13.06 cm

Solution The perimeter of a triangle is the sum of all the lengths of the sides.

$P = 10.95 + 12.1 + 13.06$ ← *Perimeter = sum of sides*
$= 36.11$ ← *Simplify*

Why Does It Work?
When we add decimals, we are really adding fractions. And when we line up decimal points and fill in zeros, we are really finding the common denominator. Consider 2.3 + 3.17 + 1.214.

$$2.3 = 2\frac{3}{10} = 2\frac{300}{1000}$$

$$3.17 = 3\frac{17}{100} = 3\frac{170}{1000}$$

$$+\ 1.214 = 1\frac{214}{1000} = \frac{1\frac{214}{1000}}{6\frac{684}{1000}}$$

Practice Problems

Add the following decimals.

See Examples 1 to 3

1.   28.83
   + 8.82

2.   4.311
   + 27.41

3.   20.2
   + 6.904

4.   72.423
   + 115.7

5. Add 75.41 + 3.767 + 47.334.
6. Add 8.918 + 41.39 + 0.199.
7. Add 5440 + 5955.67 + 6771.7 + 7026.
8. Add 12.28 + 5.37 + 2.158 + 811.
9. Larry writes checks for $19.73, $5.68, $27.15, $51.66, and $108.09. How much is the total?
10. Sally buys the following items at the store, milk, $1.59; cookies, $0.83; bread, $0.67; orange juice, $0.65; steak, $2.78. What is the total?
11. One week, Howie runs the following distances: 1.5, 2.35, 0.85, 1.8, 1.15, and 2.9 miles. What is his total for the week?
12. One month, Lynn records her gasoline purchases: 12.1, 13.7, 9, 12.25, and 8.15 gallons. How much gasoline did she buy during the month?
13. Find the perimeter of the triangle at the left.
14. Angie weighs 52.3 kilograms. Her clothes weigh 1.95 kilograms. How much does she weigh fully dressed?
15. Bob needs 0.9 gallons of paint for the kitchen, 1.3 for his bedroom, 1.15 for his child's room, 1.65 for the living room, and 0.85 for the dining room. How much paint does he need?

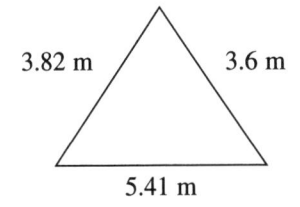

Triangle: 3.82 m, 3.6 m, 5.41 m

Practice Problem Answers

1. 37.65	2. 31.721	3. 27.104
4. 188.123	5. 126.511	6. 50.507
7. 25,193.37	8. 830.808	9. 212.31
10. 6.52	11. 10.55	12. 55.2
13. 12.83	14. 54.25	15. 5.85

The following **subtraction problems** use the same rule used for addition: *line up the decimal points.* As with addition, we are really finding the common denominator when we subtract like this.

Example 4 Subtract 10.02 − 7.593.

Solution We line up the decimal point and subtract.

$$\begin{array}{r} 10.020 \\ -\ 7.593 \\ \hline 2.427 \end{array}$$ ← *Line up decimals (fill in zero)*

← *Put decimal point below others*

Example 5 Subtract 0.00635 − 0.0009 on a hand calculator.

Solution We enter the decimals as they appear and use the − key.

PRESS	DISPLAY	COMMENT
.00635	0.00635	0.00635
⊟ .0009	0.0009	minus 0.0009

PRESS	DISPLAY	COMMENT
=	0.00545	Answer

Example 6 A store has the following sales for Monday through Thursday. Their goal for the week (Monday to Friday) is $20,000. How much more do they need to sell on Friday to make their goal?

Monday: $3975.17
Tuesday: $3611.52
Wednesday: $4023.85
Thursday: $3608.51

Solution This is a whole-and-parts problem. We have the whole ($20,000) and all the parts, except Friday. So, we subtract.

Friday = 20,000 − 3975.17 − 3611.52 − 4023.85 − 3608.51
        = 4780.95

Thus, their goal for Friday is $4780.95.

Example 7 Tina has $231.62 in her checking account. She writes a $51.78 check for a new coat. How much does she have left in her account?

Solution This is a change problem. The new balance = old − decrease. We take the original balance and subtract the amount of the check.

New = Old − Decrease
    = 231.62 − 51.78
    = $179.84

Practice Problems

Subtract the following decimals.

See Examples 4 to 7

1. 56.64
   − 1.41

2. 8.62
   − 7.013

3. 62.53
   − 7.228

4. 96.612
   − 71.36

5. Subtract 577.9 − 158.97.
6. Subtract 9.463 − 0.5352.
7. Subtract 7086 − 7.698.
8. Subtract 9400 − 5373.88.
9. Marian has $212.18 in her checking account. She writes a check for $87.53. How much does she have left in her account?
10. Ron has a board 1.73 meters long. He saws 0.8 meters off the end. How much does he have left?
11. The distance between Squirrel Valley and Raccoon Bluff is 25.3 miles. John has cycled 6.9 miles of the trip. How much does he have left?
12. Jeni has $521.53 in the bank. She withdraws $273.16 and $171.42. How much does she have left in the bank?
13. Jane buys a $39.95 broiler, but gets a $6.35 discount. What does she pay?

Practice Problem Answers

1. 55.23
2. 1.607
3. 55.302
4. 25.252
5. 418.93
6. 8.9278
7. 7078.302
8. 4026.12
9. 124.65
10. 0.93
11. 18.4
12. 76.95
13. 33.60

Since addition and subtraction are the opposites of each other, we can rewrite these examples as:

$$n - 4 = 9 \quad \text{means} \quad \begin{array}{r} n - 4 = 9 \\ + 4 + 4 \\ \hline n = 13 \end{array}$$

$$n + \frac{1}{2} = 6 \quad \text{means} \quad \begin{array}{r} n + \frac{1}{2} = 6 \\ -\frac{1}{2} -\frac{1}{2} \\ \hline n = 5\frac{1}{2} \end{array}$$

Similarly, with decimals, we add to invert subtraction, and vice versa.

Example 8 Find n: $n - 12.3 = 41.73$.

Solution Since addition is the opposite of subtraction, we add 12.3 to both sides of the equation to find n.

$$n - 12.3 = 41.73 \quad \text{means} \quad \begin{aligned} n &= 41.73 + 12.3 \\ &= 54.03 \end{aligned}$$

Example 9 Find n: $n - 0.02 = 0.735$.

Solution We invert this subtraction by adding 0.02 to both sides of the equation.

$$n - 0.02 = 0.735 \quad \text{means} \quad \begin{aligned} n &= 0.735 + 0.02 \\ &= 0.755 \end{aligned}$$

Example 10 Find n: $n + 6.29 = 17.1$.

Solution We invert this addition by subtracting 6.29 from both sides of the equation.

$$n + 6.29 = 17.1 \quad \text{means} \quad \begin{aligned} n &= 17.1 - 6.29 \\ &= 10.81 \end{aligned}$$

Example 11 What number added to 123.4 is 259.15?

Solution We let n be the number that we are looking for. The words *added* to mean +, and *is* means =.

$$\underbrace{\text{What number}}_{n} \quad \underbrace{\text{added to}}_{+} \quad \underbrace{123.4 \text{ is } 259.15?}_{123.4 = 259.15} \quad \leftarrow \textbf{\textit{Translate}}$$

Now we solve this by subtracting 123.4 from both sides of the equation:

$$n + 123.4 = 259.14 \quad \text{means} \quad \begin{aligned} n &= 259.15 - 123.4 \\ &= 135.75 \end{aligned}$$

Thus, the number is 135.75.

Practice Problems

Find n or x in the following problems.

See Examples 8 to 11

1. $n - 2 = 7.8$
2. $x - 4.8 = 11.1$
3. $n - 12.7 = 25.45$
4. $n - 28.03 = 41.75$
5. $x - 42.5 = 82$
6. $x - 0.071 = 1.52$
7. $n + 5 = 8.3$
8. $n + 2.3 = 8.1$
9. $n + 4.8 = 9.72$
10. $x + 6.05 = 10.7$
11. What number added to 307.8 is 1000?
12. What number added to 0.035 is 2.7?

Practice Problem Answers

1. 9.8
2. 15.9
3. 38.15
4. 69.78
5. 124.5
6. 1.591
7. 3.3
8. 5.8
9. 4.92
10. 4.65
11. 692.2
12. 2.665

PROBLEM SET 0.8

Perform the indicated operations.
See Examples 1 to 7

1. $2.3 + 4.69$
2. $5.92 + 47.092$
3. $43 + 59.2 + 58.01$
4. $92.92 + 47.934 + 57$
5. $579 + 46.01 + 2.129$
6. $5.029 + 29.9482 + 3.38 + 0.923$
7. $47.93 - 32.958$
8. $34.2 - 19.048$
9. $689 - 483.23$
10. $1000 - 567.473$
11. $57.099 - 12.68$
12. $10.0001 - 9.99998$

Health applications

13. A laboratory technician records the white-blood-cell count (thousands of cells per cubic millimeter of blood): 7.1; 11.3; 8.9; 7.3; 9.9; 5.9; 12.4; 5.1; 9.7; 16.8. Find the total for these counts.
14. A patient is to receive 0.05 grain of atropine sulfate. Thus far, she has received 0.035 grain of atropine sulfate. How much more of the medication does she need?

Electrical application

15. The current in a parallel circuit is the total of the currents through each branch. Find the total current in the circuit sketched below.

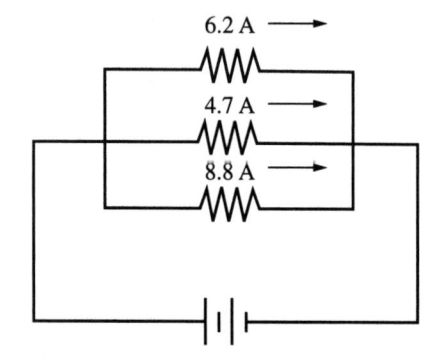

Business application

16. A company's revenue (gross) for the year was $213,571.48. Its expenses (costs) were $109,583.69. Find the net profit (revenue − cost).

17. The daily sales figures for three salespeople in one week are as follows:

Day of week	Kotowski	Dohner	Humbert
Monday	$236.93	$312.57	$274.68
Tuesday	210.57	297.47	205.79
Wednesday	296.04	310.39	346.92
Thursday	324.79	362.68	402.58
Friday	329.94	432.58	389.46

(a) Find the sales totals for each person (Add down.)

(b) Find the sales total for each day. (Add across.)

Technical applications

18. The overall length of a bushing is to be 1.25 centimeters with a tolerance of 0.012 centimeter.

(a) Find the greatest allowable length by adding the length and the tolerance.

(b) Find the smallest allowable length by subtracting the length minus the tolerance.

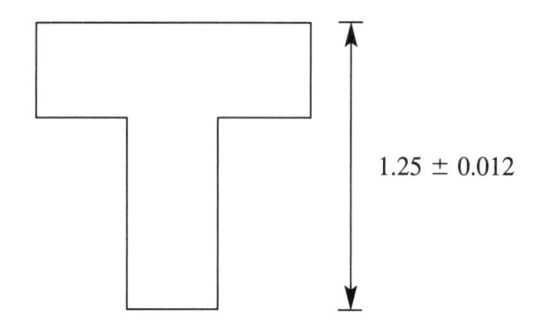

1.25 ± 0.012

Psychology application

19. While using a chi-square test, a psychologist finds that she must add the following numbers. Find the sum.

f_0^2/f_e
16.333
12.000
10.083
6.750
16.333

Consumer applications

20. The Morisons start the month with a balance in their checking account of $126.94. Use the checks and deposits given below to find their balance at the end of the month.

		Checks	Deposits
Apr 1	Starting balance		$126.94
Apr 2	Deposit		432.59
Apr 4	C & K Foods	$64.38	
Apr 6	Phone company	12.68	
Apr 9	Gas company	13.59	
Apr 11	Goniff Oil Co.	38.93	
Apr 12	MasterDebt Charge	56.82	
Apr 14	C & K Foods	59.37	
Apr 15	Deposit		432.59
Apr 18	Nonsense Shop	32.66	
Apr 21	C & K Foods	72.59	
Apr 24	Auto Finance Co	145.68	
Apr 26	Safe Form Insurance	23.69	
Apr 29	Bite Mortgage Co.	342.57	

21. In the past month, Janet has put the following amounts of gasoline in her car: 12.3, 15.2, 15, 9.8, and 11.4 gallons. Find the total.

22. Matt buys a $20 pair of pants, and he gets a $3.75 discount. How much did he pay?

23. Sandy is putting floor moldings around the room shown below.

(a) Find the missing lengths, x and y.

(b) Find the total distance around the room.

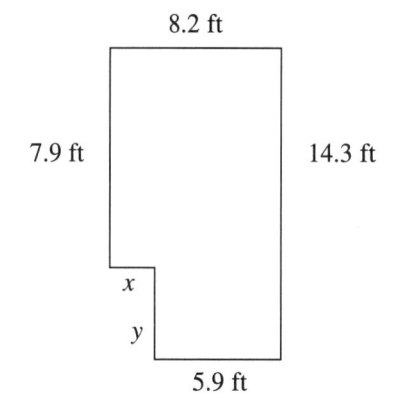

0.9 Multiplication of Decimals

Multiplying decimals is different from adding or subtracting them. Instead of lining up decimal points, we count decimal places. We start with the numbers 0.1, 0.01, 0.001, 0.0001, and so on, which we call the **decimal powers of 10.** Recall that as fractions, these are $\frac{1}{10}, \frac{1}{100}, \frac{1}{1000}, \frac{1}{10,000}$, and so on.

> To multiply a number by a decimal power of 10:
> 1. Count the number of decimal places in the decimal power of 10.
> 2. Move the decimal point of the other number that many places to the left.

Example 1 Multiply 0.01×6.23.

Solution The number 0.01 has two places, so move the decimal point two places.

0.01×6.23 ← *Count two places*

$= 006.23$ ← *Move two places left*

$= 0.0623$ ← *Fill in zeros*

Example 2 Multiply $0.001 \times 27{,}000$.

Solution $0.001 \times 27{,}000$ ← *Count three places*

$= 27000.$ ← *Move three places left*

$= 27$ ← *Simplify*

Practice Problems

Multiply the following decimals.
See Examples 1 and 2

1. 0.1×36
2. 0.01×572
3. $0.0001 \times 290{,}000$
4. 0.001×5900
5. $0.00001 \times 2{,}300{,}000$
6. 0.1×0.089

Practice Problem Answers

1. 3.6
2. 5.72
3. 29
4. 5.9
5. 23
6. 0.0089

> To multiply any two decimals:
> 1. Total the number of decimal places in both factors.
> 2. Multiply the digits as if they were whole numbers.
> 3. Move the decimal point in the answer to the left the total number of the places of the factors. (Fill in zeros, if necessary.)

0.9 MULTIPLICATION OF DECIMALS

Example 3 Multiply 8.3 × 6.9.

Solution Each factor has one decimal place, so the answer will have two decimal places.

$$\begin{array}{r} 8.3 \\ \times\, 6.9 \\ \hline 747 \\ 498 \\ \hline 5727 \end{array}$$ ← *One place*
← *One place*

← *Two places*

= 57.27

Example 4 Multiply 0.7 × 2.403 on a hand calculator.

Solution We enter the decimals as they appear and use the × key.

PRESS	DISPLAY	COMMENT
.7	0.7	0.7
☒ 2.403	2.403	times 2.403
▣	1.6821	Answer

Example 5 A school buys 4000 notebooks at $1.62 each. How much is the total cost?

Solution This is a whole-and-parts problem. Here we have 4000 equal parts, all of which cost $1.62. We could add the 4000 copies of $1.62, but it is much easier to multiply:

$$4000 \times 1.62 = 6480$$

Example 6 Karen goes to the supermarket and buys 2.78 pounds of bananas at $0.59 per pound, and 0.92 pounds of mushrooms at $2.59 per pound. How much does this cost?

Solution The word "at" indicates multiplication. We multiply to find the banana cost and the mushroom cost. Then we add to find the total cost.

Banana cost =   2.78 × 0.59 = 1.6402
Mushroom cost = 0.92 × 2.59 = 2.3828
Total Cost   4.0230

This would be rounded to the nearest penny, as $4.02. The nearest penny is always the second decimal place.

Example 7 Find the area of the rectangle shown.

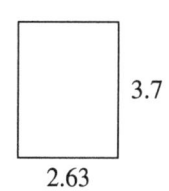

3.7

2.63

Solution The area of a rectangle is the length times the width. So we multiply 3.7 × 2.63.

$$\begin{aligned} A &= L \cdot W \\ &= (3.7) \cdot (2.63) \\ &= 9.731 \end{aligned}$$

Example 8 A patient is to receive four injections of a regular Iletin insulin solution. Each injection is 0.65 milliliter. How much total medicine is this?

Solution We find the total by multiplying 4 × 0.65.

$$\begin{array}{r} 0.65 \\ \times 4 \\ \hline 2.60 \end{array}$$

Thus the patient receives a total of 2.6 milliliters of medicine.

Why Does it Work?
When we count the decimal places in a multiplication problem, we are secretly working with fractions. Consider 3.9 × 2.73.

$$3.9 \times 2.73$$
$$= 3\frac{9}{10} \times 2\frac{73}{100} \quad \leftarrow \textit{Change to mixed number}$$
$$= \frac{39}{10} \times \frac{273}{100} \quad \leftarrow \textit{Change to improper fraction}$$
$$= \frac{10{,}647}{1000} \quad \leftarrow \textit{Multiply fractions}$$
$$= 10\frac{647}{1000} \quad \leftarrow \textit{Change to mixed number}$$
$$= 10.647 \quad \leftarrow \textit{Change to decimal}$$

Notice how we multiply 39 × 273 and multiply the denominators. When we do this as a decimal problem, we add 1 + 2 = 3 decimal places. This stands for the denominators, 10 × 100 = 1000.

Practice Problems

Multiply the following decimals.
See Examples 3 to 8

1. 1.2
   × 3.7

2. 99
   × 0.8

3. 1.25
   × 6.66

4. 0.84
   × 6.3

5. Multiply 80,000 × 5.9.
6. Multiply 0.00046 × 0.0005.
7. Multiply 3.21 × 7.905.
8. Multiply 69.2 × 1.0009.
9. A school must buy 27 gallons of milk at $1.69 a gallon. How much will this cost?
10. Lee buys 13.6 gallons of gasoline at $1.44 a gallon. How much does the gas cost him?
11. A room measures 3.2 meters by 4.6 meters. What is the area?

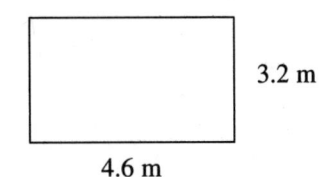

3.2 m

4.6 m

12. Fire insurance costs the Smedleys $3.25 per thousand dollars of house value. Their house is worth $28,400 (28.4 thousand). What will the insurance cost?

13. The standard car insurance is $324.76. Because Ed has a bad driving record, he must pay 1.15 times this. What will his insurance cost?

14. In a certain state, the state income tax is 0.025 times your net income. The Arbuckles' net income is $14,521.67. What is their state income tax? Round your answer to the nearest penny.

15. Jeanne makes $4.27 a hour. Last week she worked 21.5 hours. How much did she earn?

Practice Problem Answers

1. 4.44
2. 79.2
3. 8.325
4. 5.292
5. 472,000
6. 0.00000023
7. 25.37505
8. 69.26228
9. 45.63
10. 19.584 (rounded to 19.58)
11. 14.72
12. 92.30
13. 373.474 (rounded to 373.47)
14. 363.04
15. 91.805 (rounded to 91.81)

Multiplication is the opposite (or inverse) of division. As examples,

$$n \div 3 = 4 \quad \text{means} \quad n = 4 \times 3 = 12$$

$$n \div \frac{1}{2} = 10 \quad \text{means} \quad n = 10 \times \frac{1}{2} = 5$$

Similarly with decimals, multiplication is the opposite of division.

Example 9 Find n: $n \div 2 = 4.6$.

Solution Since multiplication is the opposite of division, we find n by multiplying both sides of the equation by 2.

$$n \div 2 = 4.6 \quad \text{means} \quad n = 4.6 \times 2 = 9.2$$

Example 10 Find n: $n \div 3.9 = 7.1$.

Solution We invert this division by multiplying by 3.9.

$$n \div 3.9 = 7.1 \quad \text{means} \quad n = 3.9 \times 7.1 = 27.69$$

Example 11 Find n: $\dfrac{n}{16.5} = 2.4$.

Solution Recall that $\dfrac{a}{b}$ means $a \div b$, so that $\dfrac{n}{16.5}$ means $n \div 16.5$. We solve this by multiplying both sides by 16.5 as we have done in Examples 9 and 10.

$$\dfrac{n}{16.5} = 2.4 \quad \leftarrow \textit{Given}$$

$$\cancel{16.5} \times \dfrac{n}{\cancel{16.5}} = 16.5 \times 2.4 \quad \leftarrow \textit{Muliply both sides of the equation by 16.5}$$

$$n = 39.6 \quad \leftarrow \textit{Cancel and simplify}$$

Practice Problems

Find *n* or *x* in the following problems.
See Examples 9 to 11

1. $n \div 3 = 6.5$
2. $x \div 5 = 2.4$
3. $n \div 2.6 = 7.3$
4. $n \div 1.7 = 20.3$
5. $\dfrac{x}{7} = 3.4$
6. $\dfrac{n}{5.6} = 10$
7. $\dfrac{n}{0.02} = 8$
8. $\dfrac{x}{0.5} = 4.8$

Practice Problem Answers

1. 19.5
2. 12
3. 18.98
4. 34.51
5. 23.8
6. 56
7. 0.16
8. 2.4

PROBLEM SET 0.9

Multiply the following decimals.
See Examples 1 to 8

1. 0.001×45.789
2. 0.01×5932
3. 0.1×594.38
4. 0.0001×672.85
5. 2.3×5.8
6. 6.3×2.9
7. 4.38×6.1
8. 20×5.68
9. 0.0023×5.2
10. 0.0003×0.0045
11. $18{,}000 \times 0.0003$
12. 7200×0.0000005

See Examples 9 to 11

13. Find *n*: $n \div 6 = 4.3$.
14. Find *n*: $n \div 4.2 = 6.9$.
15. Find *n*: $n \div 1.8 = 5.6$.

Business application

16. A $90 radio sells at a 25% markup. Compute this markup, 90×0.25.
17. A salesperson makes 7% commission on a $23,000 sale. Compute this commission, $0.07 \times 23{,}000$.
18. The present value of $1 (in 10 years, 8% interest) is 0.46319. Compute the present value of $200,000, which is $0.46319 \times 200{,}000$.

Health applications

19. A patient received 0.25 of a 0.25-grain ephedrine sulfate tablet. How much medication did she receive?

20. A nurse is preparing 500 milliliters of a 3% cresol solution. Compute the amount of pure liquid cresol that is needed, 500×0.03.

Technical applications

21. Below is a sketch of a countersink head rivet. Dimension X is 0.425 times diameter D. Dimension Y is 1.85 times D. Find X and Y if $D = 0.6$.

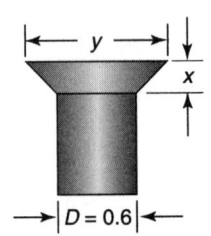

22. The tap-drill size for the screw shown below is given by $1.75 - (1.5013 \times 0.13)$. Compute this. (Multiply first; then subtract.)

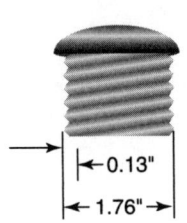

Science applications

23. The amount of heat (in Btu) needed to raise the temperature of a 2.5-pound silver bar by 7.5°F is

given by 2.5 × 7.5 × 0.056. (The number 0.056 is the specific heat capacity of silver.) Compute this.

24. The hydrostatic pressure exerted by water at a depth of 8.5 feet is given by 8.5 × 62.4. Compute this.

Psychology application

25. The average score on a learning test is 34.6. The 95% confidence range is the set of scores between
    (a) 34.6 + (1.96 × 6.1)
    (b) 34.6 − (1.96 × 6.1)

Compute these numbers. (Multiply first; then add or subtract.)

Consumer applications

26. The Hermans are buying fire insurance at $2.08 per $1000 value of the house. Their house is worth $52,000 (52 thousands). How much is their insurance premium?

27. The Bakers are buying a house. Their mortgage payments are $10.53 per thousand dollars borrowed. If they borrow $47,000 (47 thousands), how much is their monthly mortgage payment?

28. The Gorgins put 15.8 gallons of gasoline in their car. At the time, gasoline was $1.38 per gallon. How much did the gasoline cost them?

29. Because of inflation, food now costs 1.14 times what it did last year. If the Nelsons' food bill last year was $78.34 per week, what is it this year?

30. Diane is tiling a rectangular room 5.2 meters by 3.1 meters. Find the area.

31. Distance in kilometers is 1.61 times the distance in miles. The distance from Chicago to Los Angeles is 2166 miles. What is the distance in kilometers?

32. Weight in kilograms is 0.454 times the equivalent weight in pounds. Fran weighs 132 pounds. How many kilograms is this?

33. Capacity in liters is 3.78 times the equivalent capacity in gallons. A certain Nova holds 22.6 gallons of gasoline. How many liters is this?

34. The amount that a person cannot deduct as a medical deduction on his or her income tax is 0.03 times the income. If the Gleasons' income is $19,572, how much of their doctors' bills can they not deduct?

0.10 Division of Decimals

Just as addition, subtraction, and multiplication of decimals are based on addition, subtraction, and multiplication of whole numbers, so is **division of decimals** based on division of whole numbers. We start with the divisor being a whole number.

> To divide a decimal by a whole number:
> 1. Divide the numbers as if they were both whole numbers.
> 2. The decimal point for the answer goes directly above the decimal point of the dividend.

Example 1   Divide 38.29 ÷ 7.

We set up the usual division process.

Solution
```
      5.47
   7)38.29      ← Divide like whole numbers
     35
      3 2       ← Put decimal point above the point in the dividend
      2 8
        49
        49
```

Practice Problems

Divide the following decimals and whole numbers
See Example 1

1. 55.2 ÷ 8
2. 148.11 ÷ 3
3. 25.02 ÷ 9
4. 493.15 ÷ 5

Practice Problem Answers

1. 6.9
2. 49.37
3. 2.78
4. 98.63

To divide decimals and round the answer:
1. Set up the problem like long division of whole numbers.
2. Move the decimal point of the divisor (outside number) to the right until it is a whole number.
3. Move the decimal point of the dividend (inside number) the *same* number of places to the right.
4. Place the decimal point for the quotient right above the new decimal point in the dividend.
5. Divide the numbers as if they were whole numbers.
6. Carry out the division to one place beyond that desired. Using the rounding-off rules (round decimal place up or down, depending on the digit in the last place.)

Example 2 Divide 34.86 ÷ 4.2.

Solution We move the decimal point one place to the right in both numbers to make 4.2 into a whole number, 42.

$$4.2\overline{)34.86}$$ ← *Set up like long division*
← *Move decimal point of 4.2 one place right; move decimal point of 34.86 one place right*

$$42\overline{)348.6}$$ quotient 8.3
  336
  ---
  12 6
  12 6

← *Divide like whole numbers; put decimal point above joint in dividend*

Example 3 Divide 0.735 by 1.75 on a hand calculator.

Solution We enter the decimals as they appear and use the ÷ key, since "by" means division.

PRESS	DISPLAY	COMMENT
.735	0.735	0.735
÷ 1.75	1.75	Divided by 1.75
=	0.42	Answer

0.10 DIVISION OF DECIMALS

Example 4 Divide $\dfrac{0.071651}{0.0104}$ on a hand calculator and round answer to the nearest thousandth.

Solution Recall that $\dfrac{a}{b}$ means $a \div b$.

PRESS	DISPLAY	COMMENT
.0.71651	0.071651	0.071651
⊞ .0104	0.0104	Divided by 0.0104
⊟	6.889519231	Answer

Most division problems result in more decimal places than are needed. The answer here is typical. The thousandth place is the third place, so we examine the next place right.

6.889**5**19231 ← *Examine fourth place*
↑
6.890 ← *Round up, since digit is 5*

YES	NO
$0.12\overline{)2.40}$ with 20 above	$0.12\overline{)2.40}$ with 20 above (crossed out)

We move the decimal point the *same* number of places in the divisor and dividend.

Example 5 Thomas earns $400 for a 31.5-hour job. What rate of pay is this in dollars per hour (round answer to the nearest penny)?

Solution The question asks for dollars *per* hour, and "per" indicates divide (or find a rate).

Find dollars per hour ← *Given problem*
400 ÷ 31.5 ← *Translate and substitute*
12.6984127 ← *Divide*
12.70 ← *Round off*

Example 6 Sherry drives 293.1 miles on 16.2 gallons of gas. How many miles per gallon is this (to one decimal place)?

Solution To find miles *per* gallon, we divide the number of miles driven by the number of gallons of gas used. Here it is 293.1 ÷ 16.2. (The word *per* usually indicates division.)

PRESS	DISPLAY	COMMENT
293.1	293.1	Miles
÷	293.1	per

56 CHAPTER 0 THE FUNDAMENTAL OPERATIONS OF REAL NUMBERS

PRESS	DISPLAY	COMMENT
16.2	16.2	Gallon
=	18.09259259	Answer

As usual, this is too many places, so we round it to the nearest tenth: 18.1. So Sherry gets about 18.1 miles per gallon.

Example 7 Clyde Arbuckle leaves an estate of $1,239,543.71 to be split evenly among his wife and 6 children. What is each person's share (to the nearest penny)?

Solution We divide the estate by 7 since there are 7 people. The nearest penny is the second decimal place, so we carry the division three places before the round off. This is a whole-and-parts problem. We have the whole estate, and we want to split it into seven equal parts. The "split" indicates divide.

$1239543.71 \div 7 = 177077.6729$ ← *Divide*

→ 177077.67 ← *Round off*

Thus each person's share of the estate is $177,077.67.

Why Does It Work?
When we move the decimal points in a division problem, we are really multiplying both the divisor and dividend by some power of 10 that will make the division simpler. Consider $2.4 \div 0.12$. Suppose that we write this division as a fraction and multiply the numerator and denominator by 100.

$$\frac{2.4}{0.12} = \frac{2.4 \times 100}{0.12 \times 100} = \frac{240}{12} = 20$$

Now the problem is $240 \div 12$. This is the same result that we get when we move the decimal point two places:

$$0.12\overline{)2.40}\quad = 20$$

Practice Problems

Divide the following decimals.
See Examples 2 to 7

Round the answers in problem 1 to 6 to the nearest hundredth.

1. $22\overline{)10.94}$
2. $1.3\overline{)7.467}$
3. $0.04\overline{)0.9368}$
4. $87.3\overline{)95.52}$
5. $0.005\overline{)5700}$
6. $9.5\overline{)660.00}$

7. Divide $\dfrac{171}{309}$ and round your answer to three decimal places.

8. Divide $\dfrac{17}{0.00036}$ and round your answer to the nearest hundredth.

9. One week Andy worked 17.2 hours and earned $69.37. What was his hourly wage? Round your answer to the nearest cent.

10. Mary drove 225.6 miles on 11.9 gallons of gas. How many miles per gallon was this? Round your answer to the nearest tenth.

11. If a 3.9-pound roast costs $8.37, what is the price per pound? Round your answer to the nearest cent.
12. Jose runs 3.4 miles in 19.82 minutes. How many minutes per mile is this? Round your answer to two decimal places.
13. Dan Casey got 113 hits in 352 times at bat. What was his batting average (hits ÷ at bats, written as a three-place decimal)?
14. Five men have the following weights: 198.2, 182.6, 179.7, 158.4, and 216.4 pounds. Find the average weight. (Add the weights and divide by 5. Leave your answer as a one-place decimal.)
15. A disposable lighter selling for $1.39 lasted for 2984 lights. Find the cost per light as a decimal rounded to the nearest ten-thousandth of a cent. (*Hint:* First write $1.39 in cents.)
16. A clothes washing machine lasted exactly 5 years (one of them was a leap year). It cost $283.37 to buy new, plus $87.63 in repairs.
    (a) How many days was the machine owned?
    (b) How much did the machine cost to own (in cents)?
    (c) What was the average *daily* cost of owning the washer? Leave your answer in cents, rounded to the nearest tenth of a cent.
17. Lake Michigan covers 57,441 square kilometers. The U.S. portion of Lake Huron covers 23,245 square kilometers. Lake Michigan is how many times larger than the U.S. portion of Lake Huron? Leave your answer as a three-place decimal.

Practice Problem Answers

1. 0.50
2. 5.74
3. 23.42
4. 1.09
5. 1,140,000
6. 69.47
7. 0.553
8. 47,222.22
9. 4.03
10. 19.0
11. $2.15
12. 5.83
13. 0.321
14. 187.1
15. 0.0466
16. (a) 1826
    (b) 37100
    (c) 20.3
17. 2.471

Division is the opposite of multiplication. As examples (recall that dot · means multiplication).

$$6 \cdot n = 42 \quad \text{means} \quad n = 42 \div 6 = 7$$

$$\frac{2}{3} \cdot n = \frac{7}{5} \quad \text{means} \quad n = \frac{7}{5} \div \frac{2}{3}$$

$$= \frac{7}{5} \times \frac{3}{2} = \frac{21}{10}$$

Similarly with decimals, we use division to invert multiplication.

Example 8 Find n: $6 \cdot n = 0.84$.

Solution Since division is the opposite of multiplication, we divide by 6 to find n.

$$6 \cdot n = 0.84 \quad \text{means} \quad n = \frac{0.84}{6} = 0.14$$

Example 9 Find n: $1.5 \cdot n = 750$.

Solution We find n by dividing both sides of the equation by 1.5.

$1.5 \cdot n = 750$ ← *Given*

$\dfrac{\cancel{1.5} \cdot n}{\cancel{1.5}} = \dfrac{750}{1.5}$ ← *Divide both sides by 1.5*

$= 500$ ← *Divide*

Thus $n = 500$.

Example 10 What number times 0.02 is 8.4?

Solution We translate this as follows (we let n be the number we want):

What number times 0.02 is 8.4
↓ ↓ ↓ ↓ ↓
$n \quad \cdot \quad 0.02 = 8.4$ ← *Translate*

$\dfrac{n \cdot \cancel{0.02}}{\cancel{0.02}} = \dfrac{8.4}{0.02}$ ← *Divide both sides by 0.02*

$= 420$ ← *Divide*

Thus, $n = 420$.

Practice Problems

Find n or x in the following problems.
See Examples 8 to 10

1. $3 \cdot n = 2.7$
2. $20 \cdot x = 4.8$
3. $0.2 \cdot n = 50$
4. $0.07 \cdot n = 210$
5. $0.14 \cdot x = 35$
6. $2.1 \cdot n = 0.714$
7. What number times 3.6 is 0.72?
8. What number times 0.25 is 13.2?

Practice Problem Answers

1. 0.9 2. 0.24 3. 250
4. 3000 5. 250 6. 0.34
7. 0.2 8. 52.8

PROBLEM SET 0.10

Divide the following decimals.
See Examples 2 to 4

1. $164.4 \div 3$
2. $13.44 \div 8$
3. $64.74 \div 13$
4. $0.9583 \div 37$
5. $6.63 \div 1.7$
6. $0.8979 \div 7.3$
7. $3.8014 \div 0.83$
8. $2.3568 \div 0.03$
9. $\dfrac{0.7245}{20.7}$
10. $\dfrac{0.005559}{0.17}$

Divide the following decimals. Round the answers to problems 11 and 12 to the nearest tenth; problems 13 and 14 to the nearest hundredth; problems 15 and 16 to the nearest thousandth.

See Examples 2 to 7

11. $3.69 \div 0.7$
12. $\dfrac{56.8}{0.3}$
13. $0.46 \div 0.72$
14. $\dfrac{800}{1.2}$
15. $12.3 \div 0.0046$
16. $\dfrac{0.0057}{0.0009}$

Find *n* in the following problems.

See Examples 8 to 10

17. $4 \cdot n = 5.6$
18. $5 \cdot n = 87.2$
19. $0.03 \cdot n = 9$
20. $1.2 \cdot n = 0.0108$

Business applications

21. It costs a company $5.25 to produce 300 pencils. Find the cost per pencil to the nearest thousandth of a dollar.
22. A company pays its salespeople an 8% commission. The amount of sales needed to provide a $20,000 commission is given by $\dfrac{20{,}000}{0.08}$. Compute this.
23. An insurance company is studying the likelihood of men dying in their forth-fifth year. It studied 9049 men, of whom 48 died at age 45, and 9001 lived to be 46.
    (a) Compute the likelihood of a 45-year-old man living to be 46: $\dfrac{9001}{9049}$. Write the answer as a four-place decimal.
    (b) Compute the likelihood of a 45-year-old man dying before he is 46: $\dfrac{48}{9049}$. Write this as a four-place decimal.

Health applications

24. Six radiation workers reported the following radiation exposures for a certain time period: 1.3, 1.7, 0.9, 0.8, 0.7, and 1.1 rads. Find the average of these radiations (as a decimal accurate to the nearest hundredth).

25. The average adult dosage of atrophien sulfate is 0.4 milligram. The dosage for a child with a body surface area of 0.56 square meter can be computed as $\dfrac{0.56}{1.73} \times 0.4$.
    (a) Compute this by dividing first, then multiplying.
    (b) Compute this by multiplying first, then dividing.

Electrical application

26. The effective current *I* through a 100-watt light bulb is given by
    $$I = \frac{P}{V} = \frac{100 \text{ watts}}{115 \text{ volts}}$$
    (a) Compute the effective current *I*.
    (b) Compute the peak current by calculating $I_{peak} = \dfrac{I}{0.707}$.

Science applications

27. The molecular weight of oxygen (O_2) is 32. The weight of 1 liter of oxygen is given by $\dfrac{32}{22.4}$ grams. Compute this.
28. The ramp sketched below makes moving items easier. The *mechanical advantage MA* is given
    $$MA = \frac{\text{length of ramp}}{\text{height of ramp}} = \frac{4.5}{1.2}$$
    (a) Compute *MA*.
    (b) For a 300-pound box, the *effort* is $\dfrac{300}{MA}$. Compute this.

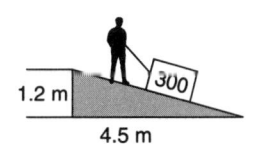

Consumer applications

29. A gallon of low-fat milk costs about $1.90. It contains about 128 grams of protein. Calculate the protein-per-dollar content of milk, $\dfrac{128}{1.90}$.

30. Frank's car gets 22.7 miles per gallon of gasohol. If gasohol costs $1.45 per gallon, compute Frank's cost per mile (in cents per mile), $\frac{145}{22.7}$.

31. Denise borrows $4552 to buy a car. The finance charge is 8% add-on for 3 years.
   (a) Compute the finance charge = 4552 × 0.08 × 3.
   (b) Compute the total debt = 4552 + finance charge.
   (c) Compute the monthly payments = $\frac{\text{total debt}}{36}$.

32. One month, Ken keeps the following records for gas mileage.

Miles	Gallons
289.2	14.6
312.7	16.1
257.9	12.3
298.6	14.9

   (a) Find the total number of miles driven.
   (b) Find the total number of gallons used.
   (c) Compute the average miles per gallon for the month.

0.11 Fraction-Decimal Conversions

Recall that the fraction $\frac{3}{4}$ means 3 divided by 4. We cannot do this division with whole numbers, but we can do it with decimals.

$$3 \div 4 = 0.75$$

We perform this division, and we get 0.75. We say that $\frac{3}{4}$ is 0.75 when written as a decimal.

To change a fraction into a decimal:
1. Divide the denominator into the numerator.
2. Round off.

Example 1 Write $\frac{5}{6}$ as a two-place decimal.

Solution Since we want two decimal places, we divide and examine the third decimal place.

PRESS	DISPLAY	COMMENT
5	5.	Numerator
÷	5.	over
6	6.	Denominator
=	0.833333333	Answer

As usual, this is too many places, so we round it to the nearest hundredth, or to two decimal places. To do this, we examine the third decimal place.

0.83**3**333333 ← *Examine the third place*
↑
0.83 ← *Round down, since digit is 3*

0.11 FRACTION-DECIMAL CONVERSIONS

Example 2 Write $\dfrac{1}{13}$ as a three-place decimal.

Solution Since we want three decimal places, we examine the fourth decimal place.

PRESS	DISPLAY	COMMENT
1	1.	Numerator
$\div$	1.	over
13	13.	Denominator
$=$	0.076923077	Answer

As usual, this is too many places, so we round it to the nearest thousandth, or to three decimal places. To do this, we examine the fourth decimal place.

$\quad$ 0.076**9**23077 $\quad \leftarrow$ *Examine fourth place*
$\qquad \uparrow$
$\quad$ 0.077 $\qquad \leftarrow$ *Round up since digit is 9*

Example 3 Write $10\dfrac{2}{3}$ as a decimal accurate to the nearest hundredth.

Solution There are two ways to solve this. One method is to divide $2 \div 3$ to convert $\dfrac{2}{3}$ to a decimal.

PRESS	DISPLAY	COMMENT
2	2.	Numerator
$\div$	2.	over
3	3.	Denominator
$=$	0.666666667	Decimal for 2/3

We round this to 0.67 and add it to 10 to get

$$10\dfrac{2}{3} = 10.67$$

If your calculator has an ab/c key, you might work it as follows.

PRESS	DISPLAY	COMMENT
10 [ab/c] 2 [ab/c] 3	10 ⌐2 ⌐3	$10\dfrac{2}{3}$
[F ↔ D]	10.66666667	Decimal of $10\dfrac{2}{3}$

The F ↔ D key is the "Fraction to Decimal" key that allows us to convert a fraction to a decimal. First you enter the fraction, then press this key.

> **Why Does It Work?**
>
> As mentioned before, this rule works because the fraction $\frac{a}{b}$ means $a \div b$. When we divide, we get the decimal equivalent.

Practice Problems

For problems 1 to 6, write each fraction as a decimal accurate to two decimal places.
See Examples 1 to 3

1. $\frac{8}{9}$ 2. $\frac{3}{7}$ 3. $\frac{7}{8}$ 4. $\frac{1}{17}$ 5. $\frac{2}{15}$ 6. $\frac{17}{3}$

For problems 7 to 12, write each fraction or mixed number as a decimal accurate to three places.

7. $\frac{5}{52}$ 8. $10\frac{5}{12}$ 9. $\frac{1}{75}$ 10. $\frac{114}{193}$ 11. $1\frac{1}{50}$ 12. $35\frac{6}{19}$

13. A board is $5\frac{7}{12}$ inches long. Write this as a decimal accurate to the nearest hundredth.

14. Ruth uses $\frac{11}{15}$ of a tank of gas on a trip. Write this as a decimal accurate to the nearest thousandth.

15. A certain lot is $\frac{1}{88}$ mile long. Write this as a decimal accurate to the nearest thousandth.

16. The Garcias' share of a town-house cooperative is $\frac{1}{120}$. Write this decimal accurate to the nearest ten-thousandth.

Practice Problem Answers

1. 0.89   2. 0.43   3. 0.88   4. 0.06   5. 0.13   6. 5.67
7. 0.096   8. 10.417   9. 0.013   10. 0.591   11. 1.02   12. 35.316
13. 5.58   14. 0.733   15. 0.011   16. 0.0083

Frequently, we wish to change a decimal into a fraction. This is the procedure that we use.

> To change a decimal to a reduced fraction:
> 1. Determine the column value of the last decimal place.
> 2. Put the original number (without any decimal point) over this column-value denominator.
> 3. Reduce the fraction, if necessary.

0.11 FRACTION-DECIMAL CONVERSIONS **63**

Example 4   Write 0.93 as a reduced fraction.

Solution   The last column is the hundredths' column.

$$0.93 \qquad \leftarrow \textit{Determine the last column value}$$
$$ \qquad \frac{1}{100}$$
$$= \frac{93}{100} \qquad \leftarrow \textit{Put number over this denominator}$$

Example 5   Write 0.025 as a reduced fraction.

Solution   The last column is the thousandth's column.

$$0.025 \qquad \leftarrow \textit{Determine last column value}$$
$$ \qquad \frac{1}{1000}$$
$$= \frac{25}{1000} \qquad \leftarrow \textit{Put number over this denominator}$$
$$= \frac{1}{40} \qquad \leftarrow \textit{Reduce fraction}$$

Example 6   Write 6.7 as a reduced fraction.

Solution   The last column is the tenth's column.

$$6.7 \qquad \leftarrow \textit{Determine last place value}$$
$$ \qquad \frac{1}{10}$$
$$= \frac{67}{10} \qquad \leftarrow \textit{Put the number over this denominator}$$
$$\text{or } 6\frac{7}{10} \qquad \leftarrow \textit{Write as a mixed number}$$

Example 7   Write 0.0059 as a reduced fraction.

Solution   The last column is the ten-thousandth's column.

$$0.0059 \qquad \leftarrow \textit{Determine last column value}$$
$$ \qquad \frac{1}{10{,}000}$$
$$= \frac{59}{10{,}000} \qquad \leftarrow \textit{Put number over this denominator}$$

Why Does It Work?

The column values of decimals are fractions: $\frac{1}{10}, \frac{1}{100}, \frac{1}{1000}$, and so on.

When we expand out a decimal, we see that the last place acts as a common denominator. As examples, consider

$$0.217 = \frac{2}{10} + \frac{1}{100} + \frac{7}{1000} = \frac{200}{1000} + \frac{10}{1000} + \frac{7}{1000} = \frac{217}{1000}$$

$$5.83 = 5 + \frac{8}{10} + \frac{3}{100} = \frac{500}{100} + \frac{80}{100} + \frac{3}{100} = \frac{583}{100} \text{ or } 5\frac{83}{100}$$

Example 8   The sponsors of a new TV show, Name That Clown, have computed that their share of the viewing market is 0.15. Write this as a reduced fraction.

Solution   The place value of the last digit is $\frac{1}{100}$. Thus we get

0.15 ← **Determine last column value**

$\frac{1}{100}$

$= \frac{15}{100}$ ← **Put number over this denominator**

$= \frac{3}{20}$ ← **Reduce**

This says that 3 of every 20 viewers watch Name That Clown.

Practice Problems

Write the following decimals as reduced fractions or mixed numbers.
See Examples 4 to 8

1. 0.7   2. 0.35   3. 0.231   4. 0.44   5. 2.1
6. 0.2   7. 7.25   8. 99.99   9. 0.98   10. 0.125
11. 6.14   12. 0.055   13. 1.12   14. 36.95   15. 0.76531
16. A car is 14.8 feet long. Write this as a mixed number.
17. Mary buys 0.64 pounds of chocolate. Write this as a reduced fraction.
18. Reggie earns $3.60 an hour. Write this as a reduced mixed number.

Practice Problem Answers

1. $\frac{7}{10}$ 2. $\frac{7}{20}$ 3. $\frac{231}{1000}$ 4. $\frac{11}{25}$ 5. $2\frac{1}{10}$ 6. $\frac{1}{5}$
7. $7\frac{1}{4}$ 8. $99\frac{99}{100}$ 9. $\frac{49}{50}$ 10. $\frac{1}{8}$ 11. $6\frac{7}{50}$ 12. $\frac{11}{200}$
13. $1\frac{3}{25}$ 14. $36\frac{19}{20}$ 15. $\frac{76,531}{100,000}$ 16. $14\frac{4}{5}$ 17. $\frac{16}{25}$ 18. $3\frac{3}{5}$

PROBLEM SET 0.11

Write the following fractions as three-place decimals.
See Examples 1 to 3

1. $\frac{6}{7}$ 2. $\frac{5}{9}$ 3. $\frac{11}{13}$
4. $\frac{32}{29}$ 5. $3\frac{1}{4}$ 6. $23\frac{4}{11}$

Write the following decimals as reduced fractions (or mixed numbers).
See Examples 4 to 8

7. 0.74   8. 0.002   9. 0.35
10. 0.0005   11. 1.2   12. 5.25

Business applications

13. A marketing survey finds that $\frac{114}{349}$ of a sample population eats cereal in the morning. Write this as a three-place decimal.

14. According to life insurance tables, the chance of a 21-year-old man dying before he reaches 22 is 0.0025. Write this as a reduced fraction.

Health applications

15. A hospital stocks 0.375-grain ephedrine sulfate ampules. Express this as a reduced fraction.

16. A patient receives $\frac{1}{4}$ of a $\frac{1}{2}$-grain codeine sulfate tablet.
    (a) How much medication did the patient receive (as a fraction of a grain)?
    (b) How much medication did the patient receive (as a decimal)?

Technical application

17. For the bolt sketched below, express each dimension as a decimal.

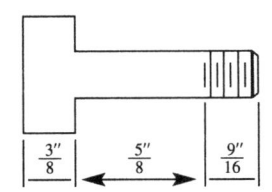

$\frac{3''}{8}$ $\frac{5''}{8}$ $\frac{9''}{16}$

Chemistry application

18. One coulomb of electricity supplied to a solution of AlCl$_3$ (aluminum chloride) deposits $\frac{9}{96,500}$ gram of aluminum. If 19,300 coulombs are supplied to the solution, $19{,}300 \times \frac{9}{96{,}500}$ grams of aluminum are deposited.
    (a) Express this as a reduced fraction.
    (b) Express this as a decimal.

Consumer applications

19. At a sale, pants were $\frac{1}{4}$ off list price.
    (a) Express this as a decimal.
    (b) How much money was saved on a $23.75 purchase?

20. Mike calculates that 0.28 of his total gross income goes to taxes. Express this as a reduced fraction.

21. Below is a sketch of a room that the Rossis are carpeting.

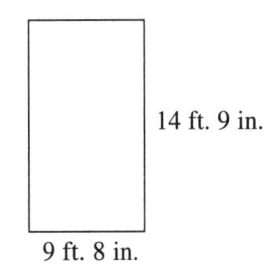

14 ft. 9 in.

9 ft. 8 in.

(a) Express each dimension as a mixed number of feet. (For example, 2 feet 7 inches is $2\frac{7}{12}$ feet.)
(b) Express each dimension as a two-place decimal.
(c) Compute the area as a mixed number and as a decimal. [Use the numbers from parts (a) and (b).]

0.12 Areas of Plane Figures

The Area of a Rectangle

The **area A of a rectangle** is the product of the length l and the width w (Figure 1), that is

$$A = lw.$$

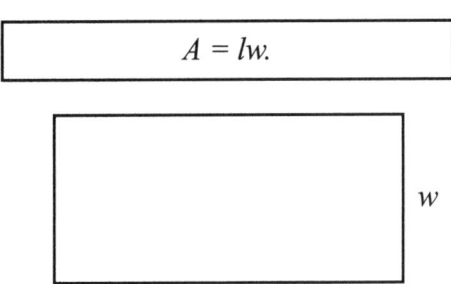

Figure 1

Example 1 Find the area of a rectangle with a length of 7 centimeters and a width of 5 centimeters.

$$A = lw$$
$$= 7 \cdot 5 = 35$$

Therefore, the area of the rectangle is 35 square centimeters.

The Area of a Square

The formula for the **area of a square** is very similar to the formula for the area of a rectangle. This is because a square is a special kind of rectangle. If the length of each side of a square is s units (Figure 2), then the area A is given by the following formula:

$$A = s^2.$$

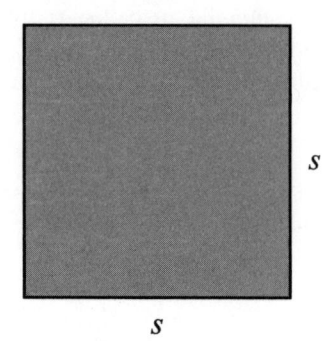

Figure 2

Example 2 Find the area of a square with sides of 4.3 meters.

Solution
$$A = s^2$$
$$= (4.3)^2 = 18.49$$

Therefore, the area of the square is 18.49 square meters.

The Area of a Parallelogram

The **area A of a parallelogram** is given by the following formula:

$$A = bh,$$

where b is the base and h is the height of the parallelogram (Figure 3).

Figure 3

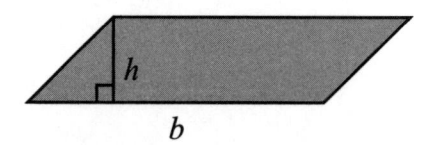

Example 3 Find the area of a parallelogram with a base of 5 feet and a height of 3 feet.

Solution The area of a parallelogram is

$$A = bh$$
$$= 5 \cdot 3 = 15.$$

Therefore, the area $A = 15$ square feet.

The Area of a Triangle

The **area A of a triangle** is equal to one-half the product of the base b and the height h (Figure 4):

$$A = \frac{1}{2}bh.$$

Figure 4

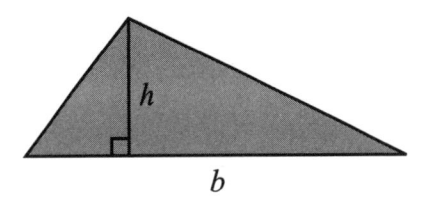

Example 4 Find the area of a triangle with a base of 10 centimeters and a height of 3.7 centimeters.

Solution The area of the triangle is

$$A = \frac{1}{2} \cdot 10 \cdot 3.7 = 18.5.$$

Therefore, the area is 18.5 square centimeters.

The Area of a Trapezoid

The **area A of a trapezoid** is equal to one-half the product of the height h and the sum of the bases b_1 and b_2 (Figure 5):

$$A = \frac{1}{2}h(b_1 + b_2)$$

Figure 5

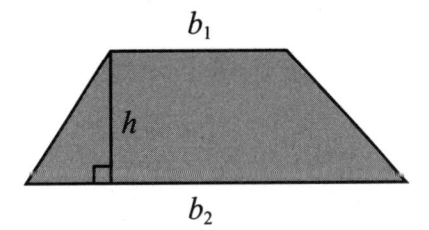

Example 5 Find the area of a trapezoid with bases of 12 inches and 8 inches, and with a height of 5 inches.

Solution We substitute $h = 5$, $b_1 = 12$, and $b_2 = 8$ in the formula

$$A = \frac{1}{2}h(b_1 + b_2),$$

and we have:

$$A = \frac{1}{2}(5)(12 + 8) = 50.$$

Therefore, the area is 50 square inches.

The Area of a Circle

The **area A of a circle** with radius r is π times r^2, in which $\pi = 3.1416\ldots$ (Figure 6):

$$A = r^2\pi.$$

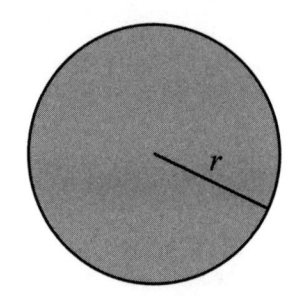

Figure 6

Example 6 Find the area of a circle with a radius of 10 inches.

Solution
$$A = \pi r^2$$
$$= \pi(10)^2 = 100\pi \approx 314.16$$

Therefore, the area of the circle is approximately 314.16 square inches.

PROBLEM SET 0.12

In problems 1–6, find the area of the rectangle with the given length l and width w.

1. $l = 5$ inches and $w = 3$ inches
2. $l = 7$ yards and $w = 2$ yards
3. $l = 6.2$ centimeters and $w = 4.1$ centimeters
4. $l = 9.3$ feet and $w = 8.5$ feet
5. $l = 7.3$ feet and $w = 5.6$ feet
6. $l = 14.3$ meters and $w = 11.6$ meters

In problems 7–12, find the area of the square with the given side s.

7. $s = 8$ centimeters
8. $s = 13$ inches
9. $s = 6.3$ feet
10. $s = 11.3$ meters
11. $s = 9.7$ yards
12. $s = 6\frac{2}{3}$ feet

In problems 13–18, find the area of the parallelogram with the given base b and height h.

13. $b = 5$ feet and $h = 3.4$ feet
14. $b = 12$ centimeters and $h = 8$ centimeters
15. $b = 11.7$ meters and $h = 5.3$ meters
16. $b = 6.3$ yards and $h = 5.8$ yards
17. $b = 4\frac{1}{3}$ inches and $h = 6$ inches
18. $b = 3\frac{1}{4}$ inches and $h = 5\frac{2}{3}$ inches

In problems 19–24, find the area of the triangle with the given base b and height h.

19. $b = 6$ feet and $h = 3$ feet
20. $b = 8$ inches and $h = 10$ inches
21. $b = 5.1$ meters and $h = 7.3$ meters
22. $b = 11.7$ yards and $h = 39.3$ yards
23. $b = 8.9$ inches and $h = 4.2$ inches
24. $b = 6\frac{2}{5}$ feet and $h = 7\frac{3}{4}$ feet

In problems 25–30, find the area of the trapezoid with the given height h and bases b_1 and b_2.

25. $h = 5$ feet, $b_1 = 7$ feet, and $b_2 = 9$ feet
26. $h = 3.2$ yards, $b_1 = 5.9$ yards, and $b_2 = 7.2$ yards
27. $h = 3.8$ inches, $b_1 = 4.7$ inches, and $b_2 = 8.2$ inches
28. $h = 11.3$ meters, $b_1 = 9.7$ meters, and $b_2 = 6.5$ meters
29. $h = 6\frac{1}{3}$ feet, $b_1 = 7\frac{1}{2}$ feet, and $b_2 = 5\frac{1}{4}$ feet
30. $h = 5\frac{5}{6}$ yards, $b_1 = 3\frac{1}{2}$ yards, and $b_2 = 4\frac{2}{3}$ yards

In problems 31–36, find the area of the circle with the given radius r.

31. $r = 5$ feet
32. $r = 7$ yards
33. $r = 3.1$ meters
34. $r = 6.3$ feet
35. $r = 4.4$ inches
36. $r = 8.2$ centimeters

0.13 Perimeters of Plane Figures

The Perimeter of a Rectangle

The **perimeter P of a rectangle** is the sum of twice the length (and twice the width w (Figure 1):

$$P = 2l + 2w.$$

Figure 1

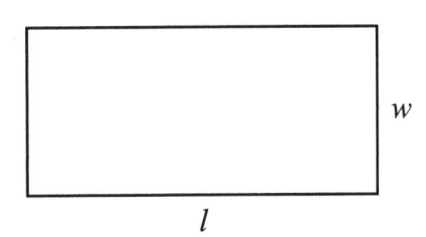

Example 1 Find the perimeter of a rectangle with a length of 8 meters and a width of 5 meters.

Solution
$$P = 2l + 2w$$
$$= 2(8) + 2(5) = 16 + 10 = 26$$

Therefore, the perimeter is 26 meters.

The Perimeter of a Square

The **perimeter P of a square** is four times the length of its side (Figure 2):

$$P = 4s.$$

Figure 2

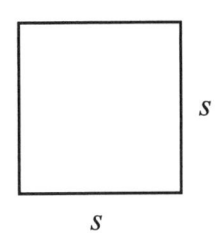

Example 2 Find the perimeter of a square with sides of 8 centimeters.

Solution
$$P = 4s$$
$$= 4(8) = 32$$

Therefore, the perimeter of the square is 32 centimeters.

The Perimeter of a Parallelogram

The **perimeter P of a parallelogram** is the sum of twice the length l and twice the width w (Figure 3):

$$P = 2l + 2w$$

Figure 3

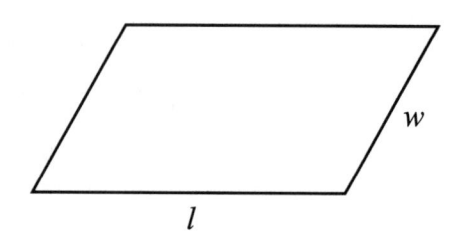

Example 3 Find the perimeter of a parallelogram with a length of 10 inches and a width of 6 inches.

Solution
$$P = 2l + 2w$$
$$= 2(10) + 2(6) = 20 + 12 = 32$$

Therefore, the perimeter of the parallelogram is 32 inches.

The Perimeter of a Triangle

The **perimeter P of a triangle** is the sum of the lengths of its sides a, b, and c (Figure 4):

$$P = a + b + c.$$

Figure 4

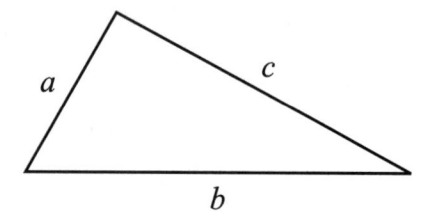

Example 4 Find the perimeter of a triangle with sides 5 inches, 6 inches, and 8 inches.

Solution
$$P = a + b + c$$
$$= 5 + 6 + 8 = 19$$

Therefore, the perimeter of the triangle is 19 inches.

The Perimeter of a Trapezoid

The **perimeter P of a trapezoid** is the sum of the lengths of the sides a, b, c, and d (Figure 5):

$$P = a + b + c + d.$$

Figure 5

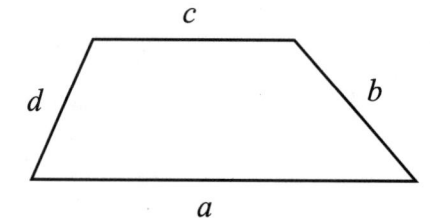

Example 5 Find the perimeter of a trapezoid with sides 8 inches, 5 inches, 6 inches, and 5 inches.

Solution
$$P = a + b + c + d$$
$$= 8 + 5 + 6 + 5 = 24$$

Therefore, the perimeter of the trapezoid is 24 inches.

The Circumference of a Circle

The **circumference C of a circle** of radius r is 2π times r (Figure 6):

$$C = 2\pi r.$$

Figure 6

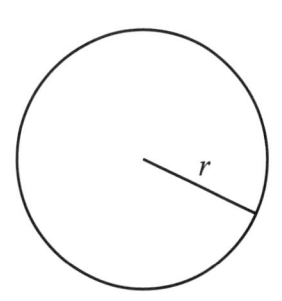

Example 6 Find the circumference of a circle with a radius of 5 inches.

$$C = 2\pi r$$
$$= 2\pi(5) = 10\pi \approx 31.42$$

Therefore, the circumference of the circle is approximately 31.42 inches.

PROBLEM SET 0.13

In problems 1–6. find the perimeter of the rectangle with the given length l and width w.

1. $l = 5$ inches and $w = 3$ inches
2. $l = 12$ yards and $w = 7$ yards
3. $l = 6.2$ centimeters and $w = 4.1$ centimeters
4. $l = 9.3$ feet and $w = 8.5$ feet
5. $l = 7.3$ feet and $w = 5.6$ feet
6. $l = 14.3$ meters and $w = 11.6$ meters

In problems 7–12. find the perimeter of the square with the given side s.

7. $s = 8$ centimeters
8. $s = 13$ inches
9. $s = 6.3$ feet
10. $s = 11.3$ meters
11. $s = 9.7$ yards
12. $s = 6\frac{2}{3}$ feet

In problems 13–18. find the perimeter of the parallelogram with the given length l and width w.

13. $l = 5$ feet and $w = 3.4$ feet
14. $l = 12$ centimeters and $w = 8$ centimeters
15. $l = 11.7$ meters and $w = 5.3$ meters
16. $l = 6.3$ yards and $w = 5.8$ yards
17. $l = 6$ inches and $w = 4\frac{1}{3}$ inches
18. $l = 5\frac{2}{3}$ inches and $w = 3\frac{1}{4}$ inches

In problems 19–24, find the perimeter of the triangle with the given sides a, b and c.

19. $a = 3$ inches, $b = 4$ inches, and $c = 5$ inches
20. $a = 5$ feet, $b = 12$ feet, and $c = 13$ feet
21. $a = 5.6$ meters, $b = 4.9$ meters, and $c = 7.2$ meters
22. $a = 10.7$ yards, $b = 13.2$ yards, and $c = 6.1$ yards

23. $a = 3\frac{1}{4}$ inches, $b = 4\frac{2}{3}$ inches, and $c = 2\frac{5}{6}$ inches
24. $a = 12\frac{1}{5}$ feet, $b = 5\frac{11}{15}$ feet, and $c = 10\frac{2}{3}$ feet

In problems 25–30, find the perimeter of the trapezoid with the given sides a, b, c, and d.

25. $a = 10$ inches, $b = 4$ inches, $c = 6$ inches, and $d = 4$ inches
26. $a = 7\frac{2}{3}$ feet, $b = 4\frac{1}{3}$ feet, $c = 3$ feet, and $d = 5$ feet
27. $a = 23.9$ meters, $b = 9.1$ meters, $c = 9.3$ meters, and $d = 8.1$ meters
28. $a = 8.2$ yards, $b = 2.8$ yards, $c = 4.7$ yards, and $d = 3.2$ yards
29. $a = \frac{1}{2}$ foot, $b = \frac{1}{4}$ foot, $c = \frac{1}{3}$ foot, and $d = \frac{1}{4}$ foot
30. $a = 3\frac{3}{7}$ inches, $b = 2\frac{1}{7}$ inches, $c = 1\frac{3}{7}$ inches, and $d = 1\frac{6}{7}$ inches

In problems 31–36. find the circumference of the circle with given radius r.

31. $r = 7$ feet
32. $r = 15$ feet
33. $r = 5.2$ inches
34. $r = 7.9$ meters
35. $r = 16.4$ centimeters
36. $r = 3\frac{1}{2}$ inches

0.14 Volumes and Surface Areas

The Volume and Surface Area of a Rectangular Solid (Box)

The **volume V of a rectangular solid** is the product of the length l, the width w, and the height h (Figure 1):

$$V = lwh.$$

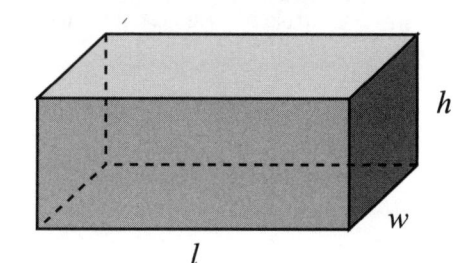

Figure 1.

The **total surface area S** of such a solid is the sum of the areas of its faces and its bases:

$$S = 2(lw + lh + wh).$$

Example 1 Find the volume and the total surface area of a rectangular box with dimensions $l = 7$, $w = 5$, and $h = 3$ centimeters.

Solution The volume is given by the formula

$$V = lwh = 7(5)(3)$$
$$= 105.$$

Therefore, the volume of the box is 105 cubic centimeters. The total surface area is given by the formula

$$S = 2[lw + lh + wh]$$
$$= 2[7(5) + 7(3) + 5(3)]$$
$$= 2(35 + 21 + 15)$$
$$= 142.$$

Therefore, the total surface area is 142 square centimeters.

The Volume and the Surface Area of a Right Circular Cylinder

The **volume V of a right circular cylinder** is the product of the height h and the area πr^2 of a base of the cylinder (Figure 2):

$$V = \pi r^2 h.$$

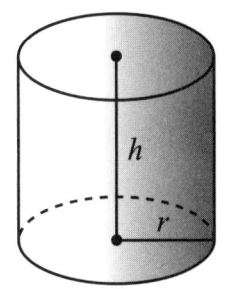

Figure 2.

The **lateral surface area LS** is the product of the height h of the cylinder and the circumference $2\pi r$ of a base:

$$LS = 2\pi rh.$$

The **total surface area S** is the sum of the area $2\pi r^2$ of the two bases and the lateral surface area $2\pi rh$ of the cylinder:

$$S = 2\pi r^2 + 2\pi rh.$$

Example 2 Find the lateral surface area, the total surface area, and the volume of a right circular cylinder with a radius of 3 centimeters and a height of 6 centimeters.

Solution
$$LS = 2\pi rh = 2\pi(3)(6)$$
$$= 36\pi$$

Thus, the lateral surface area of the cylinder is 36π square centimeters.

$$S = 2\pi r^2 + 2\pi rh$$
$$= 2\pi(3)^2 + 36\pi = 18\pi + 36\pi = 54\pi.$$

Therefore, the total surface area of the cylinder is 54π square centimeters.

$$V = \pi r^2 h = \pi(3)^2(6) = 54\pi.$$

The volume of the cylinder is 54π cubic centimeters.

The Volume and Surface Area of a Pyramid

A pyramid is a three-dimensional figure whose base is a polygon and whose lateral faces are triangles (Figure 3). If the base of a pyramid is a regular polygon and the pyramid has equal lateral edges, it is called a regular pyramid. As you can see. the pyramids shown in Figure 3a and 3b are regular. Those shown in Figure 3c and 3d are not.

74 CHAPTER 0 THE FUNDAMENTAL OPERATIONS OF REAL NUMBERS

Figure 3

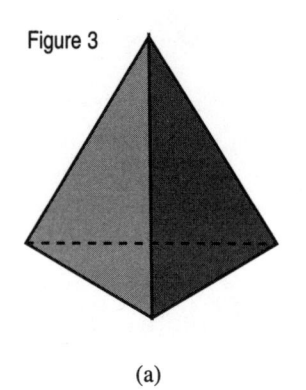

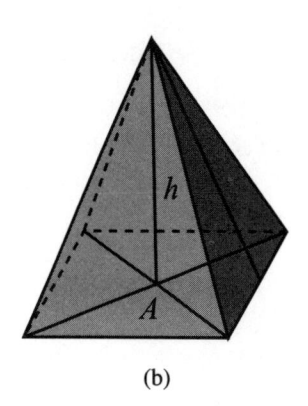

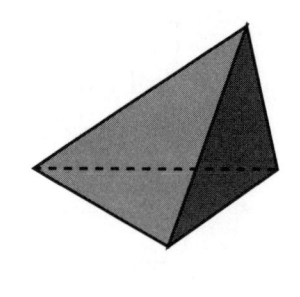

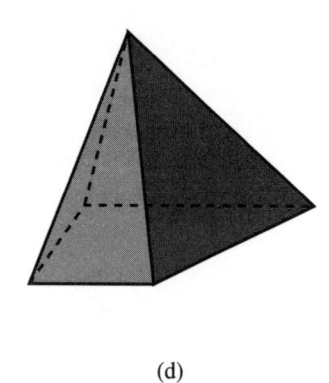

(a)　(b)　(c)　(d)

The **volume V of a regular pyramid** is one-third the product of its height h and the area A of its base (Figure 3b):

$$V = \frac{1}{3} Ah.$$

The **lateral surface area LS of a regular pyramid** is one-half the product of the perimeter P of the pyramid's base and the slant height l (the height of a triangular lateral face) (Figure 3b):

$$LS = \frac{1}{2} Pl.$$

Example 3　Find the lateral surface area and the volume of a regular pyramid whose base is a square with 12-inch sides, if the height is $2\sqrt{7}$ inches and the slant height is 8 inches (Figure 4).

Figure 4

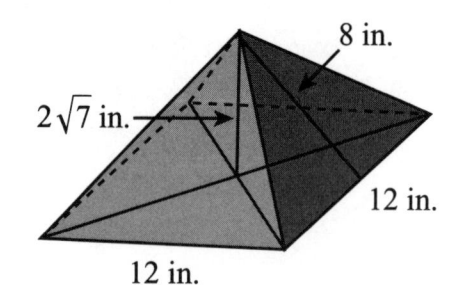

Solution　The perimeter of the pyramid's base is

$$P = 4(12) = 48.$$

Thus, the lateral surface area is given by

$$LS = \frac{1}{2} Pl$$

$$= \frac{1}{2}(48)(8) = 192.$$

Therefore, the lateral surface area of the pyramid is 192 square inches. The volume of the pyramid is

$$V = \frac{1}{3} Ah$$

$$= \frac{1}{3}(144)(2\sqrt{7}) = 96\sqrt{7}.$$

Therefore, the volume is $96\sqrt{7}$ cubic inches.

The Volume and Surface Area of a Right Circular Cone

The right circular cone resembles the pyramid. except that the base of the cone is a circle (Figure 5). The volume V and the lateral surface area LS of a right circular cone are given by the following formulas:

$$V = \frac{1}{3}\pi r^2 h \quad \text{and} \quad LS = \pi r l.$$

in which r is the radius of the base, h is the height, and l is the slant height.

Figure 5

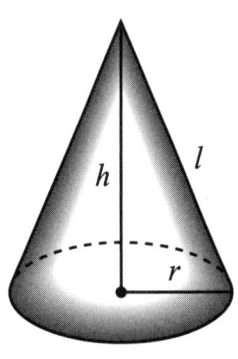

Example 4 Find the lateral surface area and the volume of a right circular cone with radius 6 inches, a height of 8 inches, and slant height of 10 inches.

Solution
$$LS = \pi r l = \pi(6)(10) = 60\pi$$

Therefore, the lateral surface area of the cone is 60π square inches.

$$V = \frac{1}{3}\pi r^2 h$$
$$= \frac{1}{3}\pi(6)^2(8)$$
$$= 96\pi.$$

Thus, the volume of the cone is 96π cubic inches.

The Volume and Surface Area of a Sphere

A sphere is the set of all points in space at a given distance, r, from a point (Figure 6). If r is the radius, then the **surface area S** and the **volume V of a sphere** are given by the following formulas:

$$S = 4\pi r^2 \quad \text{and} \quad V = \frac{4}{3}\pi r^3.$$

Figure 6

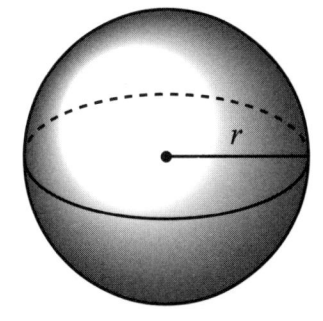

Example 5 Find the surface area and the volume of a sphere with a radius of 3 meters.

Solution
$$S = 4\pi r^2$$
$$= 4\pi(3)^2$$
$$= 36\pi$$

Therefore, the surface area of the sphere is 36π square meters.

$$V = \frac{4}{3}\pi r^3$$
$$= \frac{4}{3}\pi(3)^3$$
$$= 36\pi.$$

Thus, the volume of the sphere is 36π cubic meters.

PROBLEM SET 0.14

In problems 1–6, find the volume and the surface area of the rectangular solid (box) with the given length l, width w, and height h.

1. $l = 5$ inches, $w = 4$ inches, and $h = 3$ inches
2. $l = 7$ meters, $w = 6$ meters, and $h = 8$ meters
3. $l = 5\frac{1}{3}$ yards, $w = 3\frac{1}{4}$ yards, and $h = 4\frac{2}{3}$ yards
4. $l = 8.9$ centimeters, $w = 6.7$ centimeters, and $h = 5.3$ centimeters
5. $l = 11.9$ feet, $w = 10.3$ feet, and $h = 7.5$ feet
6. $l = 11.21$ meters, $w = 9.03$ meters, and $h = 4.17$ meters

In problems 7–12, find the volume, the lateral surface area, and the total surface area of the right circular cylinder with the given radius r and height h.

7. $r = 5$ inches and $h = 10$ inches
8. $r = 7$ yards and $h = 9$ yards
9. $r = 4.1$ meters and $h = 11.3$ meters
10. $r = 8.2$ centimeters and $h = 6.5$ centimeters
11. $r = 3.4$ feet and $h = 7.6$ feet
12. $r = 4\frac{1}{3}$ feet and $h = 5\frac{2}{3}$ feet

In problems 13–18, find the volume and the lateral surface area of the given pyramid.

13. A regular square pyramid with a base edge of 10 inches, a height of 12 inches, and a slant height of 13 inches.
14. A regular square pyramid with a base edge of 1.8 meters, a height of 1.2 meters, and a slant height of 1.5 meters.
15. A regular triangular pyramid with a base edge of 8 feet, a base area of $16\sqrt{3}$ square feet, a height of 10 feet, and a slant height of 10.3 feet.
16. A regular triangular pyramid with a base edge of 14.3 centimeters, a base area of 88.6 square centimeters, a height of 5.6 centimeters, and a slant height of 6.9 centimeters.
17. A regular pentagonal pyramid with a base edge of 12 inches, a height of 9 inches, a base area of 247.8 square inches, and a slant height of 12.2 inches.
18. A regular pentagonal pyramid with a base edge of 8 feet, a height of 15 feet, a base area of 110.1 square feet, and a slant height of 16 feet.

In problems 19–24, find the volume and the lateral surface area of the right circular cone with the given radius r, height h, and slant height l.

19. $r = 4$ inches, $h = 3$ inches, and $l = 5$ inches
20. $r = 5$ meters, $h = 12$ meters, and $l = 13$ meters
21. $r = 9.2$ feet, $h = 6.9$ feet, and $l = 11.5$ feet
22. $r = 8.5$ yards, $h = 20.4$ yards, and $l = 22.1$ yards
23. $r = 3.2$ centimeters, $h = 2.4$ centimeters, and $l = 4$ centimeters
24. $r = 3\sqrt{3}$ feet, $h = 3$ feet, and $l = 6$ feet

In problems 25–30, find the volume and the surface area of a sphere with the given radius r.

25. $r = 4$ feet
26. $r = 3.1$ centimeters
27. $r = 5.6$ meters
28. $r = 2.5$ yards
29. $r = 10$ inches
30. $r = 8.7$ meters

REVIEW PROBLEM SET

1. Find the area of Figure 1 if ABCD is a rectangle and the curve is a semicircle of radius 7 feet.

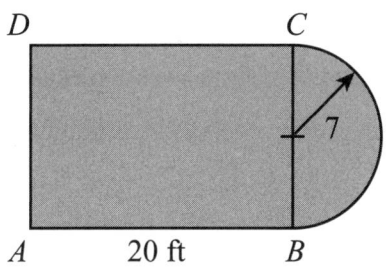

Figure 1

2. In Figure 2, find the area of △DEF, where D, E, and F are the midpoints of the sides of △ABC whose height is 10 inches and whose corresponding base is 15 inches.

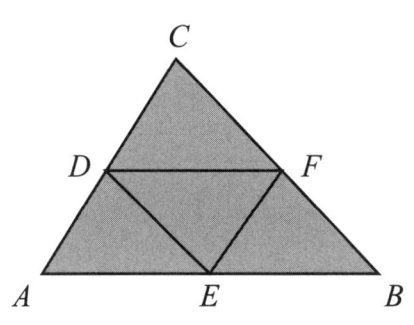

Figure 2

3. In Figure 3, find the area of the shaded region determined by circles A, B, and O, if $|\overline{AB}| = 2$. (The notation $|\overline{AB}|$ represents the length of the line segment AB.)

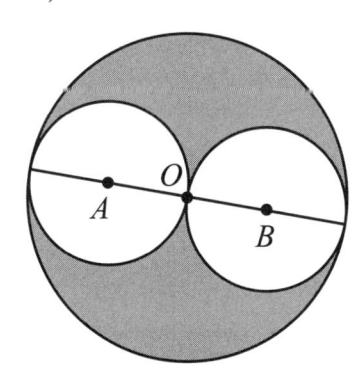

Figure 3

4. Let ABCD be a square whose area is 100 square units. What is the length of its side? What is the area of the square whose vertices are the midpoints of the sides of ABCD?

5. Let ABC be an isosceles triangle with $|\overline{AB}| = |\overline{AC}| = 13$. If $|\overline{BC}| = 10$, find the length of the perpendicular line $\overline{AD}$ from A to BC. Find the area of △ABC.

6. If a floor is 80 feet long and 20 feet wide, how many tiles are needed to cover it if each is a square 4 inches on each side?

In problems 7–12. given that ABCD is a square (Figure 4). $|\overline{DE}| = |\overline{EC}|$, $\overline{EF} \parallel \overline{CG}$, $|\overline{AB}| = 9$ inches, and $\overline{HB}/\overline{HC} = \frac{1}{2}$, find:

7. Area of square ABCD
8. Area of rectangle FHCD
9. Area of parallelogram FGCE
10. Area of quadrilateral FHCE
11. Area of rectangle ABHF
12. Area of triangle GHC

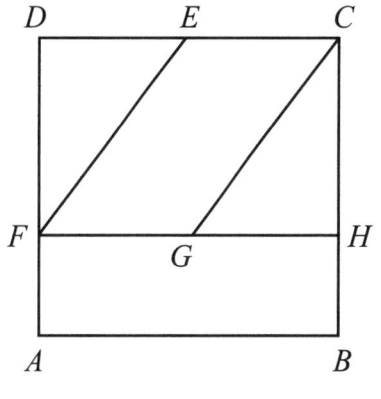

Figure 4

In problems 13–20, find the area of the given plane figure.

13. A rectangle if the base is 25 centimeters and the perimeter is 90 centimeters.
14. A rectangle if the perimeter is 50 feet and the ratio of the sides is $\frac{2}{3}$.
15. A square if the perimeter is 81 meters.
16. A parallelogram if the base is $(x - 5)$ units and the height is $(x + 5)$ units.
17. A triangle if the base is $5x$ units and the height is $4x$ units.
18. An equilateral triangle if the perimeter is 36 inches and its height is $6\sqrt{3}$ inches.
19. A trapezoid if $a = 20$ meters, $b = 40$ meters, and $h = 16$ meters.
20. A circle if the circumference is 36π centimeters.

In problems 21–26, find the volume and the surface area of each of the solids.

21. A rectangular solid (box) if $l = 6$, $w = 2$, and $h = 7$ (units in centimeters).
22. A regular triangular pyramid with a base edge of 4 meters, a base area of $4\sqrt{3}$ square meters, a height of 5 meters, and a slant height of 5.13 meters.
23. A right circular cylinder with radius of 3 inches and height of 5 inches.
24. A right circular cone with radius of 4 inches, height of 5 inches, and slant height of $\sqrt{41}$ inches.
25. A sphere with radius of 5 inches.
26. A cube whose side is 10 centimeters.

SUMMARY OF DECIMAL CONCEPTS

Important Ideas

0.7) Knowing the place value of decimals.
   Rounding off numbers.
   Comparing the size of decimals.
0.8) Adding decimals.
   Subtracting decimals.
   Knowing that addition and subtraction are opposite operations.
0.9) Multiplying decimals.
   Knowing that multiplication is the opposite of division.
0.10) Dividing decimals.
   Knowing that division is the opposite of multiplication.
0.11) Writing a fraction as a decimal.
   Writing a decimal as a fraction.

REVIEW PROBLEM SET FOR SECTIONS 0.1–0.6

In problems 1–4, write each statement in symbols.

1. Eighteen is the product of six and three.
2. Five plus nine equals fourteen.
3. Seven is not equal to two plus four.
4. Eleven minus five is six.

In problems 5–14, use the order of operations to find the value of each expression.

5. $7 + 6 \cdot 2$
6. $6 \cdot 3 - 2 \cdot 5$
7. $5 \cdot 12 - 5 \cdot 6 + 13$
8. $3 \cdot 2 + 3 \cdot 8 - 3 \cdot 4$
9. $8 \div 4 + 3 \cdot 7$
10. $64 \div 8(4 + 3)$
11. $4(3 + 2) - 10 \div 2$
12. $9 - 3(5 - 4) - 6$
13. $\dfrac{12 \div 4(5 - 3)}{10 + 2(8 - 5) - 14}$
14. $\dfrac{6(7 + 2) - 20 \div (12 - 8)}{3 + 2(6 + 5)}$

In problems 15–18, graph each set on a number line.

15. $\{-4, -1, 0, 3, 5\}$
16. $\{5, -5, 3, -3, 1, -1\}$
17. $\{2, 4, 6, 8\}$
18. $\{1, 3, 5, 7\}$

In problems 19–22, change each percent to a decimal.

19. 18.2%
20. 17.34%
21. 23.8%
22. 16.481%

In problems 23–26, change each percent to a fraction.

23. 11%
24. 31%
25. 29%
26. 16.7%

In problems 27–30, find the indicated value.

27. 15% of 300
28. 21% of 500
29. $18\frac{1}{4}$% of 1,400
30. $12\frac{3}{4}$% of 675

In problems 31–46, state the property of real numbers that is illustrated by each statement.

31. $6 + (-4) = (-4) + 6$
32. $10 + (2 + 7) = (10 + 2) + 7$
33. $13 \cdot 2 = 2 \cdot 13$
34. $(12 + 1) \cdot 5 = 12 \cdot 5 + 1 \cdot 5$
35. $8 + (7 + 11) = (8 + 7) + 11$
36. $0 + (-11) = -11$
37. $5(10 + 3) = 5 \cdot 10 + 5 \cdot 3$
38. $7 + 100 = 100 + 7$
39. $8 + 0 = 8$

40. $13 \cdot \dfrac{1}{13} = 1$
41. $11 \cdot (9 \cdot 2) = (11 \cdot 9) \cdot 2$
42. $(-19) \cdot (-13) = (-13) \cdot (-19)$
43. $1 \cdot 3 = 3$
44. $(-2) \cdot [(-3) \cdot (-5)] = [(-2) \cdot (-3)] \cdot (-5)$
45. $\dfrac{1}{7} \cdot 7 = 1$
46. $(-9) \cdot 1 = -9$

In problems 47–52, find the additive inverse and the multiplicative inverse of each number.

47. 12
48. 20
49. -17
50. -14
51. $\dfrac{5}{3}$
52. $-\dfrac{7}{9}$

In problems 53–56, simplify each expression by performing the indicated operations.

53. $3[2 + 5\{3 + (8 - 4) + 1\} - 7]$
54. $7 - 2\{5 + [6 + 2(7 - 3)]\} + 80$
55. $5 + 3\{2 + 3[1 + 3(8 - 5)] + 18\}$
56. $\{3 + 4[5 - 2(3 - 1) + 8] - 2(6 + 7)\}$

In problems 57–62, find the value of each expression.

57. $|22|$
58. $|-44|$
59. $|-18|$
60. $|81|$
61. $-|14|$
62. $-|-3|$

In problems 63–116, perform the indicated operations.

63. $13 + 12$
64. $25 + 32$
65. $(-18) + (-14)$
66. $(-28) + (-22)$
67. $(-21) + 17$
68. $29 + (-13)$
69. $21 - 8$
70. $23 - 17$
71. $31 - (-9)$
72. $39 - (-14)$
73. $-37 - 18$
74. $-42 - 16$
75. $-20 - (-21)$
76. $-73 - (-19)$
77. $13 + (-18)$
78. $100 + (-200)$
79. $(-27) + (-31) + 28$
80. $(-205) + (-95) + 100$
81. $(-15) - (-5) + 10$
82. $120 - (-30) - 50$
83. $(-3) \cdot 8$
84. $7 \cdot (-11)$
85. $(-6) \cdot (-7)$
86. $(-10) \cdot (-10)$
87. $(-1) \cdot (-2) \cdot (-8)$
88. $(-2) \cdot (-3) \cdot 5$
89. $(-2) \cdot 9 \cdot (-3)$
90. $(-1) \cdot (-5) \cdot (-2) \cdot (-1)$
91. $12 \div (-6)$
92. $(-54) \div 9$
93. $(-15) \div (-3)$
94. $32 \div (-4)$
95. $(-27) \div 9$
96. $(-100) \div (-25)$
97. $\dfrac{48}{-16}$
98. $\dfrac{-110}{-11}$
99. $\dfrac{-30}{6}$
100. $\dfrac{124}{-31}$
101. $\dfrac{-42}{-7}$
102. $\dfrac{-51}{17}$
103. $\dfrac{0}{-3}$
104. $\dfrac{-4}{0}$
105. $\dfrac{3}{13} + \dfrac{23}{13}$
106. $\dfrac{8}{19} + \dfrac{7}{19}$
107. $\dfrac{23}{16} - \dfrac{15}{16}$
108. $\dfrac{17}{25} - \dfrac{2}{25}$
109. $\dfrac{3}{12} + \dfrac{5}{16}$
110. $\dfrac{11}{30} + \dfrac{8}{35}$
111. $\dfrac{29}{32} - \dfrac{31}{40}$
112. $\dfrac{7}{90} - \dfrac{4}{27}$
113. $\dfrac{14}{25} \cdot \dfrac{10}{21}$
114. $\left(-\dfrac{11}{24}\right) \cdot \dfrac{3}{22}$
115. $\dfrac{18}{25} \div \dfrac{4}{15}$
116. $\dfrac{33}{35} \div \left(-\dfrac{11}{7}\right)$

In problems 117–126, use the order of operations to perform the indicated operations.

117. $9 + (-5) \cdot 8$
118. $17 - 4 \cdot (-3)$
119. $(-2) \cdot (-3) - (12) \div (-4)$
120. $15 \div (-3) + (-4) \cdot (-5)$
121. $-3(7 - 9) + 12 \div (10 - 12)$
122. $-8 - 2(3 - 1) - 10 \div (6 - 8)$
123. $\dfrac{4}{9} \cdot \dfrac{3}{2} + \dfrac{5}{6}$
124. $\dfrac{7}{15} \div \left(-\dfrac{14}{35}\right) - \dfrac{3}{4} \cdot \dfrac{8}{9}$
125. $\left(-\dfrac{3}{11} \div \dfrac{15}{22}\right) \cdot \dfrac{4}{3} - \dfrac{4}{45}$
126. $\dfrac{8}{25} \cdot \left(-\dfrac{5}{16}\right) + \dfrac{3}{4} \div \left(-\dfrac{9}{16}\right)$

127. An elevator started on the sixth floor and rose 27 floors. It then came down 16 floors and stopped. At which floor did the elevator stop?
128. A football team is 20 yards from making a touchdown. In the next four plays the team gains 6 yards, loses 14 yards, loses another 8 yards, and then gains 18 yards. Does the team make a touchdown?
129. If the temperature is $-10°C$ at 6 A.M. and rises at the rate of 3° per hour, what will the temperature be at 2 P.M.?
130. Pikes Peak is 4,309 meters above sea level and Death Valley is 85 meters below sea level. What is their difference in altitude?

CHAPTER 0 TEST

1. Write each statement in mathematical symbols.
   (a) Thirty-five equals the sum of five times six and five.
   (b) The product of seven and four is greater than nine.

2. Use the order of operations to find the value of each expression.
   (a) $5 \times 9 - 8 \div 2 + 7$
   (b) $\dfrac{22 + 10 \div 2}{24 \div 2 - 3}$

3. Graph the set $\{-5, -3, 0, 2, 4\}$ on a number line.
4. Change 17.3% to a decimal.
5. Change 12% to a fraction.
6. Find $5\frac{1}{2}$% of 120.
7. Find the additive inverse and multiplicative inverse of each number.
   (a) 4
   (b) $\frac{2}{3}$
8. Simplify the expression.
   $$12 + 3[2(10 - 3) - 3(5 - 4)].$$
9. Find the value of each expression.
   (a) $|-13|$
   (b) $-|-5|$
10. Perform each addition or subtraction.
    (a) $(-7) + (-8)$
    (b) $(-15) + 10$
    (c) $(-8) + 12$
    (d) $17 - (-3)$
    (e) $(-12) - (-9)$
    (f) $(-21) - (-40)$
11. Find each product or quotient.
    (a) $(-7) \cdot (-3)$
    (b) $(-3) \cdot 6$
    (c) $(-12) \div 4$
    (d) $(-21) \div (-3)$
    (e) $(-1) \cdot (-1) \cdot (-2) \cdot (-3) \cdot (-5) \cdot 2$
12. Reduce each fraction to lowest terms.
    (a) $\dfrac{60}{84}$
    (b) $\dfrac{-27}{72}$
13. Perform each addition or subtraction and simplify the result.
    (a) $\dfrac{3}{17} + \dfrac{5}{17}$
    (b) $\dfrac{11}{18} - \dfrac{5}{18}$
    (c) $\dfrac{25}{42} + \dfrac{6}{35}$
    (d) $\dfrac{10}{33} - \dfrac{8}{77}$
14. Find each product or quotient and simplify the result.
    (a) $\dfrac{12}{35} \cdot \dfrac{7}{20}$
    (b) $\left(-\dfrac{11}{30}\right) \cdot \dfrac{15}{22}$
    (c) $\dfrac{5}{16} \div \dfrac{15}{32}$
    (d) $\dfrac{18}{33} \div \left(-\dfrac{30}{11}\right)$
    (e) $\dfrac{4}{5} \cdot \left(-\dfrac{10}{7}\right) \div \dfrac{16}{21}$

15. In 2.4512, what is the value of the 5?
    (a) $\dfrac{5}{10}$
    (b) $\dfrac{5}{100}$
    (c) $\dfrac{5}{1000}$
    (d) $\dfrac{512}{10,000}$
    (e) $\dfrac{45}{100}$
16. Round 16.7169 to the nearest hundredth.
    (a) 16.72
    (b) 16.71
    (c) 16.717
    (d) 17
    (e) 16.8
17. Round 25289.3 to the nearest thousand.
    (a) 25300
    (b) 26000
    (c) 25000
    (d) 2500
    (e) 2520
18. Which of the following decimals is the largest?
    (a) 0.02
    (b) 0.022
    (c) 0.0202
    (d) 0.0222
    (e) 0.00222
19. Add 3.14
           5.8
       +13.982
    (a) 22.922
    (b) 21.882
    (c) 21.822
    (d) 11.82
    (e) 11.92
20. Subtract
         325.6
       -144.73
    (a) 279.3
    (b) 281.13
    (c) 180.93
    (d) 181.33
    (e) 180.87
21. Add $2.3 + 8.17 + 14 + 6.397$.
    (a) 22.767
    (b) 725.1
    (c) 30.867
    (d) 72.51
    (e) 7.251
22. Subtract $1000 - 231.8$.
    (a) 178.2
    (b) 879.2
    (c) 779.2
    (d) 868.2
    (e) 768.2
23. Find n: $n - 0.6 = 3.1$.
    (a) 3.7
    (b) 2.5
    (c) 18.6
    (d) 9.1
    (e) 0.51
24. Multiply 2.6×1.9.
    (a) 4.94
    (b) 49.4
    (c) 494
    (d) 4.84
    (e) 48.4

25. Multiply 31.2 × 1.003.
  (a) 4.056   (b) 0.456
  (c) 31.2936   (d) 312.936
  (e) 3129.36

26. Multiply 0.004 × 0.07.
  (a) 0.28   (b) 0.0028
  (c) 0.0028   (d) 0.00028
  (e) 0.000028

27. Divide 6.444 ÷ 1.8.
  (a) 0.347   (b) 3.47
  (c) 0.358   (d) 35.8
  (e) 3.58

28. Divide $0.07\overline{)4.531}$.
  (a) 64.73   (b) 6.47
  (c) 0.65   (d) 63.47
  (e) 6.48

29. Divide 0.397 ÷ 0.51. (Round your answer to the nearest hundredth.)
  (a) 1.28   (b) 0.77
  (c) 7.78   (d) 0.78
  (e) 0.08

30. Find n: $(0.8) \cdot n = 6.4$.
  (a) 7.2   (b) 5.6
  (c) 5.12   (d) 8
  (e) 0.08

31. Write $\frac{7}{20}$ as a decimal.
  (a) 0.286   (b) 2.86
  (c) 0.35   (d) 3.5
  (e) 0.035

32. Write $6\frac{1}{8}$ as a decimal.
  (a) 6.120   (b) 6.180
  (c) 6.125   (d) 0.613
  (e) 0.612

33. Write 0.97 as a fraction.
  (a) $\frac{100}{97}$ (b) $\frac{10}{97}$
  (c) $\frac{97}{10}$ (d) $\frac{97}{100}$
  (e) $\frac{97}{1000}$

34. Write 4.8 as a mixed number.
  (a) $4\frac{8}{100}$ (b) $4\frac{4}{5}$
  (c) $4\frac{4}{50}$ (d) $\frac{48}{50}$
  (e) $4\frac{48}{100}$

Chapter 1

EXPONENTS AND POLYNOMIALS

1.1 Multiplication with Exponents
1.2 Division with Exponents
1.3 Common Roots and Radicals
1.4 Properties of Radicals
1.5 Addition and Subtraction of Radicals
1.6 Multiplication of Radicals
1.7 Division of Radicals
1.8 Applications of Exponents and Radicals

1.1 Multiplication with Exponents

Exponents are a shorthand notation for writing repeated factors. For example,

$$3^4 = 3 \cdot 3 \cdot 3 \cdot 3 = 81$$
$$5^3 = 5 \cdot 5 \cdot 5 = 125$$
$$(-2)^5 = (-2) \cdot (-2) \cdot (-2) \cdot (-2) \cdot (-2) = -32$$
$$\left(-\frac{5}{7}\right)^2 = \left(-\frac{5}{7}\right) \cdot \left(-\frac{5}{7}\right) = \frac{25}{49}$$

and

$$a^6 = a \cdot a \cdot a \cdot a \cdot a \cdot a.$$

In general, if x is a real number and n is a positive integer, the notation x^n represents a **product,** in which the factor x occurs n times. That is,

$$x^n = \overbrace{x \cdot x \cdot x \cdot x \cdots x}^{n \text{ factors}}.$$

We say that the expression x^n is in **exponential form** and we refer to x as the **base** and n as the **exponent** or **power** to which the base is raised.

The expression x^2 is usually read "x squared," and the expression y^3 is read "y cubed." Any other power, such as t^7, is read "the 7th power of t" or "t to the 7th power."

Except section 1.4, from *Beginning Algebra, 4th edition* by M.A. Munem and W. Tschirhart. Copyright © 2000 by Kendall/Hunt Publishing Company. Reprinted by permission.

84 CHAPTER 1 EXPONENTS AND POLYNOMIALS

An exponent applies only to the base to which it is attached. We use grouping symbols (when necessary) to indicate that the base consists of more than one part. For example,

$$5x^2 = 5 \cdot x \cdot x, \qquad rs^4 = r \cdot s \cdot s \cdot s \cdot s, \qquad -4x^6 = -4 \cdot x \cdot x \cdot x \cdot x \cdot x \cdot x,$$

and

$$2y^5 = 2 \cdot y \cdot y \cdot y \cdot y \cdot y,$$

while

$$(2y^5) = (2y) \cdot (2y) \cdot (2y) \cdot (2y) \cdot (2y).$$

The right side of each of these expressions is written in an **expanded form**.

In the preceding definition of x^n, if $n = 1$, then

$$x^1 = x.$$

Thus, when no exponent is shown, it is assumed to be the number 1. For example,

$$2 = 2^1, \qquad -5 = -(5)^1, \qquad \text{and} \qquad u^3v = u^3v^1.$$

Example 1 Identify the base and the exponent in each expression, and find the value of the expression.

(a) 4^3 (b) $(\frac{1}{3})^4$ (c) $(-3)^2$ (d) -3^2 (e) $(-1)^3$ (f) -1^3

Solution (a) $4^3 = 4 \cdot 4 \cdot 4 = 64.$ The base is 4 and the exponent is 3.

(b) $\left(\frac{1}{3}\right)^4 = \frac{1}{3} \cdot \frac{1}{3} \cdot \frac{1}{3} \cdot \frac{1}{3} = \frac{1}{81}.$ The base is $\frac{1}{3}$ and the exponent is 4.

(c) $(-3)^2 = (-3) \cdot (-3) = 9.$ The base is -3 and the exponent is 2.

(d) $-3^2 = -(3^2) = -(3 \cdot 3) = -9.$ Note that because there are no parentheses, as there are in part (c), the base is 3, not -3. The exponent is 2. T

(e) $(-1)^3 = (-1) \cdot (-1) \cdot (-1) = -1.$ The base is -1 and the exponent is 3.

(f) $-1^3 = -(1^3) = -(1 \cdot 1 \cdot 1) = -1.$ The base is 1 and the exponent is 3.

Example 2 Rewrite each expression in exponential form.

(a) $5 \cdot 5 \cdot 5 \cdot 5 \cdot 5 \cdot 5$ (b) $(-4) \cdot (-4) \cdot (-4)$

(c) $x \cdot x \cdot x \cdot x \cdot y \cdot y \cdot y$ (d) $\frac{3}{7} \cdot \frac{3}{7} \cdot \frac{3}{7} \cdot \frac{3}{7}$

Solution (a) $5 \cdot 5 \cdot 5 \cdot 5 \cdot 5 \cdot 5 = 5^6$ (b) $(-4) \cdot (-4) \cdot (-4) = (-4)^3$

(c) $x \cdot x \cdot x \cdot x \cdot y \cdot y \cdot y = x^4y^3$ (d) $\frac{3}{7} \cdot \frac{3}{7} \cdot \frac{3}{7} \cdot \frac{3}{7} = (\frac{3}{7})^4$

Let us consider what happens when 2^3 and 2^2 are multiplied. We can use the definition of exponents and the properties of real numbers to write the product as

$$2^3 \cdot 2^2 = (2 \cdot 2 \cdot 2)(2 \cdot 2)$$
$$= 2 \cdot 2 \cdot 2 \cdot 2 \cdot 2$$
$$= 2^5.$$

Notice that the same result could have been obtained by adding the exponents of the factors 2^3 and 2^2, that is,

$$2^3 \cdot 2^2 = 2^{3+2}$$
$$= 2^5.$$

The exponent 5, in the product, is the sum of the exponents 3 and 2 in the factors.

The product $x^3 \cdot x^4$ can be found in a similar manner:

$$x^3 \cdot x^4 = (x \cdot x \cdot x) \cdot (x \cdot x \cdot x \cdot x)$$
$$= x \cdot x \cdot x \cdot x \cdot x \cdot x \cdot x$$
$$= x^7,$$

so that

$$x^3 \cdot x^4 = x^{3+4}$$
$$= x^7.$$

In general, we have the following property:

PROPERTY 1   Multiplication of Like Bases Property

> Let a be a real number, and let m and n be positive integers. Then
> $$a^m \cdot a^n = a^{m+n}.$$

In other words, to *multiply* two exponential expressions having a common base, we add the exponents and retain the common base.

Example 3   Use Property 1 to find the products of the following expressions. Write each answer in exponential form.

(a) $3^4 \cdot 3^7$ (b) $(-3)^6 \cdot (-3)^{12}$ (c) $x^3 \cdot x$ (d) $(-a)^3 \cdot (-a)^4$

Solution   (a) $3^4 \cdot 3^7 = 3^{4+7} = 3^{11}$

(b) $(-3)^6 \cdot (-3)^{12} = (-3)^{6+12}$
$= (-3)^{18}$
$= 3^{18}$ (why?)

(c) $x^3 \cdot x = x^3 \cdot x^1 = x^{3+1} = x^4$

(d) $(-a)^3 \cdot (-a)^4 = (-a)^{3+4}$
$= (-a)^7$
$= -a^7$ (why?)

Warning: When you try to find a product such as $4^3 \cdot 4^5$, do *not* multiply the bases to get 16^8. Be sure to *keep the common base:*

$$4^3 \cdot 4^5 = 4^{3+5}$$
$$= 4^8.$$

Now let us consider a procedure for simplifying exponential expressions such as $(5^4)^3$. The exponent 3 tells us that the base 5^4 is a factor three times:

$$(5^4)^3 = 5^4 \cdot 5^4 \cdot 5^4 = \overbrace{5 \cdot 5 \cdot 5 \cdot 5 \cdot 5 \cdot 5 \cdot 5 \cdot 5 \cdot 5 \cdot 5 \cdot 5 \cdot 5}^{12 \text{ factors}}$$
$$= 5^{12}$$

Because $3 \cdot 4 = 12$, we have

$$(5^4)^3 = 5^{4 \cdot 3}$$
$$= 5^{12}.$$

In general, we have the following property:

PROPERTY 2   Power of a Power Property

> Let a be a real number, and let m and n be positive integers. Then
> $$(a^m)^n = a^{mn}.$$

In other words, to raise a **power to a power,** we multiply the exponents and keep the base.

Example 9 Use Property 2 to simplify the following expressions. Write each answer in exponential form.

(a) $(4^2)^6$ (b) $[(-2)^3]^5$ (c) $(x^{10})^8$ (d) $[(-a)^2]^3$

Solution
(a) $(4^2)^6 = 4^{2 \cdot 6} = 4^{12}$
(b) $[(-2)^3]^5 = (-2)^{3 \cdot 5} = (-2)^{15} = -2^{15}$
(c) $(x^{10})^8 = x^{10 \cdot 8} = x^{80}$
(d) $[(-a)^2]^3 = (-a)^{2 \cdot 3} = (-a)^6 = a^6$

Now we develop a property for raising a product to a power. Consider the expression $(3y)^4$. The expression can be rewritten as follows:

$$(3y)^4 = (3y)(3y)(3y)(3y)$$
$$= 3 \cdot 3 \cdot 3 \cdot 3 \cdot y \cdot y \cdot y \cdot y \quad \text{(Why?)}$$
$$= 3^4 \cdot y^4.$$

Therefore,
$$(3y)^4 = 3^4 \cdot y^4.$$

Note that we would have obtained the same result by simply raising each factor of the base to the indicated power.

In general, we have the following property:

PROPERTY 3 Power of a Product Property

> Let a and b be real numbers, and let n be a positive integer. Then
> $$(ab)^n = a^n b^n.$$

In other words, to raise a **product to a power,** we raise each factor to the indicated power.

Example 5 Use Property 3 to rewrite the following expressions and simplify the result,

(a) $(5x)^2$ (b) $(xy)^3$ (c) $(-x)^5$ (d) $(-x)^6$ (e) $(xyz)^4$

Solution
(a) $(5x)^2 = 5^2 x^2 = 25x^2$ (b) $(xy)^3 = x^3 y^3$
(c) $(-x)^5 = [(-1)]x]^5$ (d) $(-x)^6 = [(-1)]x]^6$
$\quad = (-1)^5 x^5$ $\quad = (-1)^6 x^6$
$\quad = (-1)x^5$ $\quad = (1)x^6$
$\quad = -x^5$ $\quad = x^6$

(e) $(xyz)^4 = [x(yz)]^4$
$\quad = x^4(yz)^4 = x^4 y^4 z^4$

Example 6 Use the properties of exponents to rewrite each expression in a simpler form.

(a) $(2x^3)^4$ (b) $(2x^2y^3)^3$ (c) $(-3xy^2)^2(2x^3y)^3$

Solution
(a) $(2x^3)^4 = 2^4(x^3)^4$ (Property 3)
$\quad\quad\quad\, = 16x^{12}$ (Property 2)

(b) $(2x^2y^3)^3 = 2^3(x^2)^3(y^3)^3$ (Property 3)
$\quad\quad\quad\quad\, = 8x^6 y^9$ (Property 2)

(c) $(-3xy^2)^2(2x^3y)^3 = (-3)^2 \cdot x^2 \cdot (y^2)^2 \cdot 2^3(x^3)^3 \cdot y^3$ (Property 3)
$\quad\quad\quad\quad\quad\quad\quad = 9x^2 y^4 \cdot 8x^9 y^3$ (Property 2)
$\quad\quad\quad\quad\quad\quad\quad = 9 \cdot 8 \cdot x^{2+9} y^{4+3}$ (Property 1)
$\quad\quad\quad\quad\quad\quad\quad = 72 x^{11} y^7$

Scientific Notation

Many problems in science and business require the use of very large numbers. For example, the speed C of light is given by
$$C = 300{,}000{,}000 \text{ meters per second.}$$
The number N of heartbeats in a person's normal lifetime is given by
$$N = 2{,}200{,}000{,}000.$$
A convenient way of expressing such numbers is by scientific notation. A positive number x is written in **scientific notation** if x has the form
$$x = s \times 10^n,$$
where s is a number between 1 and 10 ($1 \leq s < 10$) and n is a positive integer.

> To change a large number (greater than or equal to 10) from ordinary decimal form to scientific notation, we move the decimal point n places to the left to obtain a number between 1 and 10. Then, we multiply this number by 10^n.

For example, to change the number 468,000 to scientific notation, we move the decimal point 5 places to the left, then we multiply this number by 10^5. That is,
$$468{,}000 = 4.68 \times 10^5.$$
Also,
$$300{,}000{,}000 = 3 \times 10^8 \quad \text{and} \quad 2{,}200{,}000{,}000 = 2.2 \times 10^9.$$

Example 7 Rewrite 487,000,000 in scientific notation.

Solution We move the decimal point 8 places to the left. Then, we multiply this number by 10^8. The number is written in scientific notation as
$$487{,}000{,}000 = 4.87 \times 10^8.$$
This procedure can be reversed whenever a number is given in scientific notation and we wish to write it in ordinary decimal form.

Example 8 Rewrite 7.85×10^4 in ordinary decimal form.

Solution We first note that $10^4 = 10{,}000$, then we multiply 7.85 by 10,000. That is,
$$7.85 \times 10^4 = 7.85 \times 10{,}000 = 78{,}500.$$
Multiplying 7.85 by 10^4 simply moves the decimal point 4 places to the right. Therefore, we can omit the middle step, and write
$$7.85 \times 10^4 = 78{,}500.$$
Many calculators automatically switch to scientific notation whenever the number is too large to be displayed in ordinary decimal form. When a number such as 4.675×10^{13} is displayed, the multiplication sign and the base do not appear and the display shows simply
$$4.675 \qquad 13.$$

PROBLEM SET 1.1

In problems 1–10, identify the base and the exponent in each expression.

1. 2^4
2. 3^{11}
3. $(-3)^5$
4. $(-7)^8$
5. -7^2
6. $-t^3$
7. x^{10}
8. $(2x)^5$
9. $-y^8$
10. $-(-7t)^2$

In problems 11–20, rewrite each expression in exponential form.

11. $3 \cdot 3 \cdot 3 \cdot 3$
12. $5 \cdot 5 \cdot 5$
13. $(-2) \cdot (-2) \cdot (-2) \cdot (-2)$
14. $(-4) \cdot (-4) \cdot (-7) \cdot (-7)$
15. $u \cdot u \cdot u \cdot u \cdot u \cdot u \cdot u$
16. $(-z) \cdot (-z) \cdot (-z) \cdot (-z) \cdot (-z) \cdot (-z)$
17. $x \cdot x \cdot x \cdot y \cdot y$
18. $-5 \cdot w \cdot w \cdot w \cdot z \cdot z \cdot z \cdot z \cdot z$
19. $-8 \cdot r \cdot r \cdot s \cdot s \cdot s + t \cdot t \cdot t \cdot t$
20. $2 \cdot u \cdot u \cdot u + 9 \cdot v \cdot v \cdot v \cdot v$

In problems 21–30, find the value of each expression.

21. 2^3
22. $(-3)^5$
23. -3^4
24. 6^3
25. $(-4)^3$
26. -7^2
27. $-(-1)^5$
28. $3^4 + 5^2$
29. $(\frac{1}{2})^6$
30. $2^3 + (-2)^3$

In problems 31–40, use Property 1 to find the products of the expressions. Write each answer in exponents form.

31. $2^3 \cdot 2^2$
32. $5^2 \cdot 5^4$
33. $3^5 \cdot 3$
34. $(-4)^3(-4)^5$
35. $(-5)^4(-5)^7$
36. $7^4 \cdot 7$
37. $m^3 m^5$
38. $y^8 y^8$
39. $(-x)^7(-x)^2$
40. $t^3 t^4 t$

In problems 41–50, use Property 2 to simplify the expressions. Write each answer in exponential form.

41. $(3^2)^3$
42. $(4^3)^8$
43. $[(-2)^4]^2$
44. $[(-5)^3]^4$
45. $(x^4)^3$
46. $(y^7)^5$
47. $[(-m)^2]^5$
48. $(u^8)^9$
49. $(z^5)^6$
50. $[(-t)^3]^3$

In problems 51–60, use Property 3 to simplify the expressions. Write each answer in exponential form.

51. $(3x)^2$
52. $(-2x)^3$
53. $(uv)^6$
54. $(rs)^5$
55. $(-4t)^3$
56. $(-5z)^2$
57. $[-2(-x)]^4$
58. $[-3(-v)]^3$
59. $(3xy)^2$
60. $(-4rst)^4$

In problems 61–66, rewrite each number in scientific notation.

61. 3,782
62. 385,000
63. 384,000
64. 5,800
65. 780,000,000
66. 431,000,000

In problems 67–72, rewrite each number in expanded form.

67. 2.1×10^2
68. 8.6×10^3
69. 7.5×10^5
70. 5.41×10^4
71. 3.12×10^7
72. 1.87×10^8

In problems 73–76, rewrite each statement so that all numbers are expressed in scientific notation.

73. The diameter of the sun is approximately 870,000 miles.
74. The sun contains about 1,489,000,000,000,000,000,000,000,000 tons of hydrogen.
75. The mass of the earth is approximately 5,980,000,000,000,000,000,000,000 tons.
76. One gram of hydrogen contains 602,300,000,000,000,000,000,000 atoms.

1.2 Division with Exponents

In Section 1.1, we saw that multiplication of two powers of the same base is accomplished by adding exponents. Now we consider how to divide two powers of the same base.

In order to find the quotient

$$\frac{a^m}{a^n},$$ where $a \neq 0$ and m and n are positive integers,

we must examine three cases: when m is greater than n, when m is equal to n, and when m is less than n.

To illustrate the first case, when m is greater than n, we consider the quotient $5^6/5^4$. This is written as

$$\frac{5^6}{5^4} = \frac{5 \cdot 5 \cdot 5 \cdot 5 \cdot 5 \cdot 5}{5 \cdot 5 \cdot 5 \cdot 5}$$
$$= 5 \cdot 5 = 5^2.$$

Note that the difference $6 - 4$ gives the new exponent 2. That is,

$$\frac{5^6}{5^4} = 5^{6-4} = 5^2.$$

Similarly, x^5/x^3 can be found by writing

$$\frac{x^5}{x^3} = \frac{x \cdot x \cdot x \cdot x \cdot x}{x \cdot x \cdot x}$$
$$= x \cdot x = x^2.$$

Again, we would have obtained the same result by simply subtracting exponents:

$$\frac{x^5}{x^3} = x^{5-3} = x^2.$$

To see what happens to the quotient a^m/a^n when $m = n$ and we try subtracting exponents, consider the quotient $2^6/2^6$. We obtain

$$\frac{2^6}{2^6} = 2^{6-6} = 2^0.$$

However,

$$\frac{2^6}{2^6} = \frac{64}{64} = 1.$$

Therefore, we must define

$$2^0 = 1.$$

In general, we have the following definition:

DEFINITION 1   **Zero Exponent**

> If a is any nonzero real number, then
> $$a^0 = 1.$$

Notice that 0^0 is not defined.

To see what happens to the quotient a^m/a^n when $m < n$, and we try subtracting exponents, consider the quotient $5^3/5^7$. We have

$$\frac{5^3}{5^7} = 5^{3-7} = 5^{-4}.$$

Since 5^{-4} has not yet been defined, we note that

$$\frac{5^3}{5^7} = \frac{5 \cdot 5 \cdot 5}{5 \cdot 5 \cdot 5 \cdot 5 \cdot 5 \cdot 5 \cdot 5}$$
$$= \frac{1}{5 \cdot 5 \cdot 5 \cdot 5} = \frac{1}{5^4}.$$

Therefore, we must define

$$5^{-4} = \frac{1}{5^4}.$$

In general we have the following:

DEFINITION 2   **Negative Integer Exponent**

> If n is a positive integer and a is any nonzero real number, then
> $$a^{-n} = \frac{1}{a^n}.$$

Example 1 Use Definition 1 or 2 to evaluate the following expressions.

(a) 3^0 (b) $(-7)^0$ (c) -7^0 (d) 4^{-3}
(e) -3^{-4} (f) $(-3)^{-4}$ (g) $(x+1)^0$ (h) y^{-4}

Solution

(a) $3^0 = 1$

(b) $(-7)^0 = 1$

(c) $-7^0 = -(7^0) = -(1) = -1$

(d) $4^{-3} = \dfrac{1}{4^3} = \dfrac{1}{64}$

(e) $-3^{-4} = -(3^{-4}) = -\dfrac{1}{3^4} = -\dfrac{1}{81}$

(f) $(-3)^{-4} = \dfrac{1}{(-3)^4} = \dfrac{1}{81}$

(g) $(x+1)^0 = 1,$ if $x + 1 \neq 0$ or $x \neq -1$

(h) $y^{-4} = \dfrac{1}{y^4}$

With the above definitions, we can now state the following property.

PROPERTY 1 Division of Like Bases Property

> If m and n are positive integers and $a \neq 0$, then
> $$\dfrac{a^m}{a^n} = a^{m-n}.$$

Thus, the quotient of two powers of the same base is obtained by subtracting exponents.

Example 2 Use Property 1 to find each quotient. Write the results using only positive exponents, and then simplify, if possible.

(a) $\dfrac{3^4}{3^2}$ (b) $\dfrac{x^4}{x^4}$ (c) $\dfrac{2^3}{2^7}$ (d) $\dfrac{(-y)^7}{(-y)^3}$ (e) $\dfrac{(-z)^2}{(-z)^5}$

Solution

(a) $\dfrac{3^4}{3^2} = 3^{4-2} = 3^2 = 9$

(b) $\dfrac{x^4}{x^4} = x^{4-4} = x^0 = 1,$ if $x \neq 0$

(c) $\dfrac{2^3}{2^7} = 2^{3-7} = 2^{-4} = \dfrac{1}{2^4} = \dfrac{1}{16}$

(d) $\dfrac{(-y)^7}{(-y)^3} = (-y)^{7-3} = (-y)^4 = y^4$

(e) $\dfrac{(-z)^2}{(-z)^5} = (-z)^{2-5} = (-z)^{-3} = \dfrac{1}{(-z)^3} = \dfrac{1}{-z^3} = -\dfrac{1}{z^3}$

Now we show how to simplify expressions in which a quotient is raised to a power. For example, consider the expression $(x/y)^4$. We can rewrite this expression as follows:

$$\left(\dfrac{x}{y}\right)^4 = \dfrac{x}{y} \cdot \dfrac{x}{y} \cdot \dfrac{x}{y} \cdot \dfrac{x}{y}$$
$$= \dfrac{x \cdot x \cdot x \cdot x}{y \cdot y \cdot y \cdot y}$$
$$= \dfrac{x^4}{y^4}.$$

We would have obtained the same result by simply raising both the numerator and the denominator of the base to the indicated power.

In general, we have the following property:

PROPERTY 2 Power of a Quotient Property

> Let a and b be real numbers where $b \neq 0$, and let n be a positive integer. Then
> $$\left(\frac{a}{b}\right)^n = \frac{a^n}{b^n}.$$

In other words, to find a **power of a quotient,** raise both the numerator and the denominator to the indicated power.

Example 3 Use Property 2 to rewrite the following expressions, and simplify the results when possible.

(a) $\left(\dfrac{3}{4}\right)^2$ (b) $\left(\dfrac{z}{y}\right)^6$ (c) $\left(\dfrac{-x}{y}\right)^3$

Solution (a) $\left(\dfrac{3}{4}\right)^2 = \dfrac{3^2}{4^2} = \dfrac{9}{16}$ (b) $\left(\dfrac{z}{y}\right)^6 = \dfrac{z^6}{y^6}$

(c) $\left(\dfrac{-x}{y}\right)^3 = \dfrac{(-x)^3}{y^3} = -\dfrac{-x^3}{y^3} = -\dfrac{x^3}{y^3}$

Example 4 Use the properties of exponents to rewrite each expression in a simpler form.

(a) $\left(\dfrac{x^4}{y^3}\right)^5$ (b) $\dfrac{(t^2)^{17}}{(t^6)^4(t^2)^5}$ (c) $\dfrac{9x^2(y^4)^3}{(3x^2y^3)^2}$

Solution (a) $\left(\dfrac{x^4}{y^3}\right)^5 = \dfrac{(x^4)^5}{(y^3)^5}$ (Property 2, Section 1.2)

$\phantom{\left(\dfrac{x^4}{y^3}\right)^5} = \dfrac{x^{20}}{y^{15}}$ (Property 2, Section 1.1)

(b) $\dfrac{(t^2)^{17}}{(t^4)^6(t^2)^5} = \dfrac{t^{34}}{t^{24}t^{10}}$ (Property 2, Section 1.1)

$\phantom{\dfrac{(t^2)^{17}}{(t^4)^6(t^2)^5}} = \dfrac{t^{34}}{t^{34}}$ (Property 1, Section 1.1)

$\phantom{\dfrac{(t^2)^{17}}{(t^4)^6(t^2)^5}} = t^0 = 1$ (Property 1, Section 1.2)

(c) $\dfrac{9x^2(y^4)^3}{(3x^2y^3)^2} = \dfrac{9x^2y^{12}}{3^2(x^2)^2(y^3)^2}$ (why?)

$\phantom{\dfrac{9x^2(y^4)^3}{(3x^2y^3)^2}} = \dfrac{9x^2y^{12}}{9x^4y^6} = x^{2-4}y^{12-6} = x^{-2}y^6 = \dfrac{1}{x^2} \cdot y^6 = \dfrac{y^6}{x^2}$

The properties of exponents also apply for exponents that are negative integers or zero.

Example 5 Use the properties of exponents to rewrite each expression in a simpler form.

(a) $x^4 \cdot x^{-3}$ (b) $(3^{-2})^{-1}$ (c) $\dfrac{y^{-2}}{y^{-5}}$

(d) $\left(\dfrac{x}{3}\right)^{-2}$ (e) $\left(\dfrac{x^{-2}}{y^{-3}}\right)^{-4}$ (f) $(x^{-3}y^2)^{-2}$

Solution

(a) $x^4 \cdot x^{-3} = x^{4+(-3)}$ (Property 1, Section 1.1)
$= x^1 = x$

(b) $(3^{-2})^{-1} = 3^{(-2)(-1)}$ (Property 2, Section 1.1)
$= 3^2 = 9$

(c) $\dfrac{y^{-2}}{y^{-5}} = y^{-2-(-5)}$ (Property 1, Section 1.2)
$= y^3$

(d) $\left(\dfrac{x}{3}\right)^{-2} = \dfrac{x^{-2}}{3^{-2}} = \dfrac{1/x^2}{1/3^2} = \dfrac{1}{x^2} \cdot \dfrac{3^2}{1}$
$= \dfrac{3^2}{x^2} = \dfrac{9}{x^2}$

(e) $\left(\dfrac{x^{-2}}{y^{-3}}\right)^{-4} = \dfrac{(x^{-2})^{-4}}{(y^{-3})^{-4}}$ (Property 2, Section 1.2)
$= \dfrac{x^{(-2)(-4)}}{y^{(-3)(-4)}}$ (why?)
$= \dfrac{x^8}{y^{12}}$

(f) $(x^{-3}y^2)^{-2} = (x^{-3})^{-2}(y^2)^{-2}$ (Property 3, Section 1.1)
$= x^{(-3)(-2)}y^{(2)(-2)}$ (why?)
$= x^6 y^{-4}$
$= x^6 \cdot \dfrac{1}{y^4} = \dfrac{x^6}{y^4}$

More on Scientific Notation

Recall from Section 1.1 that a large number x greater than 10 is written in scientific notation as
$$x = s \times 10^n$$
where $1 \leq s < 10$, and n is a positive integer. Now that the meaning of zero and negative integer exponents has been introduced, we can also write positive numbers less than 1 in scientific notation.

A positive number x is written in scientific notation, if x has the form
$$x = s \times 10^n$$
where $1 \leq s < 10$, and n is *any* integer.

> To change a small number (less than 1) from ordinary decimal form to scientific notation, we move the decimal point n places to the *right* to obtain a number between 1 and 10. Then we multiply this number by 10^{-n}.

For example, to change the number 0.00573 to scientific notation, we move the decimal point 3 places to the right. Then we multiply this number by 10^{-3}. That is,
$$0.00573 = 5.73 \times 10^{-3}.$$

The following table lists additional examples of numbers written in ordinary decimal form and in scientific notation.

Ordinary Decimal Form	Numbers in Numbers in Scientific Notation
6	6×10^0
0.35	3.5×10^{-1}
0.078	7.8×10^{-2}
0.00567	5.67×10^{-3}
0.000384	3.84×10^{-4}
0.000083	8.3×10^{-5}

The properties of exponents can be used to simplify expressions in involving numbers written in scientific notation.

Example 6 Simplify the expression and write the answer in scientific notation.
$$\frac{(6.75 \times 10^{-5}) \cdot (8.25 \times 10^{-4})}{13.50 \times 10^{-11}}.$$

Solution
$$\frac{(6.75 \times 10^{-5}) \cdot (8.25 \times 10^{-4})}{13.50 \times 10^{-11}} = \frac{(6.75) \cdot (8.25)}{13.50} \times \frac{10^{-5} \cdot 10^{-4}}{10^{-11}}$$
$$= 4.125 \times 10^2.$$

Applications Involving Exponents

Banks and loan associations use the formula
$$M = \frac{Pr}{12\left[1 - \left(1 + \frac{r}{12}\right)^{-12t}\right]}$$

to calculate the size M of a **monthly car payment** or a **monthly home mortgage payment,** where P is the amount borrowed, r is the interest rate (in decimal form), and t is the time in years.

Example 7 Suppose that a worker borrows $9,000 from a credit union at 7.9% interest for 4 years to purchase a car. What is the size of the monthly payment for the worker to purchase the car?

Solution Substituting $P = \$9,000$, $r = 0.079$, and $t = 4$ into the formula, we have
$$M = \frac{Pr}{12\left[1 - \left(1 + \frac{r}{12}\right)^{-12t}\right]} = \frac{(9,000)(0.079)}{12\left[1 - \left(1 + \frac{0.079}{12}\right)^{-12(4)}\right]}$$

Therefore, the monthly car payment is $219.29.

Example 8 An engineer borrowed $109,000 on his house at 9.5% for 30 years. What is the monthly payment on the home mortgage for the engineer?

Solution Substituting $P = \$109,000$, $r = 0.095$, and $t = 30$ in the formula, we have
$$M = \frac{Pr}{12\left[1 - \left(1 + \frac{r}{12}\right)^{-12t}\right]} = \frac{(109,000)(0.095)}{12\left[1 - \left(1 + \frac{0.095}{12}\right)^{-12(30)}\right]}$$
$$= 916.53$$

Therefore, the monthly mortgage payment is $916.53.

PROBLEM SET 1.2

In problems 1–10, write each expression with a positive exponent, then simplify.

1. 4^{-2}
2. 8^{-2}
3. 10^{-3}
4. 10^{-5}
5. $(-2)^{-3}$
6. $(-3)^{-5}$
7. -3^{-4}
8. -7^{-2}
9. $7x^{-3}$
10. $9p^{-4}$

In problems 11–18, simplify each expression.

11. 5^0
12. 9^0
13. $(-5)^0$
14. $(-9)^0$
15. -5^0
16. -9^0
17. $(3+x)^0$
18. $(-7-y)^0$

In problems 19–40, use Property 1 to rewrite each expression, and simplify the result when possible.

19. $\dfrac{2^4}{2^2}$
20. $\dfrac{3^8}{3^5}$
21. $\dfrac{(-3)^5}{(-3)^2}$
22. $\dfrac{(-5)^5}{(-5)^3}$
23. $\dfrac{z^6}{z^3}$
24. $\dfrac{s^{11}}{s^7}$
25. $\dfrac{x^4}{x^7}$
26. $\dfrac{z^4}{z^{10}}$
27. $\dfrac{t^{15}}{t^{15}}$
28. $\dfrac{(-x)^7}{(-x)^7}$
29. $\dfrac{(-u)^8}{(-u)^5}$
30. $\dfrac{(-y)^4}{(-y)^9}$
31. $\dfrac{(-z)^4}{(-z)^4}$
32. $\dfrac{w^{12}}{w^{18}}$
33. $\dfrac{r^7}{r^{11}}$
34. $\dfrac{(-v)^3}{(-v)^{11}}$
35. $\dfrac{10^{-2}}{10^{-5}}$
36. $\dfrac{x^{-1}}{x^{-1}}$
37. $\dfrac{y^{-5}}{y}$
38. $\dfrac{t^8}{t^{-3}}$
39. $\dfrac{x^{-4}}{x^{-3}}$
40. $\dfrac{a^{-3}}{a^{-8}}$

In problems 41–54, use Property 2 to rewrite each expression, and simplify the result when possible.

41. $\left(\dfrac{2}{3}\right)^3$
42. $\left(\dfrac{4}{5}\right)^2$
43. $\left(\dfrac{x}{y}\right)^4$
44. $\left(\dfrac{m}{n}\right)^3$
45. $\left(\dfrac{-u}{v}\right)^5$
46. $\left(\dfrac{-y}{z}\right)^7$
47. $\left(\dfrac{5}{-t}\right)^3$
48. $\left(\dfrac{w}{-u}\right)^9$
49. $\left(\dfrac{-z}{w}\right)^4$
50. $\left(\dfrac{-y}{x}\right)^{12}$
51. $\left(\dfrac{4}{5}\right)^{-3}$
52. $\left(\dfrac{2}{x}\right)^{-2}$
53. $\left(\dfrac{-3}{z}\right)^{-2}$
54. $\left(\dfrac{5}{-y}\right)^{-4}$

In problems 55–66, use the properties of exponents to rewrite each expression so that it contains only positive exponents and simplify.

55. $5^{-2} \cdot 5^{-3}$
56. $3^{-4} \cdot 3^2$
57. $y^{-7} \cdot y^{-2}$
58. $(-3^{-2})^{-1}$
59. $(p^{-3})^{-4}$
60. $(x^{-2}y^4)^{-2}$
61. $(-5z^{-2})^4$
62. $(4a^{-2}b^{-3})^3$
63. $(p^{-1}y^{-3})^{-2}$
64. $\left(\dfrac{x^{-2}}{y^{-1}}\right)^{-3}$
65. $\left(\dfrac{x^{-3}}{y^{-7}}\right)^{-4}$
66. $\dfrac{(x^{-3})^7}{(x^2)^{-2} \cdot (x^{-16})^2}$

C In problems 67–70, use the formula on page 81 to calculate the amount of monthly car payment M for the amount borrowed P for t years, at interest rate r if

67. $P = \$7{,}000$, $r = 7.9\%$, $t = 4$ years.
68. $P = \$16{,}000$, $r = 11.2\%$ $t = 4$ years.
69. $P = \$14{,}500$, $r = 5.6\%$, $t = 3$ years.
70. $P = \$11{,}700$, $r = 3.9\%$, $t = 5$ years.

C In problems 71–74, use the formula on page 81 to calculate the amount of a monthly house payment M on a home mortgage P for t years at interest rate r if

71. $P = \$19{,}000$, $r = 9\%$, $t = 30$ years.
72. $P = \$32{,}000$, $r = 9.5\%$, $t = 35$ years.
73. $P = \$65{,}000$, $r = 8.75\%$, $t = 15$ years.
74. $P = \$78{,}000$, $r = 8.75\%$, $t = 15$ years.

In problems 75–78, simplify the expression and write the answer in scientific notation.

75. 0.038
76. 7
77. 0.000069
78. 0.000000238

In problems 79–82, simplify the expression and write the answer in scientific notation. Round off each answer to four decimal places.

79. $\dfrac{(4.72 \times 10^{-3}) \cdot (2.31 \times 10^4)}{5.62 \times 10^3}$

80. $\dfrac{(3.91 \times 10^{-5}) \cdot (4.62 \times 10^3)}{(2.31 \times 10^{-2}) \cdot (3.9 \times 10^4)}$

81. $\dfrac{(5.81 \times 10^4) \cdot (3.37 \times 10^{-8})}{(4.52 \times 10^7) \cdot (2.42 \times 10^{-2})}$

82. $\dfrac{(5.73 \times 10^{-6}) \cdot (14.3 \times 10^{-8})}{(3.4 \times 10^2) \cdot (1.94 \times 10^4)}$

1.3 Common Roots and Radicals

Recall from Section 1.1 that

$$3^2 = 9 \quad \text{and} \quad (-3)^2 = 9.$$

We say that 9 is the square of 3, and that 9 is also the square of -3. We refer to the base of each exponential expression as a *square root* and say that 3 and -3 are square roots of 9. Similarly, we say that both 5 and -5 are square roots of 25 because

$$5^2 = 25 \quad \text{and} \quad (-5)^2 = 25.$$

More formally, we define a square root as follows:

DEFINITION 1 Square Root

> If a is a real number. we say that the real number b is a **square root** of a if $b^2 = a$.

Every positive number has two square roots, one positive and one negative. For example,

16 has 4 and -4 as its square roots, since $4^2 = 16$ and $(-4)^2 = 16$.

We refer to the positive square root as the **principal square root.** We use the symbol $\sqrt{16}$ to represent the principal square root of 16, which is 4. Thus,

$$\sqrt{16} = 4 \quad \text{since} \quad 4^2 = 16, \quad \sqrt{25} = 5 \quad \text{since} \quad 5^2 = 25,$$

and

$$\sqrt{49} = 7 \quad \text{since} \quad 7^2 = 49.$$

In general, if a and b are positive real numbers, we write:

> $$\sqrt{a} = b \text{ if and only if } b^2 = a.$$

We call b the principal square root of a.

We use the symbol $-\sqrt{16}$ to represent the negative square root of 16, which is -4. Thus,

$$\sqrt{16} - \quad 4.$$

Also,

$$-\sqrt{25} = -5.$$

Note: It is a common error to think of $\sqrt{16}$ as representing either 4 or -4; $\sqrt{16}$ means the principal square root of 16, which is only 4.

In general, if a is any positive number, then $\sqrt{a}$ represents the **positive square root** of a, and $-\sqrt{a}$ represents the negative square root of a. It follows that

$$(\sqrt{a})^2 = a \quad \text{and} \quad (-\sqrt{a})^2 = a.$$

So far we have only considered square roots of positive numbers. Note that the number 0 has only one square root, that is,

$$\sqrt{0} = 0.$$

If we attempt to find the square root of a negative number, we run into trouble. For example, if we try to find a square root of -4, we are seeking a real number $b = \sqrt{-4}$ such that $b^2 = -4$. But this is impossible, since the square of any real number is either positive or zero. Therefore, $\sqrt{-4}$ does not represent a real number.

The concept of square root can be extended to find cube roots, fourth roots, and so on.

The **cube root** of a number is one of the number's three equal factors. We use the symbol $\sqrt[3]{a}$ to represent the cube root of a. We say

$$\sqrt[3]{27} = 3 \quad \text{since} \quad 3^3 = 27,$$
$$\sqrt[3]{-8} = -2 \quad \text{since} \quad (-2)^3 = -8,$$

and

$$\sqrt[3]{-64} = -4 \quad \text{since} \quad (-4)^3 = -64.$$

In general, we write:

$$\sqrt[3]{a} = b \text{ if and only if } b^3 = a.$$

The **fourth root** (denoted by $\sqrt[4]{\ }$), the **fifth root** (denoted by $\sqrt[5]{\ }$), the **sixth root** (denoted by $\sqrt[6]{\ }$). and the **nth root** (denoted by $\sqrt[n]{\ }$) are defined in a similar manner. We say that 2 and -2 are the fourth roots of 16, since

$$2^4 = 16 \quad \text{and} \quad (-2)^4 = 16.$$

Also, the fifth root of 32 is 2, because $2^5 = 32$. Thus, if n is odd

$$\sqrt[n]{a} = b \text{ if and only if } b^n = a.$$

If we agree that a square root, a fourth root, or a sixth root is an even root: while a cube root, a fifth root, or a seventh root is an odd root, then we can make the following observations:

1. Every positive number has two even roots, one of which is positive and the other negative. For example. $25 = 5^2 = (-5)^2$.
2. Every positive number has one odd root that is positive. For example, $8 = 2^3$, but $8 \neq (-2)^3$.
3. Every negative number has one odd root that is negative. For example, $-27 = (-3)^3$, but $-27 \neq 3^3$.

When a number has both a positive and a negative root, we say that the positive root is the **principal nth root** of the number. Therefore, 10 is the principal square root of 100, and 10 is also the principal fourth root of 10,000. So when we write $\sqrt[4]{81}$ we mean the principal fourth root of 81, which is 3. Thus, $\sqrt[4]{81} = 3$. We use the symbol $-\sqrt[4]{81}$ to represent the negative fourth root of 81, which is -3; that is, $-\sqrt[4]{81} = -3$. Note that a negative number does not have an even root (because it is not defined, or it does not exist as a real number). For example,

$$\sqrt[4]{-16} \text{ is not defined.}$$

In the expression $\sqrt[n]{a}$, the symbol $\sqrt{\ }$ is called the **radical** symbol (or **radical** sign); the positive integer n is called the **index**; and the number a under the radical is called the **radicand:**

index $\longrightarrow \sqrt[n]{a} \longleftarrow$ radicand
radical symbol
or radical sign

1.3 COMMON ROOTS AND RADICALS

If the index is 2, it is customary to omit the index when the radical sign is used: thus, we write $\sqrt{9}$ rather than $\sqrt[2]{9}$. We show the index only for those values of n greater than 2.

Following is a list of the most common roots. These will be used in the remainder of the chapter.

Principal Square Roots		Principal Cube Roots	Principal Fourth Roots	Principal Fifth Roots
$\sqrt{1} = 1$	$\sqrt{49} = 7$	$\sqrt[3]{1} = 1$	$\sqrt[4]{1} = 1$	$\sqrt[5]{1} = 1$
$\sqrt{4} = 2$	$\sqrt{64} = 8$	$\sqrt[3]{8} = 2$	$\sqrt[4]{16} = 2$	$\sqrt[5]{32} = 2$
$\sqrt{9} = 3$	$\sqrt{81} = 9$	$\sqrt[3]{27} = 3$	$\sqrt[4]{81} = 3$	$\sqrt[5]{243} = 3$
$\sqrt{16} = 4$	$\sqrt{100} = 10$	$\sqrt[3]{64} = 4$	$\sqrt[4]{256} = 4$	
$\sqrt{25} = 5$	$\sqrt{121} = 11$	$\sqrt[3]{125} = 5$	$\sqrt[4]{625} = 5$	
$\sqrt{36} = 6$	$\sqrt{144} = 12$	$\sqrt[3]{216} = 6$		

In Examples 1–7, evaluate each expression.

Example 1 $\sqrt{49}$

Solution $\sqrt{49}$ is the positive number whose square is 49, that is,

$$\sqrt{49} = 7 \quad \text{since} \quad 7^2 = 49.$$

Example 2 $-\sqrt{121}$

Solution $\sqrt{121} = 11$ since $11^2 = 121$. Thus, $-\sqrt{121} = -11$.

Example 3 $\sqrt[3]{-64}$

Solution   Recall that a negative number has a negative root if the index n is odd. Therefore, $\sqrt[3]{-64} = -4$ since $(-4)^3 = -64$.

Example 4 $-\sqrt[5]{-32}$

Solution $\sqrt[5]{-32} = -2$ since $(-2)^5 = -32$. Thus, $-\sqrt[5]{-32} = -(-2) = 2$.

Example 5 $\sqrt[4]{0.0001}$

Solution $\sqrt[4]{0.0001} = 0.1$ since $(0.1)^4 = 0.0001$.

Example 6 $\sqrt{-25}$

Solution   Recall that even roots of negative numbers do not exist as real numbers. Therefore, $\sqrt{-25}$ does not exist as a real number.

Example 7 $\sqrt[8]{-1}$

Solution   Since 8 is an even index, $\sqrt[8]{-1}$ does not exist as a real number.

Sometimes the radicand contains variables. In this chapter, we assume that all variables that appear under a radical sign with an even index represent *positive numbers*.

In Examples 8–11, find the value of each expression. Assume that the variables represent positive numbers.

Example 8 $\sqrt{x^2}$

Solution We seek an expression whose square is x^2. Since x is an expression whose square is x^2, we have $\sqrt{x^2} = x$.

Example 9 $\sqrt[4]{16t^4}$

Solution Since $(2t)^4 = 2^4 t^4 = 16t^4$, then
$$\sqrt[4]{16t^4} = 2t.$$

Example 10 $-\sqrt[3]{y^6}$

Solution We begin by finding an expression whose cube is y^6:
$$(y^2)^3 = y^6, \quad \text{so} \quad \sqrt[3]{y^6} = y^2,$$
and
$$-\sqrt[3]{y^6} = -y^2.$$

Example 11 $\sqrt[5]{-u^5}$

Solution We seek an expression that will be $-u^5$ when raised to the fifth power.
$$(-u)^5 = -u^5 \quad \text{so} \quad \sqrt[5]{-u^5} = -u.$$

In general, if n is an odd positive integer, then

$$\sqrt[n]{-a} = -\sqrt[n]{a}$$

Numbers such as $\sqrt{7}$, $\sqrt{13}$, and $\sqrt{19}$ are examples of *irrational numbers* because they cannot be expressed as ratios of integers. However, irrational numbers can be approximated by decimals. The easiest way to find a decimal approximation for a square root that is an irrational number is to use a calculator with a $\sqrt{}$ key.

Example 12 **C** Use a calculator to approximate each irrational number to four decimal places.

(a) $\sqrt{7}$ (b) $\sqrt{13}$ (c) $\sqrt{119}$

Solution (a) Using a calculator that gives answers to seven decimal places, we get
$$\sqrt{7} \approx 2.6457513.$$

($\approx$ means "is approximately equal to.") In order to express the answer to four decimal places, we round off by adding 1 to the fourth digit if the fifth digit is 5 or larger. Thus,
$$\sqrt{7} \approx 2.6458$$
rounded off to four decimal places.

(b) $\sqrt{13} \approx 3.6055513$, which rounds off to
$$\sqrt{13} \approx 3.6056.$$

(c) $\sqrt{119} \approx 10.908712$. which rounds off to

$$\sqrt{119} \approx 10.9087$$

by dropping the last digits, 1 and 2.

Example 13 **C** Suppose the height h (in meters) of men of given blood type is related to their weight w (in kilograms) by the equation

$$h = \sqrt[3]{0.16w}.$$

Find the height of a basketball player who weighs 101 kilograms.

Solution We substitute 101 for w in the preceding equation, so that

$$h = \sqrt[3]{0.16(101)}$$
$$= \sqrt[3]{16.16}$$
$$= 2.5282137.$$

Therefore, the height of a basketball player who weighs 101 kilograms is approximately 2.53 meters.

PROBLEM SET 1.3

In problems 1–30, evaluate each expression.

1. $-\sqrt{9}$
2. $-\sqrt{25}$
3. $\sqrt{169}$
4. $-\sqrt{144}$
5. $-\sqrt{81}$
6. $-\sqrt{400}$
7. $\sqrt[3]{-8}$
8. $-\sqrt[3]{-27}$
9. $-\sqrt[3]{-64}$
10. $-\sqrt[3]{-125}$
11. $-\sqrt[4]{16}$
12. $-\sqrt[4]{625}$
13. $\sqrt[5]{-32}$
14. $-\sqrt[5]{-243}$
15. $\sqrt{-16}$
16. $\sqrt[6]{-64}$
17. $\sqrt[6]{-1}$
18. $\sqrt[4]{-256}$
19. $-\sqrt[11]{-1}$
20. $-\sqrt{256}$
21. $-\sqrt{225}$
22. $\sqrt{1,681}$
23. $\sqrt{289}$
24. $\sqrt{1.21}$
25. $\sqrt[4]{0.0016}$
26. $\sqrt{0.81}$
27. $\sqrt[3]{-0.125}$
28. $-\sqrt[3]{0.008}$
29. $\sqrt[5]{0.00032}$
30. $-\sqrt{2.56}$

In problems 31–42, find the value of each expression. Assume that variables represent positive numbers.

31. $\sqrt{x^8}$
32. $\sqrt{y^6}$
33. $\sqrt{z^{50}}$
34. $-\sqrt{w^6}$
35. $-\sqrt[3]{u^3}$
36. $\sqrt[5]{z^5}$
37. $\sqrt{4x^2}$
38. $-\sqrt[3]{-8y^6}$
39. $\sqrt[4]{81w^4}$
40. $\sqrt[5]{-32a^{10}}$
41. $\sqrt[3]{-27w^3t^6}$
42. $\sqrt[5]{-243c^5d^{10}}$

C In problems 43–50, use a calculator with a $\sqrt{}$ key to approximate each irrational number and round off the answer to four decimal places.

43. $\sqrt{2}$
44. $\sqrt{3}$
45. $\sqrt{5}$
46. $\sqrt{10}$
47. $\sqrt{11}$
48. $\sqrt{23}$
49. $\sqrt{113}$
50. $\sqrt{279}$

51. Ecologists estimate that the amount A (tons per year) of pollution entering the atmosphere is related to the number N of people by the equation

$$A = \sqrt[3]{N^2}.$$

How many tons of pollution per year are produced if the population is 8,000 people?

C 52. In geometry, the formula for the radius r (in centimeters) of a circle of area A (in square centimeters) (Figure 1) is given by the formula

$$r = \sqrt{\frac{A}{\pi}}.$$

Find the radius of a circle whose area is 75 square centimeters.

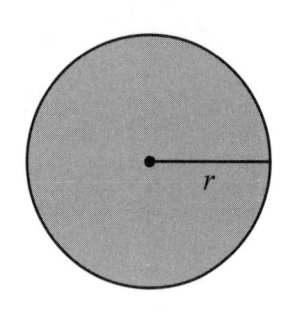

Figure 1

53. In physics, it is estimated that the time t (in seconds) that it takes a pendulum of length l (in feet) to complete one cycle (Figure 2) is given by the formula

$$t = 2\pi \sqrt{\frac{l}{32}}.$$

How long does it take a pendulum that is 2.3 feet long to complete a cycle?

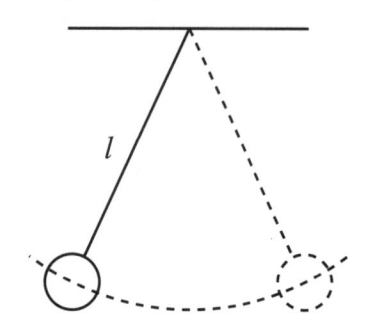

Figure 2

54. In a medical laboratory, the quantity x (in milligrams) of antigen present during an antigen-antibody reaction is related to the time (in minutes) by the equation

$$x = \frac{50 + \sqrt{60t - 300}}{10}, \quad t \geq 5$$

Find the quantity x of antigen present after one hour and 20 minutes.

55. In life sciences, the population of an endangered species is estimated to be 2,187. It is expected that the population P will decrease in coming years according to the formula

$$P = \frac{2,187}{\sqrt[8]{81t}}, \quad 0 \leq t \leq 6,$$

where t is the number of years. Determine the expected population in 4 years.

1.4 Properties of Radicals

We know from Section 1.3 that $\sqrt{36} = 6$. We also know that

$$\sqrt{4} \cdot \sqrt{9} = 2 \cdot 3 = 6.$$

Because $\sqrt{36}$ and $\sqrt{4} \cdot \sqrt{9}$ are both equal to 6, they must be equal, that is,

$$\sqrt{36} = \sqrt{4} \cdot \sqrt{9}.$$

Since it is also true that $36 = 4 \cdot 9$, we may write the last equation as

$$\sqrt{4 \cdot 9} = \sqrt{4} \cdot \sqrt{9}.$$

This illustrates the following property:

PROPERTY 1 Roots of Products

> If a and b are real numbers and n is a positive integer, then, provided that all expressions are defined,
>
> $$\sqrt[n]{ab} = \sqrt[n]{a} \cdot \sqrt[n]{b}.$$

For instance,

$$\sqrt{ab} = \sqrt{a} \cdot \sqrt{b}, \quad \sqrt[3]{ab} = \sqrt[3]{a} \cdot \sqrt[3]{b}, \quad \sqrt[4]{ab} = \sqrt[4]{a} \cdot \sqrt[4]{b},$$

$$\sqrt{6} = \sqrt{2 \cdot 3} = \sqrt{2} \cdot \sqrt{3},$$

and
$$\sqrt[3]{12} = \sqrt[3]{3 \cdot 4} = \sqrt[3]{3} \cdot \sqrt[3]{4}$$

In order to write a radical expression, such as $\sqrt{75}$, in a **simplified** form, we use the following steps:

> Step 1. Write the expression within the radical sign as a product of two expressions in which one of the factors has a perfect nth root. (That is, one factor has a power equal to the index of the radical.)
> Step 2. Apply the roots-of-products property.

For example, to simplify $\sqrt{75}$, we follow this procedure:

Step 1. Find the largest factor of 75 that is a perfect square:
$$\sqrt{75} = \sqrt{25 \cdot 3} \quad \text{(since } 25 \cdot 3 = 75 \text{ and 25 is a perfect square)}$$

Step 2. Applying Property 1, we have:
$$\sqrt{25 \cdot 3} = \sqrt{25} \cdot \sqrt{3} = 5\sqrt{3} \quad \text{(since } \sqrt{25} = 5)$$
Thus, $\sqrt{75} = 5\sqrt{3}$.

In Examples 1–4, simplify each expression.

Example 1 $\sqrt{18}$

Solution   First, we find the largest perfect square that is a factor of 18. This factor is 9, so
$$\sqrt{18} = \sqrt{9 \cdot 2} \quad \text{(since } 18 = 9 \cdot 2)$$
$$= \sqrt{9} \cdot \sqrt{2} \quad \text{(by Property 1)}$$
$$= 3\sqrt{2}. \quad \text{(since } \sqrt{9} = 3)$$

Example 2 $\sqrt[3]{16}$

Solution   Here we look for the largest perfect cube that is a factor of 16. Since $16 = 8 \cdot 2$, and $8 = 2^3$, we have:
$$\sqrt[3]{16} = \sqrt[3]{8 \cdot 2}$$
$$= \sqrt[3]{8} \cdot \sqrt[3]{2} = 2\sqrt[3]{2}$$

Example 3 $\sqrt[4]{405}$

Solution   Here we look for the largest factor of 405 that is a perfect fourth power. Since $405 = 81 \cdot 5$, and $81 = 3^4$, we have:
$$\sqrt[4]{405} = \sqrt[4]{81 \cdot 5}$$
$$= \sqrt[4]{81} \cdot \sqrt[4]{5} = 3\sqrt[4]{5}.$$

Example 4 $\sqrt[5]{-96}$

Solution $(-2)^5 = -32$, which is a factor of -96. So we have:

$$\sqrt[5]{-96} = \sqrt[5]{-32 \cdot 3}$$
$$= \sqrt[5]{-32} \cdot \sqrt[5]{3} = -2\sqrt[5]{3}.$$

If one or more variables appear as factors under the radical sign, we **simplify** the expression according to the following steps.

> Step 1. Write the algebraic expression within the radical sign as a product of two expressions such that one factor has a power equal to the index of the radical.
> Step 2. Apply the roots-of-products property.

In Examples 5–7, simplify each expression. Assume that all variables represent positive numbers.

Example 5 $\sqrt{x^3}$

Solution Here we have $x^3 = x^2 \cdot x$. We can write

$\sqrt{x^3} = \sqrt{x^2 \cdot x}$ (since $x^3 = x^2 \cdot x$ and x^2 is a perfect square)
$= \sqrt{x^2} \cdot \sqrt{x}$ (by Property 1)
$= x\sqrt{x}.$ (since $\sqrt{x^2} = x$ for $x > 0$)

Example 6 $\sqrt{9x^5}$

Solution Since $9x^5 = 9x^4 \cdot x$, we have:

$$\sqrt{9x^5} = \sqrt{9x^4 \cdot x} = \sqrt{9x^4} \cdot \sqrt{x}$$
$$= 3x^2\sqrt{x}.$$

Example 7 $\sqrt[3]{16t^4u^5}$

Solution Since $16t^4u^5 = (8t^3u^3)(2tu^2)$, we have:

$$\sqrt[3]{16t^4u^5} = \sqrt[3]{(8t^3u^3) \cdot (2tu^2)} = \sqrt[3]{8t^3u^3} \cdot \sqrt[3]{2tu^2}$$
$$= 2tu\sqrt[3]{2tu^2}.$$

A second property of radicals has to do with roots of quotients and can be stated as follows:

PROPERTY 2   Roots of Quotients

> If a and b are real numbers and n is a positive integer, then, provided that all expressions are defined,
> $$\sqrt[n]{\frac{a}{b}} = \frac{\sqrt[n]{a}}{\sqrt[n]{b}}$$

For instance,

$$\sqrt{\frac{a}{b}} = \frac{\sqrt{a}}{\sqrt{b}}, \qquad \sqrt[3]{\frac{a}{b}} = \frac{\sqrt[3]{a}}{\sqrt[3]{b}}, \qquad \text{and} \qquad \sqrt[4]{\frac{a}{b}} = \frac{\sqrt[4]{a}}{\sqrt[4]{b}}.$$

A radical expression that involves a fraction within the radical sign is said to be *simplified* if no radical symbol appears in a denominator of the fraction. For example, we simplify $\sqrt{\frac{5}{9}}$ and $\sqrt[3]{\frac{7}{8}}$ as follows:

$$\sqrt{\frac{5}{9}} = \frac{\sqrt{5}}{\sqrt{9}} = \frac{\sqrt{5}}{3}, \qquad \text{and} \qquad \sqrt[3]{\frac{7}{8}} = \frac{\sqrt[3]{7}}{\sqrt[3]{8}} = \frac{\sqrt[3]{7}}{2}$$

In Examples 8–11, simplify each expression.

Example 8 $\sqrt{\dfrac{25}{36}}$

Solution $\sqrt{\dfrac{25}{36}} = \dfrac{\sqrt{25}}{\sqrt{36}}$ (by Property 2)

$\qquad\qquad = \dfrac{5}{6}$ (since $\sqrt{25} = 5$ and $\sqrt{36} = 6$)

Example 9 $\sqrt{\dfrac{11}{49}}$

Solution $\sqrt{\dfrac{11}{49}} = \dfrac{\sqrt{11}}{\sqrt{49}} = \dfrac{\sqrt{11}}{7}$

Example 10 $\sqrt[3]{\dfrac{-5}{8}}$

Solution $\sqrt[3]{\dfrac{-5}{8}} = \dfrac{\sqrt[3]{-5}}{\sqrt[3]{8}} = -\dfrac{\sqrt[3]{5}}{2}$

Example 11 $\sqrt[4]{\dfrac{7}{16}}$

Solution $\sqrt[4]{\dfrac{7}{16}} = \dfrac{\sqrt[4]{7}}{\sqrt[4]{16}} = \dfrac{\sqrt[4]{7}}{2}$

Again, if one or more variables appear as factors under the radical sign. we simplify the expression the same way we did in the preceding examples.

In Examples 12–14, simplify each expression. Assume that all variables represent positive numbers.

Example 12 $\sqrt{\dfrac{7}{x^2}}$

Solution $\sqrt{\dfrac{7}{x^2}} = \dfrac{\sqrt{7}}{\sqrt{x^2}} = \dfrac{\sqrt{7}}{x}$

Example 13 $\sqrt[3]{\dfrac{17}{t^3}}$

Solution $\sqrt[3]{\dfrac{17}{t^3}} = \dfrac{\sqrt[3]{17}}{\sqrt[3]{t^3}} = \dfrac{\sqrt[3]{17}}{t}$

Example 14 $\sqrt[5]{\dfrac{19}{32c^{10}}}$

Solution $\sqrt[5]{\dfrac{19}{32c^{10}}} = \dfrac{\sqrt[5]{19}}{\sqrt[5]{32c^{10}}} = \dfrac{\sqrt[5]{19}}{2c^2}$

PROBLEM SET 1.4

In problems 1–40, use Property 1 to simplify each expression. Assume that all variables represent positive numbers.

1. $\sqrt{32}$
2. $\sqrt{20}$
3. $\sqrt{45}$
4. $\sqrt{27}$
5. $\sqrt{160}$
6. $\sqrt{325}$
7. $\sqrt{150}$
8. $\sqrt{125}$
9. $\sqrt[3]{72}$
10. $\sqrt[3]{108}$
11. $\sqrt[3]{320}$
12. $\sqrt[3]{-56}$
13. $\sqrt[3]{-250}$
14. $\sqrt[3]{81}$
15. $\sqrt[4]{80}$
16. $\sqrt[4]{162}$
17. $\sqrt[4]{567}$
18. $\sqrt[4]{176}$
19. $\sqrt[5]{160}$
20. $\sqrt[5]{416}$
21. $\sqrt[5]{-486}$
22. $\sqrt[5]{-729}$
23. $\sqrt{x^5}$
24. $\sqrt{y^7}$
25. $\sqrt{8u^3}$
26. $\sqrt{18v^5}$
27. $\sqrt{50x^3}$
28. $\sqrt{27t^5}$
29. $\sqrt{x^3y^5}$
30. $\sqrt{m^{21}n^7}$
31. $\sqrt{20u^5v^8}$
32. $\sqrt{75x^{27}y^3}$
33. $\sqrt[3]{16z^{10}}$
34. $\sqrt[3]{54u^{17}}$
35. $\sqrt[3]{-250x^4y^5}$
36. $\sqrt[3]{-108t^4s^7}$
37. $\sqrt[4]{48u^5v^7}$
38. $\sqrt[4]{162x^5y^7}$
39. $\sqrt[5]{-224w^6z^7}$
40. $\sqrt[5]{192u^7v^9}$

In problems 41–76, use Property 2 to simplify each expression. Assume that all variables represent positive numbers.

41. $\sqrt{\dfrac{3}{25}}$
42. $\sqrt{\dfrac{7}{16}}$
43. $\sqrt{\dfrac{5}{16}}$
44. $\sqrt{\dfrac{17}{81}}$
45. $\sqrt{\dfrac{7}{121}}$
46. $\sqrt{\dfrac{8}{49}}$
47. $\sqrt{\dfrac{49}{64}}$
48. $\sqrt{\dfrac{25}{9}}$
49. $\sqrt[3]{\dfrac{8}{125}}$
50. $\sqrt[3]{-\dfrac{27}{125}}$
51. $\sqrt[3]{\dfrac{-11}{8}}$
52. $\sqrt[3]{-\dfrac{5}{64}}$
53. $\sqrt[4]{\dfrac{5}{16}}$
54. $\sqrt[4]{\dfrac{32}{81}}$
55. $\sqrt[4]{\dfrac{23}{81}}$
56. $\sqrt[4]{\dfrac{162}{81}}$
57. $\sqrt[5]{\dfrac{19}{32}}$
58. $\sqrt[5]{-\dfrac{243}{32}}$
59. $\sqrt[5]{-\dfrac{171}{243}}$
60. $\sqrt[5]{\dfrac{486}{64}}$
61. $\sqrt{\dfrac{11}{t^2}}$
62. $\sqrt{\dfrac{x}{y^2}}$
63. $\sqrt{\dfrac{3}{4x^2}}$
64. $\sqrt{\dfrac{17}{9u^4}}$
65. $\sqrt{\dfrac{a^2}{81}}$
66. $\sqrt{\dfrac{x^2}{16y^4}}$
67. $\sqrt[3]{\dfrac{13}{27a^3}}$
68. $\sqrt[3]{-\dfrac{8}{125u^9}}$
69. $\sqrt[3]{\dfrac{-17z}{64x^3}}$
70. $\sqrt[3]{\dfrac{21}{216t^{12}}}$
71. $\sqrt[4]{\dfrac{x^9}{16z^{16}}}$
72. $\sqrt[4]{\dfrac{32}{81m^8n^4}}$
73. $\sqrt[4]{\dfrac{37}{81u^8v^{12}}}$
74. $\sqrt[4]{\dfrac{111}{256t^{20}}}$
75. $\sqrt[5]{-\dfrac{19}{32x^{10}}}$
76. $\sqrt[5]{\dfrac{64w^6}{y^{25}t^5}}$

77. In physics, the voltage V needed to operate an electrical appliance is given by

$$V = \sqrt{\frac{P}{R}},$$

where R is the resistance (in ohms) and P is the power (in watts). Determine the voltage needed to operate an electric heater that uses 1,600 watts of power and has a resistance of 25 ohms.

78. The formula for the time t (in seconds) that it takes an object to hit the ground from the top of a building h feet high is given by

$$t = \sqrt{\frac{h}{16}}.$$

How long will it take an object dropped from the top of a building 169 feet high to reach the ground?

79. In business, the annual rate of return r on an investment of P dollars that grows to become S dollars in 3 years is given by the formula

$$r = \sqrt[3]{\frac{S}{P}} - 1.$$

Suppose that you invest $12,500 in the stock market and 3 years later your stocks are worth $21,600. Find the annual rate of return on your investment.

1.5 Addition and Subtraction of Radicals

We add and subtract radical expressions in much the same way that we add and subtract polynomial expressions: We apply the distributive property and combine only *like* or *similar* terms. Two or more radical expressions are said to be *like* or *similar* if they have the same index and radicand. For instance,

$$3\sqrt{7}, \quad 5\sqrt{7}, \quad x\sqrt{7}, \quad \text{and} \quad \frac{\sqrt{7}}{3}$$

are like terms, whereas

$$5\sqrt{7} \quad \text{and} \quad 5\sqrt[7]{7} \text{ are not like terms.}$$

In Examples 1 and 2, combine like terms.

Example 1 $3\sqrt{2} + 4\sqrt{2} - \sqrt{2}$

Solution Since all the terms are similar, we apply the distributive property and combine:

$$3\sqrt{2} + 4\sqrt{2} - \sqrt{2} = (3 + 4 - 1)\sqrt{2} = 6\sqrt{2}.$$

Example 2 $7\sqrt[3]{5} - 4\sqrt[3]{5} + 2\sqrt[3]{5}$

Solution $7\sqrt[3]{5} - 4\sqrt[3]{5} + 2\sqrt[3]{5} = (7 - 4 + 2)\sqrt[3]{5} = 5\sqrt[3]{5}$

Each term, in Examples 1 and 2, contains a radical expression in simplified form. Occasionally two or more of the terms containing radical expressions do not appear to be similar but contain like terms when simplified. In this case, we write *each* expression in a simplified form and then apply the distributive property to combine any like terms.

In Examples 3–8, perform each operation. Assume that all variables represent positive numbers.

Example 3 $\sqrt{8} + \sqrt{32}$

Solution We begin by writing each term so that the radical expressions are simplified.

We have:
$$\sqrt{8} + \sqrt{32} = \sqrt{4 \cdot 2} + \sqrt{16 \cdot 2}$$
$$= \sqrt{4} \cdot \sqrt{2} + \sqrt{16} \cdot \sqrt{2} \quad \text{(by Property 1)}$$
$$= 2\sqrt{2} + 4\sqrt{2}$$
$$= (2 + 4)\sqrt{2} \quad \text{(by the distributive property)}$$
$$= 6\sqrt{2}.$$

Example 4 $3\sqrt{50} + 7\sqrt{8}$

Solution
$$3\sqrt{50} + 7\sqrt{8} = 3\sqrt{25 \cdot 2} + 7\sqrt{4 \cdot 2}$$
$$= 3\sqrt{25} \cdot \sqrt{2} + 7\sqrt{4} \cdot \sqrt{2} \quad \text{(by Property 1)}$$
$$= 3(5)\sqrt{2} + 7(2)\sqrt{2}$$
$$= 15\sqrt{2} + 14\sqrt{2}$$
$$= (15 + 14)\sqrt{2} \quad \text{(by the distributive property)}$$
$$= 29\sqrt{2}$$

Example 5 $2\sqrt[3]{16} + \sqrt[3]{54}$

Solution
$$2\sqrt[3]{16} + \sqrt[3]{54} = 2\sqrt[3]{8 \cdot 2} + \sqrt[3]{27 \cdot 2}$$
$$= 2\sqrt[3]{8} \cdot \sqrt[3]{2} + \sqrt[3]{27} \cdot \sqrt[3]{2} \quad \text{(by Property 1)}$$
$$= 2(2)\sqrt[3]{2} + 3\sqrt[3]{2}$$
$$= 4\sqrt[3]{2} + 3\sqrt[3]{2}$$
$$= (4 + 3)\sqrt[3]{2} \quad \text{(by the distributive property)}$$
$$= 7\sqrt[3]{2}$$

Example 6 $4\sqrt{12} + 5\sqrt{8} - \sqrt{50}$

Solution
$$4\sqrt{12} + 5\sqrt{8} - \sqrt{50} = 4\sqrt{4 \cdot 3} + 5\sqrt{4 \cdot 2} - \sqrt{25 \cdot 2}$$
$$= 4\sqrt{4}\sqrt{3} + 5\sqrt{4}\sqrt{2} - \sqrt{25}\sqrt{2} \quad \text{(by Property 1)}$$
$$= 4 \cdot 2\sqrt{3} + 5 \cdot 2\sqrt{2} - 5\sqrt{2}$$
$$= 8\sqrt{3} + 10\sqrt{2} - 5\sqrt{2}$$
$$= 8\sqrt{3} + (10 - 5)\sqrt{2}$$
$$= 8\sqrt{3} + 5\sqrt{2}$$

(Notice in Example 6 that we combined only the last two terms. since the radicand in the first term differed from the radicand in the other two terms.)

Example 7 $\sqrt{90y} - \sqrt{40y} + 5\sqrt{10y}$

Solution
$$\sqrt{90y} - \sqrt{40y} + 5\sqrt{10y} = \sqrt{9(10y)} - \sqrt{4(10y)} + 5\sqrt{10y}$$
$$= \sqrt{9} \cdot \sqrt{10y} - \sqrt{4} \cdot \sqrt{10y} + 5\sqrt{10y}$$
$$= 3\sqrt{10y} - 2\sqrt{10y} + 5\sqrt{10y}$$
$$= (3 - 2 + 5)\sqrt{10y}$$
$$= 6\sqrt{10y}$$

Example 8 $7\sqrt{3u^3} + u\sqrt{27u} - 5\sqrt{3u^2}$

Solution
$$7\sqrt{3u^3} + u\sqrt{27u} - 5\sqrt{3u^2} = 7\sqrt{u^2(3u)} + u\sqrt{9(3u)} - 5\sqrt{u^2 \cdot (3)}$$
$$= 7\sqrt{u^2} \cdot \sqrt{3u} + u\sqrt{9} \cdot \sqrt{3u} - 5\sqrt{u^2} \cdot \sqrt{3}$$
$$= 7u\sqrt{3u} + 3u\sqrt{3u} - 5u\sqrt{3}$$
$$= 10u\sqrt{3u} - 5u\sqrt{3}$$

PROBLEM SET 1.5

In problems 1–14, combine like terms. Assume that all variables represent positive numbers.

1. $2\sqrt{2} + 3\sqrt{2}$
2. $5\sqrt{3} - 2\sqrt{3}$
3. $7\sqrt{5} + 4\sqrt{5} - 3\sqrt{5}$
4. $8\sqrt{7} - 2\sqrt{7} + 4\sqrt{7}$
5. $12\sqrt[3]{6} - 7\sqrt[3]{6} + 5\sqrt[3]{6}$
6. $-11\sqrt[3]{4} + 9\sqrt[3]{4} - 3\sqrt[3]{4}$
7. $15\sqrt{x} + 3\sqrt{x} - 8\sqrt{x}$
8. $21\sqrt{t} - 16\sqrt{t} - 2\sqrt{t}$
9. $-7\sqrt[3]{y} - 11\sqrt[3]{y} + 9\sqrt[3]{y}$
10. $17\sqrt[3]{u} + 13\sqrt[3]{u} - 12\sqrt[3]{u}$
11. $8\sqrt{z} - 3\sqrt{w} + 2\sqrt{z}$
12. $-23\sqrt{v} + 17\sqrt{w} + 13\sqrt{w}$
13. $15\sqrt[3]{x^2} + 5\sqrt[3]{x^2} - \sqrt[3]{x}$
14. $19\sqrt[3]{z^2} + 10\sqrt{z} - 9\sqrt[3]{z^2}$

In problems 15–46, perform each operation by simplifying the radical expression and combining like terms. Assume that all variables represent positive numbers.

15. $\sqrt{8} + \sqrt{18}$
16. $\sqrt{18} + \sqrt{50}$
17. $\sqrt{72} - \sqrt{2}$
18. $\sqrt{50} - \sqrt{128}$
19. $5\sqrt{125} + 6\sqrt{45}$
20. $7\sqrt{48} - 6\sqrt{12}$
21. $\sqrt{98} - \sqrt{50}$
22. $4\sqrt{18} - 2\sqrt{72}$
23. $2\sqrt{108} - 3\sqrt{27}$
24. $\sqrt{108} - 2\sqrt{48}$
25. $\sqrt{18} - \sqrt{8} + \sqrt{32}$
26. $\sqrt{20} + \sqrt{45} - \sqrt{80}$
27. $\sqrt{50} + \sqrt{72} - \sqrt{48}$
28. $\sqrt{75} - \sqrt{20} - \sqrt{12}$
29. $\sqrt[3]{16} + \sqrt[3]{54}$
30. $\sqrt[3]{250} - \sqrt[3]{432}$
31. $2\sqrt[3]{3} + \sqrt[3]{24} + 5\sqrt[3]{192}$
32. $6\sqrt[3]{108} + 5\sqrt[3]{256} - 7\sqrt[3]{32}$
33. $\sqrt{8x} + \sqrt{32x}$
34. $\sqrt{50t} - \sqrt{32t}$
35. $\sqrt{20y} + \sqrt{45y} - \sqrt{80y}$
36. $\sqrt{242u} - \sqrt{50u} - \sqrt{72u}$
37. $\sqrt[3]{54x^2} - \sqrt[3]{16x^2}$
38. $\sqrt[3]{81z^2y} + \sqrt[3]{24z^2y}$
39. $\sqrt{y^3} + \sqrt{25y^3} + \sqrt{9y}$
40. $3x\sqrt{x^7} - 2x\sqrt{x^5} + 3\sqrt{x^3}$
41. $7\sqrt{75t^3} + 9\sqrt{48t^3}$
42. $8\sqrt{200z^5} - 6\sqrt{162z^5}$
43. $2\sqrt{2x^3} + 5\sqrt{18x^3} - \sqrt{32x^3}$
44. $10\sqrt{27u^5} + 12\sqrt{48u^5} - 5\sqrt{u^5}$
45. $5\sqrt[3]{108m^4} - 2\sqrt[3]{32m^4}$
46. $3\sqrt[3]{250x^7} + 4\sqrt[3]{432x^7}$

1.6 Multiplication of Radicals

We multiply radical expressions by using the distributive property and Property 1 of Section 1.4 written in the reverse order:

$$\sqrt[n]{a} \cdot \sqrt[n]{b} = \sqrt[n]{ab}.$$

Once again, we combine like terms and simplify whenever possible.

In Examples 1–9, find each product and simplify. Assume that all variables represent positive numbers.

Example 1 $\sqrt{18} \cdot \sqrt{2}$

Solution $\sqrt{18} \cdot \sqrt{2} = \sqrt{18 \cdot 2} = \sqrt{36} = 6$

Example 2 $\sqrt[3]{3c^2d^5} \cdot \sqrt[3]{9c^4d^7}$

Solution
$$\sqrt[3]{3c^2d^5} \cdot \sqrt[3]{9c^4d^7} = \sqrt[3]{3c^2d^5 \cdot 9c^4d^7}$$
$$= \sqrt[3]{27c^6d^{12}}$$
$$= 3c^2d^4$$

Example 3 $\sqrt{3}(\sqrt{5} + 2\sqrt{3})$

Solution
$$\sqrt{3}(\sqrt{5} + 2\sqrt{3}) = \sqrt{3} \cdot \sqrt{5} + \sqrt{3} \cdot 2\sqrt{3}$$
$$= \sqrt{15} + 2\sqrt{3} \cdot \sqrt{3}$$
$$= \sqrt{15} + 2(3)$$
$$= \sqrt{15} + 6$$

Example 4 $\sqrt{6}(\sqrt{8} + \sqrt{18})$

Solution
$$\sqrt{6}(\sqrt{8} + \sqrt{18}) = \sqrt{6} \cdot \sqrt{8} + \sqrt{6} \cdot \sqrt{18}$$
$$= \sqrt{48} + \sqrt{108}$$
$$= \sqrt{16 \cdot 3} + \sqrt{36 \cdot 3}$$
$$= \sqrt{16} \cdot \sqrt{3} + \sqrt{36} \cdot \sqrt{3}$$
$$= 4\sqrt{3} + 6\sqrt{3}$$
$$= 10\sqrt{3}$$

Example 5 $(\sqrt{10} + \sqrt{2})(\sqrt{10} - \sqrt{2})$

Solution Notice that this product is in the form $(a + b)(a - b)$ with $a = \sqrt{10}$ and $b = \sqrt{2}$. Recall that $(a + b)(a - b) = a^2 - b^2$. Thus, we have:

$$(\sqrt{10} + \sqrt{2})(\sqrt{10} - \sqrt{2}) = (\sqrt{10})^2 - (\sqrt{2})^2$$
$$= 10 - 2$$
$$= 8.$$

Example 6 $(\sqrt{5} + 2\sqrt{3})^2$

Solution This product is in the form $(a + b)^2$ with $a = \sqrt{5}$ and $b = 2\sqrt{3}$. Recall that $(a + b)^2 = a^2 + 2ab + b^2$. Thus, we have:

$$(\sqrt{5} + 2\sqrt{3})^2 = (\sqrt{5})^2 + 2(\sqrt{5}) \cdot (2\sqrt{3}) + (2\sqrt{3})^2$$
$$= 5 + 4\sqrt{15} + 4(3)$$
$$= 5 + 4\sqrt{15} + 12$$
$$= 17 + 4\sqrt{15}.$$

Example 7 $(\sqrt{3} - \sqrt{2})(2\sqrt{3} + \sqrt{2})$

Solution We apply the distributive property:

$$(\sqrt{3} - \sqrt{2})(2\sqrt{3} + \sqrt{2}) = 2\sqrt{3}\sqrt{3} + \sqrt{3}\sqrt{2} - 2\sqrt{2}\sqrt{3} - \sqrt{2}\sqrt{2}$$
$$= 2(3) + \sqrt{6} - 2\sqrt{6} - 2$$
$$= 6 - 2 + \sqrt{6} - 2\sqrt{6}$$
$$= 4 - \sqrt{6}.$$

Example 8 $(\sqrt{x} + y)(\sqrt{x} + 2y)$

Solution We apply the distributive property:

$$(\sqrt{x} + y)(\sqrt{x} + 2y) = \sqrt{x} \cdot \sqrt{x} + 2y\sqrt{x} + y\sqrt{x} + 2y^2$$
$$= x + 3y\sqrt{x} + 2y^2.$$

Example 9 $(\sqrt{x} + 2\sqrt{y})^2$

Solution Using $(a + b)^2 = a^2 + 2ab + b^2$ with $a = \sqrt{x}$ and $b = 2\sqrt{y}$, we have:

$$(\sqrt{x} + 2\sqrt{y})^2 = (\sqrt{x})^2 + 2(\sqrt{x})(2\sqrt{y}) + (2\sqrt{y})^2$$
$$= x + 4\sqrt{x}\sqrt{y} + 4y$$
$$= x + 4\sqrt{xy} + 4y.$$

PROBLEM SET 1.6

In problems 1–40, find each product and simplify. Assume that all variables represent positive numbers.

1. $\sqrt{2}\sqrt{32}$
2. $\sqrt{12}\sqrt{3}$
3. $\sqrt{10}\sqrt{5}$
4. $\sqrt{8}\sqrt{6}$
5. $\sqrt[3]{5}\sqrt[3]{25}$
6. $\sqrt[3]{6}\sqrt[3]{36}$
7. $\sqrt{2x}\sqrt{8x}$
8. $\sqrt{20v^3}\sqrt{5v}$
9. $\sqrt[3]{4m^2n}\sqrt[3]{16mn^2}$
10. $\sqrt[3]{12x^4y^2}\sqrt[3]{18x^2y^4}$
11. $(3\sqrt{2})^2$
12. $(5\sqrt{x})^2$
13. $(7\sqrt{y})^2$
14. $(-2\sqrt[3]{3})^3$
15. $\sqrt{3}(\sqrt{2} + \sqrt{3})$
16. $\sqrt{5}(\sqrt{2} - \sqrt{5})$
17. $\sqrt{2}(\sqrt{18} - 2\sqrt{2})$
18. $\sqrt{7}(3\sqrt{7} + 1)$
19. $\sqrt{x}(3\sqrt{x} + 5)$
20. $\sqrt{t}(2 - 4\sqrt{t})$
21. $(\sqrt{7} + \sqrt{3})(\sqrt{7} - \sqrt{3})$
22. $(\sqrt{11} + \sqrt{5})(\sqrt{11} - \sqrt{5})$
23. $(\sqrt{14} - \sqrt{6})(\sqrt{14} + \sqrt{6})$

24. $(\sqrt{19} - \sqrt{3})(\sqrt{19} + \sqrt{3})$
25. $(3\sqrt{5} - 4)(3\sqrt{5} + 4)$
26. $(7\sqrt{2} + \sqrt{7})(7\sqrt{2} - \sqrt{7})$
27. $(\sqrt{u} - \sqrt{v})(\sqrt{u} + \sqrt{v})$
28. $(2\sqrt{x} - \sqrt{y})(2\sqrt{x} + \sqrt{y})$
29. $(\sqrt{3} + 5)^2$
30. $(\sqrt{7} + \sqrt{3})^2$
31. $(2\sqrt{5} - 1)^2$
32. $(4\sqrt{2} - 3\sqrt{3})^2$
33. $(\sqrt{m} + \sqrt{n})^2$
34. $(\sqrt{x} - 2\sqrt{y})^2$
35. $(\sqrt{x} - 3\sqrt{y})^2$
36. $(5\sqrt{u} + 2\sqrt{v})^2$
37. $(\sqrt{5} + 2\sqrt{3})(3\sqrt{5} - \sqrt{3})$
38. $(2\sqrt{7} - 3\sqrt{5})(\sqrt{7} + 4\sqrt{5})$
39. $(2\sqrt{x} - 3\sqrt{y})(4\sqrt{x} + \sqrt{y})$
40. $(\sqrt{m} + 4\sqrt{n})(5\sqrt{m} - 2\sqrt{n})$

1.7 Division of Radicals

We divide radical expressions by using Property 2 on page 301 written in the reverse order:

$$\frac{\sqrt[n]{a}}{\sqrt[n]{b}} = \sqrt[n]{\frac{a}{b}}.$$

In Examples 1–3, perform each division. Assume that all variables represent positive numbers.

Example 1 $\quad \dfrac{\sqrt{27}}{\sqrt{3}}$

Solution $\quad \dfrac{\sqrt{27}}{\sqrt{3}} = \sqrt{\dfrac{27}{3}} = \sqrt{9} = 3$

Example 2 $\quad \dfrac{\sqrt{75x^3}}{\sqrt{3x}}$

Solution $\quad \dfrac{\sqrt{75x^3}}{\sqrt{3x}} = \sqrt{\dfrac{75x^3}{3x}} = \sqrt{25x^2} = 5x$

Example 3 $\quad \dfrac{\sqrt[3]{32u^7}}{\sqrt[3]{4uv^3}}$

Solution $\quad \dfrac{\sqrt[3]{32u^7}}{\sqrt[3]{4uv^3}} = \sqrt[3]{\dfrac{32u^7}{4uv^3}} = \sqrt[3]{\dfrac{8u^6}{v^3}} = \dfrac{2u^2}{v}$

Rationalizing Denominators

Radicals may appear in the denominator of a fraction, such as $\frac{1}{\sqrt{7}}$. It is usually easier to work with fractions if their denominators are free of radicals. To rewrite a fraction so that there are no radicals in the denominator, we multiply the numerator and the denominator by a rationalizing factor for the denominator. That is, we find a factor such that the product of that factor and the denominator of the given radical expression is free of radicals. For instance, in the product

$$\sqrt{7} \cdot \sqrt{7}$$

$\sqrt{7}$ is a rationalizing factor for $\sqrt{7}$. Also, since

$$(\sqrt{5} - \sqrt{2})(\sqrt{5} - \sqrt{2}) = (\sqrt{5})^2 - (\sqrt{2})^2 = 5 - 2 = 3,$$

the expressions $\sqrt{5} + \sqrt{2}$ and $\sqrt{5} - \sqrt{2}$ are rationalizing factors for each other. The process of rewriting a fraction so that there are no radicals in the denominator is called **rationalizing the denominator.**

In Examples 4–11, rationalize the denominator of each fraction. Assume that all variables represent positive numbers.

Example 4 $\quad \dfrac{2}{\sqrt{3}}$

Solution Here $\sqrt{3}$ is a rationalizing factor for the denominator. We multiply the numerator and the denominator of the fraction by $\sqrt{3}$:

$$\frac{2}{\sqrt{3}} = \frac{2\sqrt{3}}{\sqrt{3} \cdot \sqrt{3}}$$

$$= \frac{2\sqrt{3}}{3}.$$

Example 5 $\quad \dfrac{\sqrt{8}}{\sqrt{10}}$

Solution The rationalizing factor for the denominator is 10, so we have:

$$\frac{\sqrt{8}}{\sqrt{10}} = \frac{\sqrt{8}\sqrt{10}}{\sqrt{10}\sqrt{10}} = \frac{\sqrt{80}}{10}$$

$$= \frac{\sqrt{16 \cdot 5}}{10} = \frac{\sqrt{16} \cdot \sqrt{5}}{10}$$

$$= \frac{4\sqrt{5}}{10} = \frac{2\sqrt{5}}{5}.$$

Example 6 $\quad \dfrac{3}{\sqrt{x}}$

Solution The rationalizing factor for the denominator is $\sqrt{x}$, so we have:

$$\frac{3}{\sqrt{x}} = \frac{3\sqrt{x}}{\sqrt{x}\sqrt{x}} = \frac{3\sqrt{x}}{x}.$$

Example 7 $\quad \dfrac{1}{\sqrt{3} - 1}$

Solution In this case, the rationalizing factor for the denominator is $\sqrt{3} + 1$, since

$$(\sqrt{3} - 1)(\sqrt{3} + 1) = (\sqrt{3})^2 - 1^2 = 3 - 1 = 2.$$

We multiply the numerator and the denominator by $\sqrt{3} + 1$:

$$\frac{5}{\sqrt{3} - 1} = \frac{5(\sqrt{3} + 1)}{(\sqrt{3} - 1)(\sqrt{3} + 1)} = \frac{5\sqrt{3} + 5}{(\sqrt{3})^2 - 1^2} = \frac{5\sqrt{3} + 5}{3 - 1}$$

$$= \frac{5\sqrt{3} + 5}{2}.$$

Example 8 $\dfrac{15}{\sqrt{7}+\sqrt{3}}$

Solution In this case, the rationalizing factor for the denominator is $\sqrt{7}-\sqrt{3}$, since
$$(\sqrt{7}+\sqrt{3})(\sqrt{7}-\sqrt{3}) = (\sqrt{7})^2 - (\sqrt{3})^2 = 7 - 3 = 4.$$
We multiply the numerator and the denominator by $\sqrt{7}-\sqrt{3}$:
$$\dfrac{15}{\sqrt{7}+\sqrt{3}} = \dfrac{15(\sqrt{7}-\sqrt{3})}{(\sqrt{7}+\sqrt{3})(\sqrt{7}-\sqrt{3})} = \dfrac{15(\sqrt{7}-\sqrt{3})}{(\sqrt{7})^2-(\sqrt{3})^2}$$
$$= \dfrac{15(\sqrt{7}-\sqrt{3})}{4}.$$

Example 9 $\dfrac{\sqrt{5}+\sqrt{2}}{\sqrt{5}-\sqrt{2}}$

Solution The rationalizing factor for the denominator is $\sqrt{5}+\sqrt{2}$, so we have:
$$\dfrac{\sqrt{5}+\sqrt{2}}{\sqrt{5}-\sqrt{2}} = \dfrac{(\sqrt{5}+\sqrt{2})(\sqrt{5}+\sqrt{2})}{(\sqrt{5}-\sqrt{2})(\sqrt{5}+\sqrt{2})} = \dfrac{(\sqrt{5})^2+2\sqrt{2}\sqrt{5}+(\sqrt{2})^2}{(\sqrt{5})^2-(\sqrt{2})^2}$$
$$= \dfrac{5+2\sqrt{10}+2}{5-2}$$
$$= \dfrac{7+2\sqrt{10}}{3}.$$

Example 10 $\dfrac{3}{\sqrt{x}-1}$

Solution The rationalizing factor for the denominator is $\sqrt{x}+1$, so we have:
$$\dfrac{3}{\sqrt{x}-1} = \dfrac{3(\sqrt{x}+1)}{(\sqrt{x}-1)(\sqrt{x}+1)} = \dfrac{3\sqrt{x}+3}{(\sqrt{x})^2-(1)^2} = \dfrac{3\sqrt{x}+3}{x-1}.$$

Example 11 $\dfrac{\sqrt{t}}{2\sqrt{t}+3}$

Solution The rationalizing factor for the denominator is $2\sqrt{t}-3$, so we have:
$$\dfrac{\sqrt{t}}{2\sqrt{t}+3} = \dfrac{\sqrt{t}(2\sqrt{t}-3)}{(2\sqrt{t}+3)(2\sqrt{t}-3)} = \dfrac{2\sqrt{t}\sqrt{t}-3\sqrt{t}}{(2\sqrt{t})^2-(3)^2} = \dfrac{2t-3\sqrt{t}}{4t-9}.$$

PROBLEM SET 1.7

In problems 1–14, perform each division and simplify. Assume that all variables represent positive numbers.

1. $\dfrac{\sqrt{18}}{\sqrt{2}}$

2. $\dfrac{\sqrt{48}}{\sqrt{3}}$

3. $\dfrac{\sqrt{20}}{\sqrt{5}}$

4. $\dfrac{\sqrt{28}}{\sqrt{7}}$

5. $\dfrac{\sqrt[3]{16}}{\sqrt[3]{2}}$

6. $\dfrac{\sqrt[3]{-375}}{\sqrt[3]{3}}$

7. $\dfrac{\sqrt{48x^5}}{\sqrt{3x}}$

8. $\dfrac{\sqrt{80t^3}}{\sqrt{5t}}$

9. $\dfrac{\sqrt{50u^7v^5}}{\sqrt{2uv}}$

10. $\dfrac{\sqrt{147y^9z}}{\sqrt{3yz}}$

11. $\dfrac{\sqrt{500x^5y}}{\sqrt{5x^3y}}$ 12. $\dfrac{\sqrt{99m^{12}n^5}}{\sqrt{11m^4n^3}}$ 23. $\dfrac{9}{\sqrt{2x}}$ 24. $\dfrac{4\sqrt{u}}{\sqrt{3u}}$

13. $\dfrac{\sqrt[3]{54t^{10}s}}{\sqrt[3]{2ts}}$ 14. $\dfrac{\sqrt[3]{-500u^{11}v^{19}}}{\sqrt[3]{4u^2v}}$ 25. $\dfrac{10}{\sqrt{5}-1}$ 26. $\dfrac{7}{\sqrt{7}-1}$

In problems 15–46, rationalize the denominator of each fraction and simplify. Assume that all variables represent positive numbers and that no denominator is zero.

27. $\dfrac{13}{2\sqrt{3}+2}$ 28. $\dfrac{\sqrt{2}}{3\sqrt{2}+1}$

15. $\dfrac{5}{\sqrt{3}}$ 16. $\dfrac{4}{\sqrt{5}}$ 29. $\dfrac{4}{\sqrt{t}+1}$ 30. $\dfrac{8}{\sqrt{u}-3}$

17. $\dfrac{\sqrt{2}}{\sqrt{7}}$ 18. $\dfrac{\sqrt{3}}{\sqrt{2}}$ 31. $\dfrac{\sqrt{x}}{\sqrt{x}-2}$ 32. $\dfrac{2\sqrt{y}}{\sqrt{y}+4}$

19. $\dfrac{\sqrt{10}}{\sqrt{6}}$ 20. $\dfrac{\sqrt{20}}{\sqrt{15}}$ 33. $\dfrac{6}{\sqrt{7}-\sqrt{5}}$ 34. $\dfrac{9}{\sqrt{3}-\sqrt{5}}$

21. $\dfrac{10}{\sqrt{t}}$ 22. $\dfrac{11}{\sqrt{z}}$

1.8 Applications of Exponents and Radicals

Many applications of algebra require us to find the roots of numbers. For example, finding one side of a right triangle when two sides are given, and finding the distance between two points in the plane, are problems that can be solved with radical expressions.

The Pythagorean Theorem

The lengths of the sides of a right triangle are related in the following way:

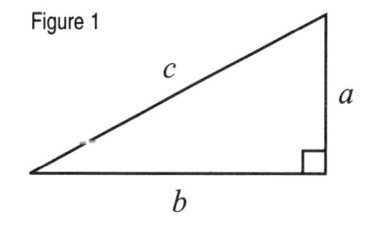

Figure 1

If a and b are the lengths of the perpendicular sides (also called the *legs*) of a right triangle and c is the length of the **hypotenuse** (Figure 1), then
$$c^2 = a^2 + b^2.$$

This result is known as the *Pythagorean theorem*.

Example 1 If the legs of a right triangle have lengths 5 and 12, find the length of the hypotenuse.

Solution See Figure 2. To find c, we use $c^2 = a^2 + b^2$, so that
$$c^2 = 5^2 + 12^2 = 25 + 144 = 169.$$

Figure 2

Taking the square root of each side of the equation, we have:
$$c = \sqrt{169} = 13.$$

114 CHAPTER 1 EXPONENTS AND POLYNOMIALS

Figure 3

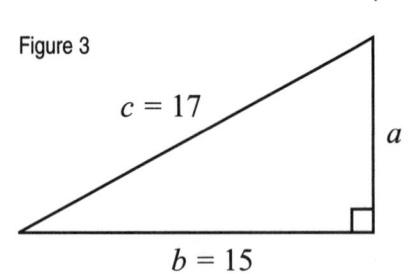

Example 2 If the length of the hypotenuse of a right triangle is 17 and the length of one leg is 15, what is the length of the other leg?

See Figure 3. To find a, we use $c^2 = a^2 + b^2$ or $a^2 = c^2 - b^2$, so that we have:

$$a^2 = (17)^2 - (15)^2 = 289 - 225 = 64$$

or

$$a = \sqrt{64} = 8.$$

Example 3 A cornfield is bounded by three straight roads that form a right triangle whose legs are 300 meters and 400 meters long. Find the length of the longest side of the cornfield.

See Figure 4. To find c, we use $c^2 = a^2 + b^2$, so that we have:

$$c^2 = (300)^2 + (400)^2$$
$$= 90{,}000 + 160{,}000$$
$$= 250{,}000$$

or

$$c = \sqrt{250{,}000}$$
$$= 500.$$

Therefore, the longest side of the cornfield is 500 meters.

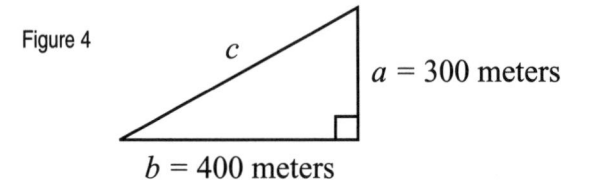

Figure 4

The Distance Formula

In the xy plane, the distance between any two points, $P_1 = (x_1, y_1)$ and $P_2 = (x_2, y_2)$, is the length of the line segment determined by the points (Figure 5). The following formula gives the distance d between $P_1 = (x_1, y_1)$ and $P_2 = (x_2, y_2)$:

$$d = \sqrt{(x_2 - x_1)^2 + (y_2 - y_1)^2}.$$

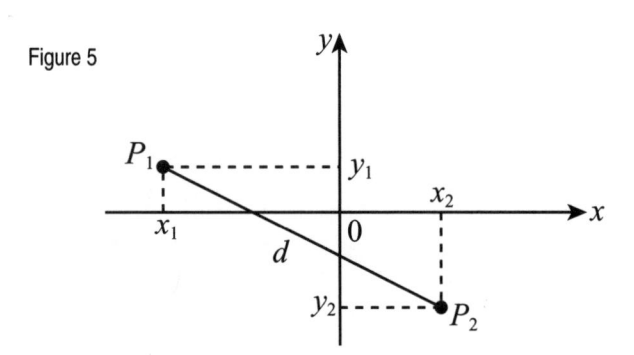

Figure 5

In Examples 4 and 5, find the distance d between the points P_1 and P_2.

Example 4 $P_1 = (-2, -1)$ and $P_2 = (2, -4)$.

Solution Using the distance formula

$$d = \sqrt{(x_2 - x_1)^2 + (y_2 - y_1)^2}$$

we have:

$$d = \sqrt{[2 - (-2)]^2 + [-4 - (-1)]^2},$$
$$= \sqrt{4^2 + (-3)^2}$$
$$= \sqrt{16 + 9}$$
$$= \sqrt{25} = 5.$$

Example 5 $P_1 = (-1, -2)$ and $P_2 = (3, 4)$.

Solution Substituting the coordinates of P_1 and P_2 in the formula

$$d = \sqrt{(x_2 - x_1)^2 + (y_2 - y_1)^2},$$

we have

$$d = \sqrt{[3 - (-1)]^2 + [4 - (-2)]^2}$$
$$= \sqrt{4^2 + 6^2}$$
$$= \sqrt{16 + 36}$$
$$= \sqrt{52}$$
$$= \sqrt{4 \cdot 13}$$
$$= 2\sqrt{13}.$$

PROBLEM SET 1.8

In problems 1–10, use the Pythagorean theorem to find the unknown side of a right triangle (Figure 1 on page 113), where c represents the hypotenuse and a and b represent the legs.

1. $a = 3, b = 4, c = ?$
2. $a = 10, b = 24, c = ?$
3. $c = 13, a = 5, b = ?$
4. $c = 25, b = 20, a = ?$
5. $a = 9, b = 12, c = ?$
6. $a = 16, b = 30, c = ?$
7. $c = 34, b = 30, a = ?$
8. $c = 39, a = 36, b = ?$
9. $a = 10, b = 24, c = ?$
10. $a = 30, b = 40, c = ?$

In problems 11–20, use the Pythagorean theorem and a calculator to find the unknown side of the right triangle. Round off the answer to two decimal places.

11. $a = 7, b = 9, c = ?$
12. $a = 12, b = 15, c = ?$
13. $a = 11.3, b = 13.4, c = ?$
14. $a = 21.7, b = 28.1, c = ?$
15. $c = 25.8, a = 17.5, b = ?$
16. $c = 317, a = 278, b = ?$
17. $c = 35.7, b = 19.3, a = ?$
18. $c = 41.6, b = 33.3, a = ?$
19. $a = 123, b = 179, c = ?$
20. $c = 17.6, a = 9.8, b = ?$

In problems 21–36, use the distance formula to find the distance d between the points P_1 and P_2.

21. $P_1 = (1, 1), P_2 = (4, 5)$
22. $P_1 = (0, 0), P_2 = (-5, -12)$
23. $P_1 = (-2, 3), P_2 = (3, -9)$
24. $P_1 = (3, 4), P_2 = (6, 8)$
25. $P_1 = (5, -1), P_2 = (-10, -7)$

26. $P_1 = (-5, 7)$, $P_2 = (3, -8)$
27. $P_1 = (-4, 5)$, $P_2 = (2, -3)$
28. $P_1 = (-2, -6)$, $P_2 = (7, 6)$
29. $P_1 = (7, 2)$, $P_2 = (4, -4)$
30. $P_1 = (8, -5)$, $P_2 = (4, 3)$
31. $P_1 = (0, 4)$, $P_2 = (2, 0)$
32. $P_1 = (-3, -5)$, $P_2 = (2, 5)$
33. $P_1 = (3, 7)$, $P_2 = (-1, 2)$
34. $P_1 = (9, 10)$, $P_2 = (7, 1)$
35. $P_1 = (-2, -3)$, $P_2 = (-8, 9)$
36. $P_1 = (-8, 3)$, $P_2 = (3, -7)$

In problems 37–42, use the Pythagorean theorem to solve the following problems.

37. A taut wire is stretched from the top of an antenna to the ground at a point 32 feet from the antenna's base. The wire forms a straight line 68 feet long. How high is the antenna?
38. Find the length of the diagonal of a rectangle if the length of one side of the rectangle is 120 meters and the length of the other side is 50 meters.
39. Two buildings are 100 feet apart. If one building is 275 feet tall and the other is 200 feet tall, what is the shortest length of cable needed to reach from the top edge of one building to the top edge of the other, connecting the walls that are 100 feet apart (Figure 6)?

Figure 6

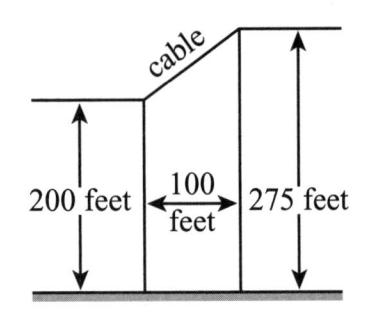

40. Two vertical poles, one 10 feet high and the other 30 feet high, are 48 feet apart. A rope is to be stretched from the top of one pole to a point on the ground halfway between the poles and then to the top of the other pole. How long must the rope be?
41. A ladder 30 feet long is placed against a building so that the foot of the ladder is 8 feet from the building. How far up the building does the ladder reach?
42. Two cars leave from the same point and drive along two straight roads that form a right angle. If one car averages 40 miles per hour and the other averages 50 miles per hour, how far apart are the two cars after 2 hours?

REVIEW PROBLEM SET

In problems 1–20, evaluate each expression. Assume that all variables represent positive numbers.

1. $\sqrt{100}$
2. $\sqrt{121}$
3. $-\sqrt{64}$
4. $-\sqrt{625}$
5. $\sqrt[3]{125}$
6. $\sqrt[3]{216}$
7. $\sqrt[4]{81}$
8. $\sqrt[4]{625}$
9. $\sqrt[5]{-243}$
10. $-\sqrt[5]{-1}$
11. $\sqrt{z^6}$
12. $\sqrt{25u^4}$
13. $\sqrt{49w^8}$
14. $\sqrt{16x^2y^4}$
15. $\sqrt[3]{-27m^9n^{30}}$
16. $\sqrt[3]{-125t^{12}s^{15}}$
17. $\sqrt[4]{16x^{16}y^{24}}$
18. $\sqrt[4]{256z^{32}w^{16}}$
19. $\sqrt[5]{32t^{35}s^{25}}$
20. $\sqrt[5]{243m^{40}n^{30}p^{15}}$

C In problems 21–26, use a calculator to approximate each irrational number and round off the answer to four decimal places.

21. $\sqrt{8}$
22. $\sqrt{12}$
23. $\sqrt{15}$
24. $\sqrt{21}$
25. $\sqrt{171}$
26. $\sqrt{379}$

In problems 27–48, use Property 1 of Section 1.4 to simplify each expression. Assume that all variables represent positive numbers.

27. $\sqrt{63}$
28. $\sqrt{48}$
29. $\sqrt{108}$
30. $\sqrt{700}$
31. $\sqrt[3]{135}$
32. $\sqrt[3]{-72}$
33. $\sqrt[3]{-192}$
34. $\sqrt[3]{375}$
35. $\sqrt[4]{32}$
36. $\sqrt[4]{243}$
37. $\sqrt[5]{224}$
38. $\sqrt[5]{-192}$
39. $\sqrt{z^9}$
40. $\sqrt{w^{11}}$
41. $\sqrt{32x^5}$
42. $\sqrt{75y^7}$
43. $\sqrt{125u^3v^{11}}$
44. $\sqrt{192z^{13}y^{17}}$
45. $\sqrt[3]{-108m^{10}n^9}$
46. $\sqrt[3]{320x^{14}z^{20}}$
47. $\sqrt[4]{48t^{15}s^{34}}$
48. $\sqrt[5]{-256u^{27}v^{34}}$

In problems 49–64, use Property 2 of Section 1.4 to simplify each expression. Assume that all variables represent positive numbers.

49. $\sqrt{\dfrac{11}{9}}$

50. $\sqrt{\dfrac{23}{4}}$

51. $\sqrt{\dfrac{16}{25}}$

52. $\sqrt{\dfrac{81}{64}}$

53. $\sqrt[3]{\dfrac{10}{27}}$

54. $\sqrt[3]{\dfrac{-29}{64}}$

55. $\sqrt[3]{\dfrac{-13}{125}}$

56. $\sqrt[4]{\dfrac{12}{81}}$

57. $\sqrt{\dfrac{15}{4u^4}}$

58. $\sqrt{\dfrac{20}{49t^6}}$

59. $\sqrt{\dfrac{8x^2}{25y^4}}$

60. $\sqrt{\dfrac{125t}{64t^3}}$

61. $\sqrt[3]{\dfrac{10}{27u^6v^9}}$

62. $\sqrt[3]{\dfrac{-5z^5}{216w^{12}}}$

63. $\sqrt[4]{\dfrac{7}{16x^8y^4}}$

64. $\sqrt[5]{\dfrac{29s^{10}}{243t^{15}}}$

In problems 65–70, combine like terms.

65. $3\sqrt{7} + 5\sqrt{7}$

66. $10\sqrt{11} + 8\sqrt{11}$

67. $8\sqrt{3} - 2\sqrt{3} + 4\sqrt{3}$

68. $7\sqrt[3]{4} + 4\sqrt[3]{4} - 6\sqrt[3]{4}$

69. $12\sqrt[3]{2t} + 9\sqrt[3]{2t} - 11\sqrt[3]{2t}$

70. $15\sqrt{3x} - 4\sqrt{3y} + 5\sqrt{3y}$

In problems 71–80, perform each operation by simplifying each radical and combining like terms. Assume that all variables represent positive numbers.

71. $2\sqrt{50} + 50\sqrt{2}$

72. $6\sqrt{98} - 4\sqrt{72}$

73. $\sqrt{8} + \sqrt{18} + \sqrt{98}$

74. $\sqrt{12} + \sqrt{48} - \sqrt{75}$

75. $\sqrt[3]{54} - \sqrt[3]{16}$

76. $5\sqrt[3]{40} + 2\sqrt[3]{135}$

77. $\sqrt{80z} + \sqrt{20z} - \sqrt{45z}$

78. $\sqrt{28t} - \sqrt{63t} + \sqrt{175t}$

79. $-5\sqrt[3]{32x^4} - 2\sqrt[3]{108x^4}$

80. $10\sqrt[3]{-81u^3v^5} + 6\sqrt[3]{375u^3v^5}$

In problems 81–92, find each product and simplify. Assume that all variables represent positive numbers.

81. $\sqrt{6}\sqrt{30}$

82. $\sqrt{12}\sqrt{36}$

83. $\sqrt{3x}\sqrt{15x^3}$

84. $\sqrt{7u^4}\sqrt{14u}$

85. $\sqrt{5}(\sqrt{15} + \sqrt{5})$

86. $\sqrt{t}(\sqrt{3t} - 2\sqrt{t})$

87. $(\sqrt{z} - 3)(\sqrt{z} + 3)$

88. $(\sqrt{4y} + \sqrt{x})(\sqrt{4y} - \sqrt{x})$

89. $(\sqrt{7} + \sqrt{2})^2$

90. $(3\sqrt{m} - 2\sqrt{n})^2$

91. $(2\sqrt{x} + \sqrt{y})(\sqrt{x} + 3\sqrt{y})$

92. $(3\sqrt{5} - 4\sqrt{2})(4\sqrt{5} + 3\sqrt{2})$

In problems 93–100, find each quotient and simplify. Assume that all variables represent positive numbers.

93. $\dfrac{\sqrt{54}}{\sqrt{3}}$

94. $\dfrac{\sqrt{320}}{\sqrt{5}}$

95. $\dfrac{\sqrt[3]{108}}{\sqrt[3]{4}}$

96. $\dfrac{\sqrt[3]{-250}}{\sqrt[3]{2}}$

97. $\dfrac{\sqrt{48x^5}}{\sqrt{3x}}$

98. $\dfrac{\sqrt{63t^7}}{\sqrt{7t^3}}$

99. $\dfrac{\sqrt[3]{-162u^5v^7}}{\sqrt[3]{6u^2v}}$

100. $\dfrac{\sqrt[3]{40zy}}{\sqrt[3]{-5zy^4}}$

In problems 101–110, rationalize the denominator of each fraction. Assume that all variables represent positive numbers and that no denominators are zero.

101. $\dfrac{13}{\sqrt{7}}$

102. $\dfrac{9}{\sqrt{11}}$

103. $\dfrac{8}{\sqrt{x}}$

104. $\dfrac{10u}{\sqrt{3u}}$

105. $\dfrac{4}{\sqrt{7} + 2}$

106. $\dfrac{6}{\sqrt{11} - \sqrt{3}}$

107. $\dfrac{3z}{\sqrt{z} - 1}$

108. $\dfrac{nm}{\sqrt{m} + \sqrt{n}}$

109. $\dfrac{\sqrt{5} + 2\sqrt{2}}{3\sqrt{5} - \sqrt{2}}$

110. $\dfrac{\sqrt{x} - 2\sqrt{y}}{4\sqrt{x} + 3\sqrt{y}}$

In problems 111–124, solve each radical equation and check your solutions.

111. $\sqrt{2x} = 10$

112. $\sqrt{3z} = 9$

113. $\sqrt{8x + 3} = 7$

114. $\sqrt{4t - 4} = 5$

115. $\sqrt{5t - 5} = 1$

116. $\sqrt{2m + 6} = 9$

117. $\sqrt{4z + 3} = 5$

118. $\sqrt{3x + 15} + 5 = 8$

119. $\sqrt{x + 2} + 4 = x$

120. $\sqrt{z - 1} + z = 7$

121. $\sqrt{3t - 11} + t = 13$

122. $x - \sqrt{4x + 4} = 2$

123. $\sqrt{4y + 41} - 3y = 1$

124. $\sqrt{5x + 49} - 2x = 2$

In problems 125–130, use the Pythagorean theorem to find the unknown side of the right triangle, where c represents the hypotenuse and a and b represent the legs.

125. $a = 15, b = 20, c = ?$

126. $a = 24, b = 45, c = ?$

127. $c = 52, a = 48, b = ?$

128. $c = 30, b = 24, a = ?$

129. $a = 7, b = 9, c = ?$

130. $c = 21, b = 11, a = ?$

In problems 131–136, use the distance formula to find the distance between the given points.

131. $P_1 = (-2, -3), P_2 = (2, 0)$

132. $P_1 = (5, 7), P_2 = (-7, 2)$

133. $P_1 = (6, 10), P_2 = (-2, -5)$

134. $P_1 = (-3, 4), P_2 = (3, -4)$

135. $P_1 = (-5, 3), P_2 = (6, 7)$

136. $P_1 = (8, -1), P_2 = (3, -7)$

In problems 137–140, use the Pythagorean theorem to solve the following problems.

137. A carpenter wishes to brace a wall 10 feet tall by fixing one end of a supporting rod to the top of the wall. and nailing the other end to the floor at a point 5 feet from the base of the wall. How long a supporting rod does he need?

138. A catcher on a baseball team stands on home plate and throws a ball to the second baseman who is standing on second base. If the distance between bases is 90 feet, how far did the catcher throw the ball?

139. A hiker walks around a large rectangular field, starting at one corner and finishing at the opposite corner. If the field is 3 miles long and 2 miles wide, how many miles could the hiker have saved by walking directly from corner to corner?

140. James lives on an island that is 1.5 miles from a straight beach and his girlfriend lives 2.3 miles up the beach. What is the shortest distance that James will have to row his boat in order to pull up to the dock of his girlfriend's house?

Chapter 2

POLYNOMIALS AND FACTORING

2.1 Polynomials
2.2 Polynomial Notation
2.3 Addition and Subtraction of Polynomials
2.4 Multiplication of Polynomials
2.5 The FOIL Method and Special Products
2.6 Division of Polynomials
2.7 Factoring Integers
2.8 Common Factors and Factoring by Grouping
2.9 Factoring the Difference of Two Squares and Perfect-Square Trinomials
2.10 Factoring Trinomials of the Form $x^2 + bx + c$
2.11 Factoring Trinomials of the Form $ax^2 + bx + c, a \neq 1$

2.1 Polynomials

Algebraic expressions such as

$$7, \quad 3x, \quad 3y + 1, \quad \tfrac{5}{7}t - \tfrac{2}{3}, \quad \text{and} \quad u^3 - 7u^2 + \tfrac{2}{5}u - 11$$

are examples of polynomials. More formally, we have the following definition:

Polynomials

> A **polynomial** is an algebraic expression that consists of the sum (or difference) of terms. Each term in a polynomial is either a constant or the product of a constant and a positive integer power of a variable.

In other words, all exponents of the variables of a polynomial are nonnegative integers, and there can be no division by a variable. (No variable can appear in the denominator of a fraction.) For example, $3x^2 + 5x + 6$ is a polynomial, whereas $5y^3 + 3y^2 + (2/y) + 6$ is *not* a polynomial because there is a division by the variable y: $(2/y = 2 \div y)$.

Three special kinds of polynomial are:

> (i) A **monomial** is a polynomial with only one term.
> (ii) A **binomial** is a polynomial with exactly two terms.
> (iii) A **trinomial** is a polynomial with exactly three terms.

From *Beginning Algebra, 4th edition* by M.A. Munem and W. Tschirhart. Copyright © 2000 by Kendall/Hunt Publishing Company. Reprinted by permission.

Example 1 Determine if the given expression is a polynomial. If it is, state whether it is a monomial, a binomial, or a trinomial.

(a) $3x^2 + \frac{2}{3}x$ (b) $-7x$ (c) $4y^3 - 2y + \frac{3}{5}$

(d) $3u^2 + 7u + \frac{1}{u}$ (e) $t^4 - \frac{2}{3}t^3 + \frac{6}{t^2} - \frac{1}{2}t + 4$ (f) $3x^2 + 3x - 17$

Solution
(a) A polynomial with variable x; a binomial.
(b) A polynomial with variable x; a monomial.
(c) A polynomial with variable y; a trinomial.
(d) Not a polynomial because of the division by the variable u.
(e) Not a polynomial because of the division by the variable t.
(f) A polynomial with a variable x; a trinomial.

Any factor of a product can be considered to be a coefficient of the product of the remaining factors. For instance, in the product xy, x is the coefficient of y, and y is the coefficient of x. The constant factors of the terms of a polynomial are called the **numerical coefficients** (or simply **coefficients**) of the polynomial. For example,

the coefficients of the polynomial $2x^2 + 3x + 7$ are 2, 3, and 7,
the coefficients of the polynomial $3t^2 - 7t - 8$ are 3, -7, and -8,

and

the coefficients of the polynomial $y^2 - y + 17$ are 1, -1, and 17.

Each term of a polynomial has a degree:

DEFINITION 2 **Degree**

> The **degree of a term** containing one variable is the exponent of the variable. The *degree* of a nonzero constant (such as 7) is 0. The number 0 has no *degree* assigned to it.
> The **degree** of a polynomial in one variable is the highest degree of any term in the polynomial.

For example,

the numerical coefficient of the term $2x$ is 2 and degree is 1,
the numerical coefficient of the term $-4y^3$ is -4 and the degree is 3,

and

the numerical coefficients of the polynomial $3x^2 - 5x + 7$ are 3, -5, and 7 and the degree is 2.

The *degree* of a term containing more than one variable is the sum of the exponents of the variable factors. For example,

the degree of the term $9x^3y^4$ is 7.

The *degree* of a polynomial that contains *more than one* variable is the degree of the term with the highest degree in the polynomial. For instance,

the degree of the polynomial $8x^3y^2 - y^4 + 3xy^2 - 2x^3 + 7xy$ is 5.

Example 2 Determine the degree and the numerical coefficients of each polynomial.

(a) $-3x^2$ (b) $6z + 7$
(c) $\frac{2}{7}y^6 - 5y^2 + 8$ (d) $-u^3 + 6u^2 - 5u + 7$
(e) $5x^2y^3 - 3x^4 + 8x^2 + 11$ (f) 12

Solution
(a) The degree is 2, and the numerical coefficient is -3.
(b) The degree is 1, and the numerical coefficients are 6 and 7.

(c) The degree is 6, and the numerical coefficients are $\frac{2}{7}$, -5, and 8.
(d) The degree is 3, and the numerical coefficients are -1, 6, -5, and 7.
(e) The degree is 5, and the numerical coefficients are 5, -3, 8, and 11.
(f) The degree is 0, and the numerical coefficient is 12.

Evaluating Polynomials

If we replace the variables in a polynomial with specific numbers, we can **evaluate** the polynomial for those values of the variable. For instance, if a specific number (say, 3) is used for x in the polynomial $12x + 13$, the polynomial becomes an arithmetic expression:

$$12x + 13 = 12(3) + 13 = 36 + 13 = 49.$$

If $x = -2$, the polynomial becomes

$$12x + 13 = 12(-2) + 13 = -24 + 13 = -11,$$

and if $x = 0$, the polynomial becomes

$$12x + 13 = 12(0) + 13 = 0 + 13 = 13.$$

We have already learned the order of operations for evaluating arithmetic expressions. Since then we have learned a new operation—raising to a power. When arithmetic operations and the power operation appear together in an expression, the power operation should be performed first, unless grouping symbols indicate otherwise.

We can state the order of operations for evaluating a polynomial as follows:

> (i) Raising to a power
> (ii) Multiplication and division
> (iii) Addition and subtraction

For example, to evaluate the polynomial

$$4x^3 + 2x^2 + 5x + 3 \quad \text{for} \quad x = -2.$$

first, we substitute -2 for x to get

$$4(-2)^3 + 2(-2)^2 + 5(-2) + 3.$$

Then, we raise -2 to the given powers, so that we have

$$4(-8) + 2(4) + 5(-2) + 3.$$

Next, we multiply from left to right to obtain

$$-32 + 8 - 10 + 3.$$

Finally, we add and subtract to get -31.

Example 3 Evaluate $y^3 + 3y - 5$ for $y = -3$.

Solution We replace y by -3:

$$\begin{aligned} y^3 + 3y - 5 &= (-3)^3 + 3(-3) - 5 \\ &= -27 - 9 - 5 \\ &= -41. \end{aligned}$$

Example 4 Evaluate $3u^2 + 2uv - 12$ for $u = -1$ and $v = 5$.

Solution We replace u by -1 and v by 5:

$$\begin{aligned} 3u^2 + 2uv - 12 &= 3(-1)^2 + 2(-1)(5) - 12 \\ &= 3 - 10 - 12 \\ &= -19. \end{aligned}$$

PROBLEM SET 2.1

In problems 1–16, determine whether the given expression is a polynomial. Then identify each polynomial as a monomial, binomial, trinomial, or none of these.

1. 7
2. -30
3. $-3x$
4. $4xy$
5. $2t + 13$
6. $8 - 3s$
7. $3u - \dfrac{5}{u}$
8. $\dfrac{4}{x^2} + 3x + \dfrac{1}{2}$
9. $y^2 + 5y + 2$
10. $z^5 + z^3 + z$
11. $x^3 - 5x^2 + 3x - 25$
12. $2y^7 - 3y^4 + 4y^3 + 9$
13. $\frac{1}{2}t^2 + \frac{3}{2}t$
14. $3x^2y + 2xy^2 + 11$
15. $\dfrac{5}{u^3} + \dfrac{1}{u} + 3$
16. $5v - 3v^2 + \dfrac{7}{v}$

In problems 17–30, determine the degree and list the numerical coefficients of each polynomial.

17. $5y + 7$
18. $-3z + 11$
19. $x^2 + 7x - 3$
20. $s^3 - s + 1$
21. $z^2 - 7z + 3$
22. $3u^2 + u + \frac{1}{2}$
23. 5
24. 0
25. $\frac{1}{2}t^3 - \frac{1}{3}t^2 + \frac{2}{5}t - \frac{3}{2}$
26. $3 - z + z^2 - z^3$
27. $4x - 7x^7 + 3x^5$
28. $x^4z + x^2z^3 - 3xz + 5$
29. $3x^2y^3 - 2xy^2 + 11$
30. $3u^2 - 7uv + v^2$

In problems 31–50, evaluate the polynomials for the given values of the variables.

31. $3t + 8, t = 3$
32. $7 - 2s, s = -5$
33. $-8x + 11, x = 5$
34. $\frac{1}{2}u + \frac{3}{2}, u = -7$
35. $z^2 + 5, z = 3$
36. $-y^2 + 5, y = 3$
37. $x^2 + x + 4, x = 2$
38. $2v^2 - 3v + 7, v = -2$
39. $y^2 + 2y - 5, y = -1$
40. $-5w^2 - 3w + 11, w = 5$
41. $2t^2 - 7t - 10, t = -2$
42. $2x^3 - 3x^2 + 5x - 4, x = -3$
43. $\frac{1}{2}x^3 + \frac{1}{4}x^2 + 3x - 5, x = 4$
44. $\frac{1}{3}s^4 - \frac{1}{2}s^3 + \frac{5}{6}s + 1, s = -1$
45. $7u^2 - 3uv + v^2, u = 1$ and $v = 2$
46. $4x^2y - 3xy^2 + xy - 2, x = -1$ and $y = 1$
47. $3z^2 + 4zy - y^2, z = -1$ and $y = 2$
48. $w^4 + w^2 + w + 5, w = -2$
49. $-5t^2 - ts + s^3, t = -2$ and $s = -1$
50. $3x^4 + 5x^3y - 3xy^2 + 11, x = -1$ and $y = 0$

51. It has been estimated that the number N of hours of sleep needed by a growing child of age A is given by the expression $17 - \frac{1}{2}A$. How many hours of sleep are needed by a child who is:
    (a) 8 years old (b) 12 years old (c) 16 years old

52. Medical studies have shown that blood pressure is related to age. The normal blood pressure (in millimeters of mercury) for a person who is A years old is given by the expression
$$\dfrac{A(A+5)}{100} + 107.$$
What is the normal blood pressure for a person who is:
    (a) 20 years old (b) 40 years old (c) 50 years old

53. If a taxicab fare is $2 for the first mile and $0.75 for each additional mile, the cost for a trip of n miles is given by the expression $0.75(n - 1) + 2$. Find the cost for a trip of 11 miles.

54. The volume of a square prism is given by the expression lw^2. Find the volume of a square prism if $l = 15$ meters and $w = 8$ meters.

55. The cost of producing x calculators is given by $0.5x^2 + 10x + 25$. Find the cost of producing (a) 15 calculators (b) 25 calculators.

56. A ball is thrown vertically upward from one foot above the ground with an initial speed of 64 feet per second. The height of the ball after t seconds is $-16t^2 + 64t + 1$. Find the height after (a) 2 seconds (b) 3 seconds.

2.2 Polynomial Notation

When evaluating polynomials, it is sometimes expedient to use a specific notation known as polynomial notation. For example, the polynomial $7x - 2$ is introduced using polynomial notation as

$P(x) = 7x - 2$ **and is read "P of x equals 7 times x minus 2."**

The polynomial can then be evaluated for specific values of x as follows:

$$P(x) = 7x - 2$$
$$P(3) = 7(3) - 2$$
$$P(3) = 21 - 2$$
$$P(3) = 19$$

Therefore, when $P(x)$ is evaluated for $x = 3$, $P(3) = 19$.

Find $P(-4)$ **and** $P(0)$.

Solution: $P(-4) = 7(-4) - 2$ and   Solution: $P(0) = 7(0) - 2$
$P(-4) = -28 - 2$ $P(0) = 0 - 2$
$P(-4) = -30$ $P(0) = -2$

We simply substitute any indicated value for the variable and evaluate the polynomial.

Other letters of the alphabet such as Q, R, H, etc. are also commonly used to indicate different polynomials that may also be under consideration.

For example,

a. If $P(x) = x^2 - 2x + 3$, find $P(4)$.
   solution: $P(4) = 4^2 - 2(4) + 3$
   $P(4) = 16 - 8 + 3$
   $P(4) = 11$

b. If $Q(x) = x + 5$, find $Q(-2)$.
   solution: $Q(-2) = -2 + 5$
   $Q(-2) = 3$

c. Given: $Q(x) = x + 5$
   $P(x) = x^2 - 2x + 3$
   $R(x) = Q(x) + P(x)$

Find: $R(-1)$.

solution A: $R(-1) = Q(-1) + P(-1)$
$R(-1) = [-1 + 5] + [(-1)^2 - 2(-1) + 3]$
$R(-1) = -1 + 5 + 1 + 2 + 3$
$R(-1) = 10$

Or we could evaluate $Q(-1)$ and $P(-1)$ separately.

solution B: $Q(x) = x + 5$ $P(x) = x^2 - 2x + 3$
$Q(-1) = -1 + 5$ $P(-1) = (-1)^2 - 2(-1) + 3$
$Q(-1) = 4$ $P(-1) = 1 + 2 + 3$
                                                                $P(-1) = 6$

Then $R(-1) = Q(-1) + P(-1)$ becomes
$R(-1) = 4 + 6$
$R(-1) = 10$

Application to life:

Business: The costs incurred to produce a product and the revenue realized from the sale of that product can sometimes be expressed as polynomials. For example, the Nakatoch T-Shirt Company bought the equipment needed to airbrush t-shirts.

The one-time equipment expense was $500.

The company bought the t-shirts for $3.00 each, airbrushed each shirt, and then sold each t-shirt for $15.00. The cost polynomial would be expressed as $C(x) = 500 + 3x$ where x represents the number of t-shirts bought.

The revenue polynomial would be expressed as $R(x) = 15x$ where x represents the number of t-shirts sold. The company could have a loss or a profit on these t-shirts depending on the number bought and sold. If we assume that every t-shirt bought by the company is designed and sold, we could write a profit polynomial that would show total revenue minus total cost: $P(x) = R(x) - C(x)$.

Problem: A. Find the total cost of producing 50 t-shirts.
B. Find the total revenue from the sale of 50 t-shirts.
C. Find the profit or loss on the production and sale of 50 t-shirts.

Solution   A. $C(x) = 500 + 3x$ dollars
$C(50) = 500 + 3(50)$
$= 500 + 150$
$C(50) = 650$ dollars

B. $R(x) = 15x$ dollars
$R(50) = 15(50)$
$R(50) = 750$ dollars

C. $P(x) = R(x) - C(x)$ dollars
$P(50) = 750 - 650$ dollars
$P(50) = 100$ dollars profit

Therefore, on the production and sale of 50 t-shirts, the Nakatosh T-Shirt Company made a profit of $100.00.

Extension: What would be their profit if they could produce and sell 100 t-shirts? What would be their profit or loss if they only produced and sold 25 t-shirts?

PROBLEM SET 2.2

Given $P(x) = 7 + x$
$Q(x) = -2x + 6$
$S(x) = -x^2 + 4x - 9$

1. Find $P(0), P(1), P(2),$ and $P(3)$.
2. Find $Q(-2), Q(-1), Q(0),$ and $Q(1)$.
3. Find $S(-1), S(0), S(1),$ and $S(2)$.
4. Find $P(3) + Q(-2) - S(0)$.
5. Given $R(x) = P(x) - Q(x)$, find $R(2), R(0)$ and $R(-3)$.

Given $P(x) = 0.4 + 1.5x$
$Q(x) = -0.25x + 6.1$
$S(x) = -3.3x^2 + 0.4x - 2.08$

6. Find $P(0), P(1), P(2),$ and $P(3)$.
7. Find $Q(-2.3), Q(-1.3), Q(0),$ and $Q(1.3)$.
8. Find $S(-1.15), S(0.15), S(1.15),$ and $S(2.15)$.
9. Find $P(3) + Q(-2.3) - S(0.15)$.
10. Given $R(x) = P(x) + Q(x)$, find $R(2), R(0)$ and $R(3)$.

Given $P(x) = \frac{3}{4} + \frac{1}{2}x$
$Q(x) = 3\frac{1}{3}x - 2\frac{5}{6}$
$s(x) = \frac{3}{8}x^2 - 4\frac{1}{4}x - 9\frac{1}{2}$

11. Find $P(0), P(1), P(2),$ and $P(3)$.
12. Find $Q(-2), Q(-1), Q(0),$ and $Q(1)$.
13. Find $S(-1), S(0), S(1),$ and $S(2)$.
14. Find $P(3) + Q(-2) - S(0)$.
15. Given $R(x) = P(x) - Q(x)$, find $R(2), R(0)$ and $R(3)$.

Applications:

16. Find the profit or loss on the production and sale of 150 t-shirts using the polynomial equations mentioned in the example in the text.
17. Find the profit or loss on the production and sale of 40 t-shirts using the polynomial equations mentioned in the example in the text.

2.3 Addition and Subtraction of Polynomials

You have worked many problems involving addition and subtraction of real numbers. Since polynomials are expressions that represent real numbers, the commutative, associative, and distributive properties hold for polynomials as well as for real numbers. That is, if $P, Q,$ and R represent polynomials, then:

(i) $P + Q = Q + P$
(ii) $P + (Q + R) = (P + Q) + R$
(iii) (a) $P(Q + R) = PQ + PR$ (b) $(P + Q)R = PR + QR$

Suppose that you want to add the monomials $5x^2$ and $2x^2$. By the distributive property, you have

$$5x^2 + 2x^2 = (5 + 2)x^2 = 7x^2.$$

The terms $5x^2$ and $2x^2$ are like or similar terms. In general:

> Two or more terms are **like terms** or **similar terms** if they contain the same power (or powers) of the variable (or variables).

Although the variable factors of like terms are the same, like terms may have different (numerical) coefficients. For example, $3x^2$ and $-7x^2$ are like terms, and so are $5y^7z^3$ and $-\frac{11}{3}y^7z^3$. However, $2x^2$ and $3x$ are not like terms because the exponents of the variables are not the same.

Two or more like terms can be combined by addition or subtraction into a single term. The distributive property is the key to this process.

Example 1 Combine like terms in each expression.

(a) $7x + 8x$ (b) $3x^2 - 6x^2$
(c) $4y^3 + 9y^3 + 2y^3$ (d) $-3z^5 - (-6z^5)$
(e) $12x^2y + 8x^2y + 3x^2y^2$ (f) $3y + 2t + 4w$

Solution (a) We apply the distributive property:
$$7x + 8x = (7+8)x = 15x.$$

(b) Here we use the fact that $a - b = a + (-b)$. Then we apply the distributive property to obtain.
$$3x^2 - 6x^2 = 3x^2 + (-6x^2) = [3 + (-6)]x^2 = -3x^2.$$

(c) $4y^3 + 9y^3 + 2y^3 = (4 + 9 + 2)y^3 = 15y^3$
(d) $-3z^5 - (-6z^5) = -3z^5 + 6z^5 = (-3 + 6)z^5 = 3z^5$
(e) The only two like terms in the expression are the first two terms. We combine these two terms,
$$12x^2y + 8x^2 + 3x^2y^2 = (12 + 8)x^2y + 3x^2y^2$$
$$= 20x^2y + 3x^2y^2.$$

(f) Because there are no like terms in the given expression, no combining can be done.

We can add and subtract polynomials by combining like terms. For example, to perform the addition.
$$(5x^2 + 3x + 2) + (3x^2 + 2x + 7),$$
we can combine like terms as follows:
$$(5x^2 + 3x + 2) + (3x^2 + 2x + 7) = 5x^2 + 3x + 2 + 3x^2 + 2x + 7$$
$$= (5x^2 + 3x^2) + (3x + 2x) + (2 + 7)$$
$$= 8x^2 + 5x + 9.$$

Example 2 Perform the addition
$$(4x^3 + 6x^2 - 3x - 7) + (5x - 2x^2 + x^3 - 2).$$

Solution We remove the parentheses, rearrange the terms, and combine like terms:
$$(4x^3 + 6x^2 - 3x - 7) + (5x - 2x^2 + x^3 - 2)$$
$$= 4x^3 + 6x^2 - 3x - 7 + 5x - 2x^2 + x^3 - 2$$
$$= 4x^3 + x^3 + 6x^2 - 2x^2 - 3x + 5x - 7 - 2$$
$$= (4 + 1)x^3 + (6 - 2)x^2 + (-3 + 5)x + (-7 - 2)$$
$$= 5x^3 + 4x^2 + 2x - 9.$$

To subtract $y^2 - 2y + 3$ from $3y^2 + 5y - 7$, we extend the definition of subtraction to polynomials:

$$P - Q = P + (-Q).$$

Thus,

$$(3y^2 + 5y - 7) - (y^2 - 2y + 3) = (3y^2 + 5y - 7) + [-(y^2 - 2y + 3)].$$

Now we form the opposite (negative) of $y^2 - 2y + 3$ simply by changing the sign of each term. That is,

$$-(y^2 - 2y + 3) = -y^2 + 2y - 3$$

so that

$$\begin{aligned}(3y^2 + 5y - 7) - (y^2 - 2y + 3) &= 3y^2 + 5y - 7 - y^2 + 2y - 3 \\ &= 3y^2 - y^2 + 5y + 2y - 7 - 3 \\ &= (3 - 1)y^2 + (5 + 2)y + (-7 - 3) \\ &= 2y^2 + 7y - 10.\end{aligned}$$

Example 3 Perform the subtraction

$$(2y^3 - y^2 + 3y + 5) - (11y^3 + 2y^2 - 7y - 2).$$

Solution

$$\begin{aligned}(2y^3 - y^2 + 3y + 5) &- (11y^3 + 2y^2 - 7y - 2) \\ &= (2y^3 - y^2 + 3y + 5) + [-(11y^3 + 2y^2 - 7y - 2)] \\ &= 2y^3 - y^2 + 3y + 5 - 11y^3 - 2y^2 + 7y + 2 \\ &= (2 - 11)y^3 + (-1 - 2)y^2 + (3 + 7)y + (5 + 2) \\ &= -9y^3 - 3y^2 + 10y + 7.\end{aligned}$$

It is often helpful to use a "vertical scheme" to rearrange like terms when we are adding or subtracting polynomials. In this procedure, like terms are arranged in vertical columns so that it is easy to add or subtract the (numerical) coefficients. For example, the polynomials $5x^2 + 3x + 2$ and $3x^2 + 2x + 7$ can be added as follows:

$$\begin{array}{r}5x^2 + 3x + 2 \\ (+)\ \underline{3x^2 + 2x + 7} \\ 8x^2 + 5x + 9.\end{array}$$

In Examples 4–7, perform each operation by using the vertical scheme.

Example 4 $(-8x^5 + 2x^2 + 4 + 5x^3) + (10x^5 + 7x^4 - x^3 - 3x)$

Solution We line up like terms and add coefficients:

$$\begin{array}{r}-8x^5 + 0x^4 + 5x^3 + 2x^2 + 0x + 4 \\ (+)\ \underline{10x^5 + 7x^4 - x^3 + 0x^2 - 3x + 0} \\ 2x^5 + 7x^4 + 4x^3 + 2x^2 - 3x + 4.\end{array}$$

Note that missing terms are written with zero coefficients

Example 5 $(x^2 - 2x + 1) + (3x^2 + 4x - 3) + (5x^2 - 3x + 4)$

Solution We arrange like terms in columns and combine:

$$\begin{array}{r}x^2 - 2x + 1 \\ 3x^2 + 4x - 3 \\ (+)\ \underline{5x^2 - 3x + 4} \\ 9x^2 - x + 2.\end{array}$$

Example 6 $(6x^2 - x - 3) - (4x^2 - 3x + 5)$

Solution We arrange like terms in columns:

$$\begin{array}{r} 6x^2 - x - 3 \\ (-)\ \underline{4x^2 - 3x + 5} \end{array}$$

We form the opposite (negative) of the bottom polynomial [that is, $-(4x^2 - 3x + 5) = -4x^2 + 3x - 5$] and add:

$$\begin{array}{r} 6x^2 - x - 3 \\ (+)\ \underline{-4x^2 + 3x - 5} \\ 2x^2 + 2x - 8. \end{array}$$

Example 7 Subtract $5x^2 - 3x + 6$ from the sum of $2x^2 - 5x + 7$ and $x^2 + 3x - 10$.

Solution First, we perform the addition of $2x^2 - 5x + 7$ and $x^2 + 3x - 10$:

$$\begin{array}{r} 2x^2 - 5x + 7 \\ (+)\ \underline{x^2 + 3x - 10} \\ 3x^2 - 2x - 3. \end{array}$$

Then we subtract $5x^2 - 3x + 6$ from this sum:

$$\begin{array}{rcr} 3x^2 - 2x - 3 & & 3x^2 - 2x - 3 \\ (-)\ \underline{5x^2 - 3x + 6} & \text{or} & (+)\ \underline{-5x^2 + 3x - 6} \\ & & -2x^2 + x - 9. \end{array}$$

If polynomials contain more than one variable, addition and subtraction are still performed by combining like terms.

Example 8 Subtract $2xy^2 - 3x^2y + 7$ from $-5x^2y + 6xy^2 - 11$.

Solution Arranging like terms in columns, we have

$$\begin{array}{r} -5x^2y + 6xy^2 - 11 \\ (+)\ \underline{-3x^2y + 2xy^2 + 7}. \end{array}$$

Now we form the opposite (negative) of the bottom polynomial and add:

$$\begin{array}{r} -5x^2y + 6xy^2 - 11 \\ (+)\ \underline{3x^2y - 2xy^2 - 7} \\ -2x^2y + 4xy^2 - 18. \end{array}$$

PROBLEM SET 2.3

In problems 1–12, combine like terms.

1. $14x^2 + 7x^2$
2. $7t^2 + (-3)t^2$
3. $4s^4 + 3s^4 + 8s^4$
4. $-11m^2 + (-9m^2)$
5. $8y^3 - 3y^3$
6. $-15u^2 - (-4u^2)$
7. $-5z^5 - (-3z^5)$
8. $6w^4 + 4w^4 - 3w^4$
9. $7x^3y + 5x^3y - 4x^3y$
10. $8st^2 - 5st^2 - (-3st^2)$
11. $(5u^2v^3 - 7u^3v^2 + 6u^2v^3)$
12. $-4z^5w^2 - z^2w^5 - (-8z^2w^5)$

In problems 13–22, perform the additions or subtractions of the given polynomials by parentheses and combining like terms.

13. $(3t^2 + 4) + (5t^2 - 8)$
14. $(7x^3 + 3) - (2x^3 + 1)$
15. $(3z^2 - 5z + 2) + (6z - 8 + 2z^2)$
16. $(5u^3 - 4u^2 + 2u - 7) + (3u^3 + 5u^2 - 6u + 8)$
17. $(3s^2 + 7s + 11) - (s^2 - 3s + 4)$
18. $(3x^2 - 7x^3 + 5 + 8x) - (15 - 3x + 3x^2 - 6x^3)$
19. $(11z^3 - 8z^2 + z - 3) - (9z^3 - 4z + 7)$
20. $(4t - 3t^2 - 10 + 5r^3) + (15 - 5t + 3t^2)$

21. $(8x^3y - 4x^2y^2 + 5xy^3) + (2xy^3 - 3x^2y^2 + 5x^3y)$
22. $(7wz^3 + 5w^2z - 8wz) - (9w^2z + wz - 6wz^3 - 5)$

In problems 23–38, perform the additions or subtractions of the given polynomials by using a vertical scheme.

23. $(3x^2 + 5x - 7) + (4x^2 + 2x + 12)$
24. $(13z^2 - 11z + 12) + (-4z^2 + 7z - 3)$
25. $(5s^3 - 4s^2 + 6s - 2) + (3s^3 - 2s^2 - s - 4)$
26. $(7y^2 - 8y + y^3 - 6) + (11y - 2y^2 + 8 - 2y^3)$
27. $(4z^2 - 3z - 11) - (z^2 + 2z - 7)$
28. $(-t^2 + t - 1) - (t^2 - t + 3)$
29. $(8u^3 + 2u^2 - 5u + 6) - (-2u^3 + u^2 - 5u - 6)$
30. $(4x^5 - 7x^3 + 8x - 2) - (3x^4 + 4x^3 - 3x^2 + 5)$
31. $(5y - 3y^2 + 13 - 5y^3) + (4y + 7y^2 - 2y^3 - 8)$
32. $(6z^4 - 3z^3 - 14 + 2z) + (15 - 2z^4 + 3z^2 - 2z^3)$
33. $(10m^3n - 8m^2n^2 - 9mn^3) - (-5m^3n + 4mn^3 - 8m^2n^2)$
34. $(3y^3z^2 + 5y^2z^3 - 8yz + 4) - (2y^2z^3 - 5y^3z^2 + 2yz)$
35. $(-4u^4v^2 + 7u^2v - 9uv^5) + (6u^2v + 8u^4v^2 - u^4v^4 - 7uv^5)$
36. $(2s^3t - 3st^2 + st - 3) - (5st + 6st^2 - 2s^3t)$
37. $(15x^3y^2 - 8x^2y + 4xy^2) - (7x^2y - xy^2 + 12x^3y^2 - 2)$
38. $(-20wz^3 + 16w^2z^2 - 18w^3z - 14) + (14w^2z^2 + 10wz^3 - 12w^3z - 16)$

In problems 39–50, perform the indicated operations.

39. $(2x^2 - 3x + 11) + (x^2 + 5x - 4) + (3x^2 - 7x + 1)$
40. $(2y^2 + 7y - 8) + (-y^2 + 3y - 2) + (3y^2 - 5y + 7)$
41. $(8t^2 - 7t + 6) + (3t^2 + 2t - 4) - (7t^2 + t + 5)$
42. $(9m^2 + 5m - 8) - (-m^2 + 3m - 1) - (2m^2 - 6m + 4)$
43. $(s^3 + s^2 - s + 1) + (2s^3 - 3s^2 + 2s + 3) + (3s^3 + 5s^2 - 7s + 6)$
44. $(y^3 - 3y^2 + 6y - 8) + (-2y^3 + 7y - 2 + y^2) + (1 - 3y + 3y^2 - y^3)$
45. $(4z^4 - 7z^2 + 6) - (z^4 + z^2 - 2) - (2z^4 - 3z^2 + 5)$
46. $(-t^3 + 2t^2 + t) - (4t^2 - t + t^3) - (-3t + t^2 - 2t^3)$
47. $(5n^2 - 4n - 1) - (-2n^2 + n + 6) + (3n^2 - 2n - 2)$
48. $(6x^3 - 2x + 1) - (5x^2 - x - 8) + (x^3 - 2x^2 + 3x)$
49. $(11x^2y^3 - 8x^3y + 2xy^2) - (7x^3y - 3xy^2 + 8x^2y^3) - (xy^2 + x^2y^3 - 2x^3y)$
50. $(9mn^2 + m^2n - 2mn) - (-2m^2n + 3mn + 5mn^2) + (6mn + mn^2 - 7m^2n)$
51. Subtract the sum of $2x^2 + 3x - 4$ and $x^2 + 7x - 4$ from $5x^2 - 8x - 11$.
52. Subtract $3t^3 - 8t^2 + 5t - 3$ from $6t^3 - 8t - 11$ and add the result to $2t^3 - 4t^2 + 10t - 13$.
53. Find the sum of $8y^2 - 3y + 7$ and $-4y^2 + 4y - 3$, and subtract the result from $7 - 13y - 10y^2$.
54. From what polynomial must $3t^2 - 2t + 1$ be subtracted so that the difference is $t^2 + 5t - 2$?
55. To what polynomial must $5x^3 - 3x^2 + 4x - 2$ be added to produce a sum of zero?
56. What polynomial must be added to $4u^2 - 5u$ to produce a polynomial equal to the sum of $2u^2 + 6u - 8$ and $-3u^2 + u - 7$?
57. The sides of a triangle are $5x + 3$, $3x - 7$, and $7x + 1$ units (Figure 1). Find an expression for the perimeter of the triangle in terms if x.

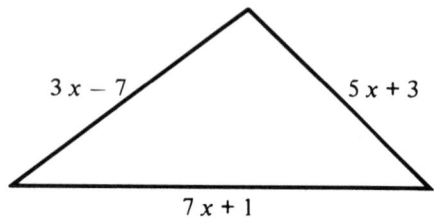

Figure 1

58. The length of a rectangle is $5x - 9$ units, and the width is $2x + 3$ units (Figure 2). Write an expression for its perimeter in terms of x.

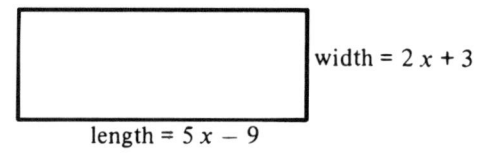

width = $2x + 3$

length = $5x - 9$

Figure 2

2.4 Multiplication of Polynomials

To find products of polynomials, we extend the properties of real numbers to polynomials. For example, suppose that P, Q, and R are polynomials, Then

(i) Commutative property for multiplication:
$$PQ = QP.$$
(ii) Associative property for multiplication:
$$P(QR) = (PQ)R.$$

2.4 MULTIPLICATION OF POLYNOMIALS

Then we apply the properties of exponents. For example, to multiply the monomials $5x^3$ and $4x^4$, we proceed as follows:

$(5x^3) \cdot (4x^4) = (5 \cdot 4) \cdot (x^3 \cdot x^4)$ (We used the commutative and associative properties.)

$\qquad\qquad\qquad = 20x^{3+4} = 20x^7$ (We multiplied coefficients and added exponents.)

In general, to multiply monomials, we follow two basic steps:

Step 1. Rewrite the product using the commutative and associative properties.

Step 2. Simplify the result by multiplying coefficients and adding exponents of like bases.

Example 1   Find the products of the following monomials and simplify.

(a) $(9x^4)(-4x)$ (b) $\left(\dfrac{-2}{3}x^3y^4\right)(9xy^2)$ (c) $(-2xy^2)(-3x^2y)(-5y^3)$

Solution   Regrouping the coefficients and the variables together, we have:

(a) $(9x^4)(-4x) = 9 \cdot (-4)(x^4 \cdot x) = -36x^5$

(b) $\left(\dfrac{-2}{3}x^3y^4\right)(9xy^2) = \left(-\dfrac{2}{3}\right) \cdot 9(x^3 \cdot x)(y^4 \cdot y^2) = -6x^4y^6$

(c) $(-2xy^2)(-3x^2y)(-5y^3) = (-2) \cdot (-3) \cdot (-5)(x \cdot x^2)(y^2 \cdot y \cdot y^3)$
$\qquad\qquad\qquad\qquad\qquad = -30x^3y^6$

To multiply a monomial by a polynomial, we apply the distributive property.

PROPERTY   Distributive Property

If P, Q, and R are polynomials, then

(i) $P(Q + R) = PQ + PR$ (ii) $(P + Q)R = PR + QR$

For example, to find the product $4x^3(3x + 2)$, we first multiply the expression $4x^3$ by each term inside the parentheses:

$$4x^3(3x + 2) = (4x^3)(3x) + (4x^3)(2).$$

We now follow the preceding procedure of multiplying monomials to simplify each term:

$$(4x^3)(3x) + (4x^3)(2) = (4 \cdot 3)(x^3 \cdot x) + (4 \cdot 2)x^3 = 12x^4 + 8x^3.$$

The procedure is further illustrated by the following example.

Example 2   Find each product and simplify.

(a) $2x(3x^3 + 5x)$ (b) $-4y^3(-3y^2 + y - 2)$ (c) $(-4 - 5t^2)(-3t^3)$

Solution   (a) We use the distributive property:

$$2x(3x^3 + 5x) = (2x)(3x^3) + (2x)(5x)$$
$$= (2 \cdot 3)(x \cdot x^3) + (2 \cdot 5)(x \cdot x)$$
$$= 6x^4 + 10x^2.$$

(b) We multiply the expression $-4y^3$ by each term inside the parentheses:

$$-4y^3(-3y^2 + y - 2) = (-4y^3)(-3y^2) + (-4y^3)(y) + (-4y^3)(-2)$$
$$= 12y^5 - 4y^4 + 8y^3.$$

(c) We apply part (ii) of the distributive property:

$$(-4 - 5t^2)(-3t^3) = (-4)(-3t^3) + (-5t^2)(-3t^3)$$
$$= 12t^3 + 15t^5.$$

To multiply one polynomial by another, we use the distributive property to reduce the given multiplication to a multiplication of monomials. For example, to multiply $x + 3$ by $x + 4$, we think of $x + 4$ as one term, call it u for now, and then apply the distributive property:

$$\begin{aligned}(x + 3)(x + 4) &= (x + 3)u \\ &= xu + 3u \\ &= x(x + 4) + 3(x + 4) \quad &\text{(We replaced } u \text{ by } x + 4.) \\ &= x^2 + 4x + 3x + 12 \quad &\text{(We used the distributive property.)} \\ &= x^2 + 7x + 12. \quad &\text{(We combined like terms.)}\end{aligned}$$

The distributive property is always the basis for multiplying polynomials. However, a somewhat shorter procedure than writing out all the steps just shown is to do as follows:

> To multiply two polynomials, multiply each term of one polynomial by each term of the other polynomial, and then simplify the results by combining like terms.

Example 3 Find each product.
(a) $(2x + 1)(x - 3)$
(b) $(t + 2)(t^2 - 2t + 7)$

Solution
(a) We multiply each term of $2x + 1$ by each term of $x - 3$:

$$(2x + 1)(x - 3) = (2x)(x) + (2x)(-3) + (1)(x) + (1)(-3)$$
$$= 2x^2 - 6x + x - 3$$
$$= 2x^2 - 5x - 3.$$

(b) $(t + 2)(t^2 - 2t + 7) = (t)(t^2) + (t)(-2t) + (t)(7) + (2)(t^2) + (2)(-2t) + (2)(7)$
$$= t^3 - 2t^2 + 7t + 2t^2 - 4t + 14$$
$$= t^3 + 3t + 14.$$

When we multiply polynomials (just as when we add and subtract them) it may be helpful to use the vertical scheme to arrange the terms. For example, to find the product of $x + 2$ and $x^3 - x^2 + 3x - 4$, we arrange the polynomials in a vertical scheme as follows:

$$\begin{array}{r} x^3 - x^2 + 3x - 4 \\ (\times)\ x + 2 \\ \hline x^4 - x^3 + 3x^2 - 4x \\ (+)\ 2x^3 - 2x^2 + 6x - 8 \\ \hline x^4 + x^3 + x^2 + 2x - 8 \end{array}$$

The vertical scheme helps us to arrange the partial products so that all like terms are in the same column, ready for the final step of addition.

Example 4 Use the vertical scheme to find the product $(3x - 4)(2x^2 + 3x + 5)$.

Solution

$$
\begin{array}{r}
2x^2 + 3x + 5 \\
(\times)\ 3x - 4 \\
\hline
6x^3 + 9x^2 + 15x \\
(+)\ \ -8x^2 - 12x - 20 \\
\hline
6x^3 + x^2 + 3x - 20.
\end{array}
$$

Polynomials with more than one variable are multiplied in the same way as polynomials with only one variable. When you multiply polynomials with more than one variable, it is a good practice to write the variables in each term in alphabetical order. This makes it easier to spot like terms.

In Examples 5 and 6, find each product.

Example 5 $(3x + 2y)(x - 3y)$

Solution We multiply each term of $3x + 2y$ by every term of $x - 3y$:
$$(3x + 2y)(x - 3y) = (3x)(x) + (3x)(-3y) + (2y)(x) + (2y)(-3y)$$
$$= 3x^2 - 9xy + 2xy - 6y^2$$
$$= 3x^2 - 7xy - 6y^2.$$

Example 6 $(u^2 - 5uv + 2v^2)(u^2 + 4uv + v^2)$

Solution We use the vertical scheme:

$$
\begin{array}{r}
u^2 - 5uv + 2v^2 \\
(\times)\ u^2 + 4uv + v^2 \\
\hline
u^4 - 5u^3v + 2u^2v^2 \\
4u^3v - 20u^2v^2 + 8uv^3 \\
(+)\ \ u^2v^2 - 5uv^3 + 2v^4 \\
\hline
u^4 - u^3v - 17u^2v^2 + 3uv^3 + 2v^4.
\end{array}
$$

PROBLEM SET 2.4

In problems 1–10, find the products of the monomials.

1. $(2x^5)(4x^2)$
2. $(7y^3)(-2y)$
3. $(-3t^3)(2t)$
4. $(4z^5)(3z^4)$
5. $(5u^2v)(3uv^4)$
6. $(-9yz)(-7y^2z^3)$
7. $(4x^3)(2x)(-5x^4)$
8. $(2t^2)(-2t^3)(3t)$
9. $(\tfrac{1}{3}mn^2)(-3n^3)(-2m^3)$
10. $(\tfrac{2}{3}x^4)(-6x)(-\tfrac{1}{4}x^5)$

In problems 11–24, find the products of the monomials and polynomials, and simplify.

11. $3y(2y + 1)$
12. $-4x(x^2 - 1)$
13. $5s(3s - 2)$
14. $5t^2(-t + 3)$
15. $-3z(4z - 6)$
16. $-2y^4(7y - 3)$
17. $2x(x^2 - x - 1)$
18. $3p(4p^2 + p - 5)$
19. $-4ab^2(-a^2b + 2ab - 3b^2)$
20. $-8xy(2x^2y - 3xy^2 + x2y^3)$
21. $(3m + 2n)5mn$
22. $(-3w^2 + 4w)(-2w)$
23. $(x^2y - 2xy + 4xy^2)(-x^3y^2)$
24. $(4m^2 + 3n^2 - 8p^2)(5mp)$

In problems 25–50, use a vertical scheme to find the product of each pair of polynomials.

25. $(2x + 1)(x + 3)$
26. $(-3y + 1)(y - 4)$
27. $(3u - 5)(u + 2)$
28. $(2y + 3)(5y - 2)$
29. $(-2t + 3)(t - 7)$
30. $(4m - 1)(-3m + 5)$
31. $(4z^2 - 1)(z^2 + 1)$
32. $(2v^2 + 5)(v^2 - 1)$
33. $(5x + 2y)(3x - 4y)$
34. $(-4m + 3p)(-2m + 5p)$
35. $(-6r + s)(-r + 3s)$
36. $(2vw - u)(3vw + 4u)$
37. $(2x - 1)(x^2 - 3x + 4)$
38. $(3y + 1)(2y^2 - 5y + 7)$

39. $(z + 3)(4z^3 - 2z^2 + z - 1)$
40. $(3p - 2)(-5p^3 + 6p^2 - 4p + 7)$
41. $(x + 2y)(3x^2 - xy + 4y^2)$
42. $(5t - s)(2t^2 + 3ts - 4s^2)$
43. $(-m + 3n)(-4m^2 + 7mn - 2n^2)$
44. $(4y - z)(5y^2 - 3yz - 6z^2)$
45. $(x^2 + 2x + 1)(2x^2 - 4x + 3)$
46. $(3p^2 - 5p + 2)(4p^2 + p - 7)$
47. $(2x^2 + 3xy - y^2)(3x^2 - xy + 2y^2)$
48. $(w^2 + 2wz + 3z^2)(w - wz + z)$
49. $(3y^2 + 4y - 1)(5y^3 - 3y^2 + 3y - 2)$
50. $(4u^2 - 3uv + v^2)(2u^3 + 4u^2v - 3uv^2 - 5v^3)$
51. The length of a rectangle is $4x + 3$ units and the width is $3x - 4$ units. Find an expression for the area of the rectangle in terms of x.
52. John has x dimes, and 6 fewer nickels than dimes. He also has 3 times as many quarters as nickels. Find an expression in terms of x that represents the total amount of money (in cents) John has.
53. The base of a triangle is $3x + 5$ units and the height is $2x + 3$ units (Figure 1). Use the formula $A = \frac{1}{2}bh$, where b is the base and h is the height, to find an expression for the area of the triangle in terms of x.

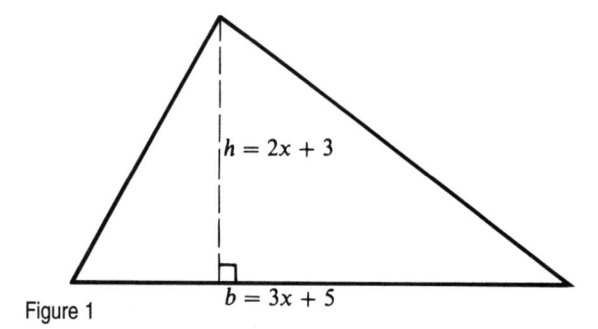

Figure 1

54. The radius of the base of a cylinder is $3x - 3$ and the height is $5x + 3$ units (Figure 2). Use the formula $A = 2\pi r^2 + 2\pi rh$, where r is the radius and h is the height of the cylinder, to find an expression for the total area of the cylinder in terms of x.

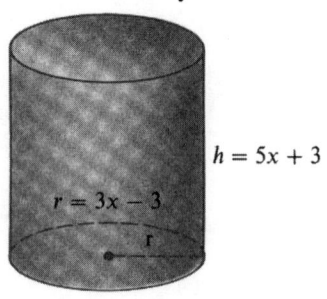

Figure 2

2.5 The FOIL Method and Special Products

In Section 2.4, we showed that the product of two polynomials is found by multiplying each term in one polynomial by each term in the other. When we are multiplying two binomials, we can use what is known as the **FOIL** method to guide us in doing this. Consider, for example, the product of the binomials

$$3x + 5 \quad \text{and} \quad 2x - 7.$$

You can find this product by multiplying each term of $3x + 5$ by each term of $2x - 7$, that is,

$$(3x + 5)(2x - 7) = (3x)(2x) + (3x)(-7) + (5)(2x) + (5)(-7)$$
$$= 6x^2 - 21x + 10x - 35$$
$$= 6x^2 - 11x - 35.$$

Look carefully at this example. Notice that the first term of the product is obtained by multiplying the first terms of the binomials:

$$(3x)(2x) = 6x^2 \quad \text{(first term of the product)}$$

The middle term of the product is found by adding the product of the two outside terms and the product of the two inside terms of the binomials:

$$(3x)(-7) = -21x \quad \text{(product of outside terms of the binomials)}$$
$$(5)(2x) = 10x \quad \text{(product of inside terms of the binomials)}$$
$$= -11x \quad \text{(middle term of the product)}$$

2.5 THE FOIL METHOD AND SPECIAL PRODUCTS

The last term of the product is obtained by multiplying the last terms of the binomials:

$(5)(-7) = -35$ (last term of the product)

Therefore, we can indicate the terms of the product with the letters F-O-I-L as follows:

$$(3x + 5)(2x - 7) = (3x)(2x) + (3x)(-7) + (5)(2x) + (5)(-7)$$

$$\uparrow \qquad \uparrow \qquad \uparrow \qquad \uparrow$$
$$\text{F} \qquad \text{O} \qquad \text{I} \qquad \text{L}$$
(first   (outside   (inside   (last
binomial binomial   binomial binomial
terms)   terms)   terms)   terms)

$$= 6x^2 - 21x + 10x - 35$$
$$= 6x^2 - 11x - 35.$$

The following three-step diagram helps to illustrate how the FOIL method works. To find the product $3x + 5$ and $2x - 7$, we have

first term: $(3x + 5)(2x - 7) = 6x^2 - 11x - 35$
$(3x)(2x)$

middle term: $(3x + 5)(2x - 7) = 6x^2 \quad - \quad 11x - 35$
$(3x)(-7) + (5)(2x)$

last term: $(3x + 5)(2x - 7) = 6x^2 - 11x - 35$
$(5)(-7)$

Therefore,

$$(3x + 5)(2x - 7) = 6x^2 - 11x - 35.$$

In Example 1-3, Use the FOIL method to find each product.

Example 1 $(x + 2)(2x + 1)$

Solution

first term: $(x + 2)(2x + 1) = 2x^2 + \_\_ + \_\_$
$(x)(2x)$

middle term: $(x + 2)(2x + 1) = \quad 2x^2 \quad + \quad 5x + \_\_$
$(1)(x) + (2)(2x)$

last term: $(x + 2)(2x + 1) = 2x^2 + 5x + 2$
$(2)(1)$

Therefore, $(x + 2)(2x + 1) = 2x^2 + 5x + 2$.

Example 2 $(3x + 2)(4x - 5)$

Solution

first term: $(3x + 2)(4x - 5) = 12x^2 + \_\_ + \_\_$
$(3x)(4x)$

middle term: $(3x + 2)(4x - 5) = \quad 12x^2 \quad + \quad -7x + \_\_$
$(3x)(-5) + (2)(4x)$

last term: $(3x + 2)(4x - 5) = 12x^2 - 7x + -10$

$(2)(-5)$

Therefore, $(3x + 2)(4x - 5) = 12x^2 - 7x - 10$.

After you have had some practice, you will be able to multiply binomials without showing any intermediate steps.

Example 3 $(2x + 3y)(3x - 5y)$

We can (mentally) obtain this product, using the following scheme:

$$(2x + 3y)(3x - 5y) = 6x^2 - xy - 15y^2$$

first term: $(2x)(3x) = 6x^2$

middle term: $(2x)(-5y) + (3y)(3x) = -xy$

third term: $(3y)(-5y) = -15y^2$

Special Products

Certain products occur frequently in algebra and in applications of algebra. When we come across these special products, we have an opportunity to use time-saving shortcuts in multiplication. Some examples follow.

Consider the problem of squaring the binomial $a + b$:

$$(a + b)^2 = (a + b)(a + b) = a^2 + ab + ba + b^2$$
$$= a^2 + 2ab + b^2.$$

SPECIAL PRODUCT 1   We state this product as follows:

> Let a and b be real numbers. Then
> $$(a + b)^2 = a^2 + 2ab + b^2.$$

In other words, *the square of the sum of two terms is equal to the square of the first term plus twice the product of the two terms plus the square of the last term.*

Warning: The expression $(a + b)^2$ is not equal to $a^2 + b^2$.

We obtain a similar result for $(a - b)^2$:

$$(a - b)^2 = (a - b)(a - b) = a^2 - ab - ba + b^2$$
$$= a^2 - 2ab + b^2.$$

Therefore, we have:

SPECIAL PRODUCT 2

> Let a and b be real numbers. Then
> $$(a - b)^2 = a^2 - 2ab + b^2.$$

In other words, *the square of the difference of two terms is equal to the square of the first term minus twice the product of the two terms plus the square of the last term.*

Example 4   Use special products to find the square of each binomial.

(a) $(t + 4)^2$ (b) $(2u - 5)^2$ (c) $(8x + 2y)^2$ (d) $(4m^3n^2 - \frac{1}{2})^2$

Solution   (a) We substitute t for a and 4 for b in $(a + b)^2 = a^2 + 2ab + b^2$ and have
$$(t + 4)^2 = t^2 + 2(t)(4) + 4^2 = t^2 + 8t + 16.$$
(b) We substitute $2u$ for a and 5 for b in $(a - b)^2 = a^2 - 2ab + b^2$ and have
$$(2u - 5)^2 = (2u)^2 - 2(2u)(5) + 5^2 = 4u^2 - 20u + 25.$$
(c) We apply Special Product 1 directly:
$$(8x + 2y)^2 = (8x)^2 + 2(8x)(2y) + (2y)^2$$
$$= 64x^2 + 32xy + 4y^2.$$
(d) We apply Special Product 2 directly:
$$(4m^3n^2 - \tfrac{1}{2})^2 = (4m^3n^2)^2 - 2(4m^3n^2)(\tfrac{1}{2}) + (-\tfrac{1}{2})^2$$
$$= 16m^6n^4 - 4m^3n^2 + \tfrac{1}{4}.$$

Another special product that occurs frequently in mathematics and in applications has the form $(a + b)(a - b)$:
$$(a + b)(a - b) = a^2 - ab + ba - b^2$$
$$= a^2 - b^2.$$

We state the product as follows:

SPECIAL PRODUCT 3   Let a and b be real numbers. Then
$$(a + b)(a - b) = a^2 - b^2.$$

In other words, *when multiplying two binomials that differ only in the sign between their terms, subtract the square of the last term from the square of the first term.*

Example 5   Use Special Product 3 to perform the multiplication.
(a) $(c + 3)(c - 3)$ (b) $(2w + 1)(2w - 1)$
(c) $(\tfrac{1}{5} - 2z)(\tfrac{1}{5} + 2z)$ (d) $(3x^2 - 2y^3)(3x^2 + 2y^3)$

Solution   (a) We substitute c for a and 3 for b in $(a + b)(a - b) = a^2 - b^2$:
$$(c + 3)(c - 3) = c^2 - 3^2$$
$$= c^2 - 9.$$
(b) Let $a = 2w$ and $b = 1$ in $(a + b)(a - b) = a^2 - b^2$:
$$(2w + 1)(2w - 1) = (2w)^2 - 1^2$$
$$= 4w^2 - 1.$$
(c) We apply Special Product 3 directly:
$$(\tfrac{1}{5} - 2z)(\tfrac{1}{5} + 2z) = (\tfrac{1}{5})^2 - (2z)^2$$
$$= \tfrac{1}{25} - 4z^2.$$
(d) We apply Special Product 3 directly:
$$(3x^2 - 2y^3)(3x^2 + 2y^3) = (3x^2)^2 - (2y^3)^2$$
$$= 9x^4 - 4y^6.$$

PROBLEM SET 2.5

In problems 1–40, find the products of the following binomials by the FOIL method.

1. $(x + 1)(x + 2)$
2. $(y + 2)(y + 6)$
3. $(z - 3)(z + 5)$
4. $(t - 2)(t - 4)$
5. $(y - 10)(y - 11)$
6. $(z + 7)(z - 9)$
7. $(v + 7)(v - 8)$
8. $(x + 5y)(x + 3y)$
9. $(z + 11)(z - 11)$
10. $(y - 7x)(y + 7x)$
11. $(2x + 3)(x + 1)$
12. $(3r - 5)(r - 5)$
13. $(4y - 5)(2y + 1)$
14. $(2v - 9u)(3v - 7u)$
15. $(7s - 3)(s - 2)$
16. $(3x - 1)(7x + 2)$

17. $(6u + 5)(u - 4)$
18. $(8m - 3n)(3m + 2n)$
19. $(2x + 1)(2x + 1)$
20. $(5x + 2y)(5x + 2y)$
21. $(3y - z)(y + 2z)$
22. $(4s - 3)(4s + 3)$
23. $(2m - 5n)(2m - 5n)$
24. $(11y - 2z)(3y + z)$
25. $(3x + 2y)(2x - 3y)$
26. $(15 + 7u)(4 + 2u)$
27. $(6t - 7)(9t + 8)$
28. $(12x - 9y)(9x + 7y)$
29. $(10y - 3z)(5y - 2z)$
30. $(8w - 7z)(6w + 5z)$
31. $(x + 7)(2x + 1)$
32. $(t + 4)(t + 5)$
33. $(2z + 3)(z + 4)$
34. $(3s - 1)(2s - 7)$
35. $(5y - 8)(4y - 3)$
36. $(8p + 9)(2p - 5)$
37. $(7m - 5n)(5m + 7n)$
38. $(11x - 4y)(10x + 3y)$
39. $(3x + 2y)(6x - 5y)$
40. $(13w + 4z)(3w + 5z)$

In problems 41–66, use Special Product 2 find the square of each binomial.

41. $(x + 5)^2$
42. $(z + 7)^2$
43. $(y - 2)^2$
44. $(p - 11)^2$
45. $(m + 6)^2$
46. $(x + 12)^2$
47. $(2x + 7)^2$
48. $(3z - 8)^2$
49. $(3z - 4)^2$
50. $(5 - 4t)^2$
51. $(11 + 2t)^2$
52. $(12u + v)^2$
53. $(5x - y)^2$
54. $(7y - 6z)^2$
55. $(4t + 3s)^2$
56. $(13 + 3s)^2$
57. $(7u - 3v)^2$
58. $(9y - 5z)^2$
59. $(6w + 10z)^2$
60. $(8x - 5y)^2$
61. $(2x + \frac{1}{2})^2$
62. $(6z - \frac{2}{3}y)^2$
63. $(3mn - p)^2$
64. $(4uv^2 + 1)^2$
65. $(5x^2y - 2z)^2$
66. $(8xy + 9z)^2$

In problems 67–86, use Special Product 3 to find the following products.

67. $(x + 4)(x - 4)$
68. $(z - 6)(z + 6)$
69. $(u + 10)(u - 10)$
70. $(11 - y)(11 + y)$
71. $(z - 8)(z + 8)$
72. $(p + 9)(p - 9)$
73. $(3x - y)(3x + y)$
74. $(v + 4u)(v - 4u)$
75. $(2t + 7)(2t - 7)$
76. $(8x + 7y)(8x - 7y)$
77. $(6x - 5y)(6x + 5y)$
78. $(5 - 2rs)(5 + 2rs)$
79. $(3m + 11n)(3m - 11n)$
80. $(20p - 13q)(20p + 13q)$
81. $(3x + \frac{1}{4})(3x - \frac{1}{4})$
82. $(\frac{2}{3} - t)(\frac{2}{3} + t)$
83. $(5t^2 - 3s^2)(5t^2 + 3s^2)$
84. $(6x^3 - y^2)(6x^3 + y^2)$
85. $(7x^2y + 4z^3)(7x^2y - 4z^3)$
86. $(8m^2 - 3np)(8m^2 + 3np)$

87. If the length of a side of a square is $7x + 11$ units, find an expression for its area in terms of x.

88. The radius of a circle is $3x + 2$ units (Figure 1). Find an expression for its area in terms of x.

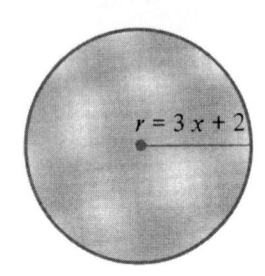

$r = 3x + 2$

Figure 1

2.6 Division of Polynomials

We use the division of like bases property of exponents (section 1.2) to divide one monomial by another. Thus, division of monomials will result in division of coefficients and subtraction of exponents with like bases. For example, to divide $6x^6y^3$ by $3x^4y^2$, we have

$$\frac{6x^6y^3}{3x^4y^2} = \left(\frac{6}{3}\right)\left(\frac{x^6}{x^4}\right)\left(\frac{y^3}{y^2}\right)$$
$$= 2x^{6-4}y^{3-2} = 2x^2y.$$

To divide a polynomial by a monomial, we divide each term of the polynomial by the monomial. For example, to divide $x^3 + 5x^2 + 3x$ by x, we write

$$\frac{x^3 + 5x^2 + 3x}{x} = \frac{x^3}{x} + \frac{5x^2}{x} + \frac{3x}{x}$$
$$= x^2 + 5x + 3.$$

The following examples illustrate this procedure.

2.6 DIVISION OF POLYNOMIALS

Example 1 Divide $x^4 + x^5$ by x^2.

Solution $\dfrac{x^4 + x^5}{x^2} = \dfrac{x^4}{x^2} + \dfrac{x^5}{x^2} = x^2 + x^3.$

Example 2 Divide $-7x^4 + 3x^3 - 5x^2 + 2x$ by x^2.

Solution $\dfrac{-7x^4 + 3x^3 - 5x^2 + 2x}{x^2} = \dfrac{-7x^4}{x^2} + \dfrac{3x^3}{x^2} - \dfrac{5x^2}{x^2} + \dfrac{2x}{x^2}$

$= -7x^2 + 3x - 5 + \dfrac{2}{x}.$

Example 3 Divide $-4u^5v^6 - 12u^4v^5 + 8u^3v^4$ by $2uv^3$.

Solution $\dfrac{-4u^5v^6 - 12u^4v^5 + 8u^3v^4}{2uv^3} = \dfrac{-4u^5v^6}{2uv^3} - \dfrac{12u^4v^5}{2uv^3} + \dfrac{8u^3v^4}{2uv^3}$

$= -2u^4v^3 - 6u^3v^2 + 4u^2v.$

Warning: When you divide a polynomial by a monomial, you should avoid making the mistake of dividing only one term by the monomial, because

$$\dfrac{4x^3 + x}{x} \neq \dfrac{4x^3}{x} + x = 4x^2 + x.$$

The correct way is to divide each term of the polynomial by the monomial:

$$\dfrac{4x^3 + x}{x} = \dfrac{4x^3}{x} + \dfrac{x}{x} = 4x^2 + 1.$$

To divide a polynomial by another polynomial, we use a method similar to the long-division method of arithmetic. For example, consider the division $6{,}741 \div 21$. We usually do this long division in the following way:

```
                    321   ← quotient
     divisor → 21)6,741   ← dividend
                  - 63      = 21 · 3
                    44
                  - 42      = 21 · 2
                    21
                  - 21      = 21 · 1
                     0    ← remainder
```

The result of this calculation can be expressed as

$$6{,}741 = 21(321) + 0$$

In other words:

---
dividend = divisor × quotient + remainder.

We can also perform this division by changing the dividend 6,741 and the divisor 21 to the following forms:

$$21 = 2 \cdot 10 + 1$$

and

$$6{,}741 = 6 \cdot 10^3 + 7 \cdot 10^2 + 4 \cdot 10 + 1.$$

Then we divide:

$$\begin{array}{r} 3\cdot 10^2 + 2\cdot 10 + 1 \leftarrow \text{quotient} \\ \text{divisor} \rightarrow 2\cdot 10 + 1 \overline{) 6\cdot 10^3 + 7\cdot 10^2 + 4\cdot 10 + 1} \\ \underline{6\cdot 10^3 + 3\cdot 10^2} \qquad\qquad [= (2\cdot 10 + 1)(3\cdot 10^2)] \\ 4\cdot 10^2 + 4\cdot 10 \\ \underline{4\cdot 10^2 + 2\cdot 10} \qquad\qquad [= (2\cdot 10 + 1)(2\cdot 10)] \\ 2\cdot 10 + 1 \\ \underline{2\cdot 10 + 1} \quad [= (2\cdot 10 + 1)(1)] \\ 0 \leftarrow \text{remainder} \end{array}$$

Note that the expanded forms of the numbers in this example are polynomials in which each term has a base of 10 (rather than a variable base).

We can change this problem to a similar one by changing base 10 to base x. We then have the long division of

$$6x^3 + 7x^2 + 4x + 1 \quad \text{by} \quad 2x + 1:$$

$$\boxed{\dfrac{6x^3}{2x} = 3x^2} \qquad \boxed{\dfrac{4x^2}{2x} = 2x} \qquad \boxed{\dfrac{2x}{2x} = 1}$$

$$\begin{array}{r} 3x^2 + 2x + 1 \leftarrow \text{quotient} \\ 2x + 1 \overline{) 6x^3 + 7x^2 + 4x + 1} \\ \text{subtract} \rightarrow \underline{6x^3 + 3x^2} \qquad\qquad [= (2x + 1)(3x^2)] \\ 4x^2 + 4x \\ \text{subtract} \rightarrow \underline{4x^2 + 2x} \qquad\qquad [= (2x + 1)(2x)] \\ 2x + 1 \\ \text{subtract} \rightarrow \underline{2x + 1} \quad [= (2x + 1)(1)] \\ 0 \leftarrow \text{remainder} \end{array}$$

This example illustrates the following step-by-step procedure for dividing one polynomial by another:

Procedure for Long Division

> Step 1. Arrange both polynomials in descending powers of one variable, and write the missing terms of the dividend with zero coefficients.
> Step 2. Find the first term of the quotient by dividing the first term of the dividend by the first term of the divisor.
> Step 3. Multiply the quotient term obtained in step 2 by the entire divisor.
> Step 4. Subtract the product obtained in step 3 from the dividend, and bring down the next term of the original dividend to form a new dividend.
> Step 5. Find the next term of the quotient by dividing the first term of the divisor into the first term of the new dividend.
> Step 6. Repeat steps 3, 4, and 5 until the remainder in step 4 is 0, or until the degree of the remainder is less than the degree of the divisor.
> Step 7. Check the calculation: Does
>
> divisor × quotient + remainder = dividend?

We use the step-by-step procedure to work the following examples:

Example 4 Divide $2x^2 - 1 + x^3 - 2x$ by $-1 + x$.

Solution Step 1. We arrange both polynomials in descending powers of x:

$$\text{divisor} \to x - 1 \overline{)x^3 + 2x^2 - 2x - 1.} \leftarrow \text{dividend}$$

Step 2. We divide x^3, the first term of the dividend, by x, the first term of the divisor, to obtain

$$\frac{x^3}{x} = x^2.$$

Step 3. We multiply x^2, the first quotient term, by $x - 1$, the divisor:

$$x^2(x - 1) = x^3 - x^2.$$

Step 4. We place the product $x^3 - x^2$ under the dividend and subtract to obtain a new dividend, so that

$$\begin{array}{r} x^2 \\ x - 1 \overline{)x^3 + 2x^2 - 2x - 1.} \\ \text{subtract} \to \underline{x^3 - x^2} \\ 3x^2 - 2x. \leftarrow \text{new dividend} \end{array}$$

Step 5. We divide x, the first term of the divisor, into $3x^2$, the first term of the new dividend, to obtain

$$\frac{3x^2}{x} = 3x.$$

Thus, we have

$$\begin{array}{r} x^2 + 3x \\ x - 1 \overline{)x^3 + 2x^2 - 2x - 1.} \\ \underline{x^3 - x^2} \\ 3x^2 - 2x. \end{array}$$

Step 6. We repeat steps 3, 4, and 5 until we obtain

$$\frac{x}{x} = 1$$

$$\begin{array}{r} x^2 + 3x + 1 \\ x - 1 \overline{)x^3 + 2x^2 - 2x - 1} \\ \underline{x^3 - x^2} \\ 3x^2 - 2x \\ \text{subtract} \to \underline{3x^2 - 3x} \quad [= 3x(x - 1)] \\ x - 1 \\ \text{subtract} \to \underline{x - 1} \quad [= 1(x - 1)] \\ 0 \leftarrow \text{remainder} \end{array}$$

Therefore, the quotient is $x^2 + 3x + 1$ and the remainder is 0.

Step 7. To check, we calculate:

$$\begin{aligned} \text{divisor} \times \text{quotient} + \text{remainder} &= (x - 1)(x^2 + 3x + 1) + 0 \\ &= x^3 + 2x^2 - 2x - 1 \\ &= \text{dividend}. \end{aligned}$$

Example 5 Divide $2t^4 + 5 - 3t + 3t^3$ by $2t - 1$.

Solution First, arrange both polynomials in descending powers of t, and write the missing term of the dividend with zero coefficient to obtain

$$\text{divisor} \to 2t - 1 \overline{)2t^4 + 3t^3 - 0t^2 - 3t + 5} \leftarrow \text{dividend}$$

The remaining steps are shown as follows:

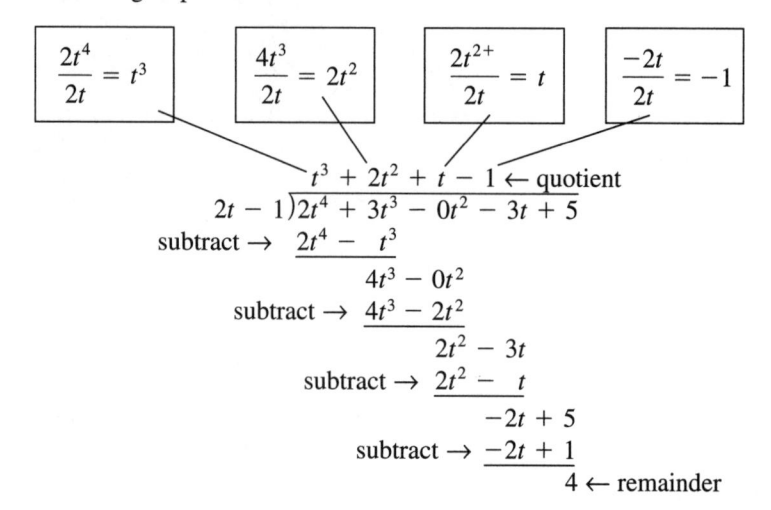

$$\boxed{\frac{2t^4}{2t} = t^3} \quad \boxed{\frac{4t^3}{2t} = 2t^2} \quad \boxed{\frac{2t^2}{2t} = t} \quad \boxed{\frac{-2t}{2t} = -1}$$

$$\begin{array}{r} t^3 + 2t^2 + t - 1 \leftarrow \text{quotient} \\ 2t - 1 \overline{) 2t^4 + 3t^3 - 0t^2 - 3t + 5} \\ \text{subtract} \to \underline{2t^4 - t^3} \\ 4t^3 - 0t^2 \\ \text{subtract} \to \underline{4t^3 - 2t^2} \\ 2t^2 - 3t \\ \text{subtract} \to \underline{2t^2 - t} \\ -2t + 5 \\ \text{subtract} \to \underline{-2t + 1} \\ 4 \leftarrow \text{remainder} \end{array}$$

Therefore, the quotient is $t^3 + 2t^2 + t - 1$ and the remainder is 4.
To check, we calculate:

$$\begin{aligned} \text{divisor} \times \text{quotient} + \text{remainder} &= (2t - 1)(t^2 + 2t^2 + t - 1) + 4 \\ &= (2t^4 + 3t^3 - 3t + 1) + 4 \\ &= (2t^4 + 3t^3 - 3t + 5) \\ &= \text{dividend}. \end{aligned}$$

Example 6 Divide $-3w^3 + 2w^4 + 5w^2 + 2w + 7$ by $-w + w^2 + 1$.

Solution First, we arrange both polynomials in descending powers of w, and follow the procedure in the previous examples to obtain

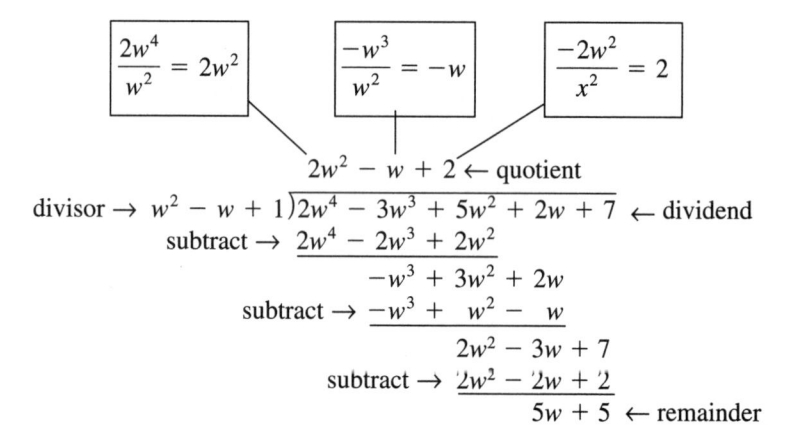

$$\boxed{\frac{2w^4}{w^2} = 2w^2} \quad \boxed{\frac{-w^3}{w^2} = -w} \quad \boxed{\frac{-2w^2}{x^2} = 2}$$

$$\begin{array}{r} 2w^2 - w + 2 \leftarrow \text{quotient} \\ \text{divisor} \to w^2 - w + 1 \overline{) 2w^4 - 3w^3 + 5w^2 + 2w + 7} \leftarrow \text{dividend} \\ \text{subtract} \to \underline{2w^4 - 2w^3 + 2w^2} \\ -w^3 + 3w^2 + 2w \\ \text{subtract} \to \underline{-w^3 + w^2 - w} \\ 2w^2 - 3w + 7 \\ \text{subtract} \to \underline{2w^2 - 2w + 2} \\ 5w + 5 \leftarrow \text{remainder} \end{array}$$

Therefore, the quotient is $2w^2 - w + 2$ and the remainder is $5w + 5$.
To check, we calculate

$$\begin{aligned} \text{divisor} \times \text{quotient} + \text{remainder} &= (w^2 - w + 1)(2w^2 - w + 2) + 5w + 5 \\ &= (2w^4 - 3w^3 + 5w^2 - 3w + 2) + 5w + 5 \\ &= 2w^4 - 3w^3 + 5w^2 + 2w + 7 \\ &= \text{dividend}. \end{aligned}$$

Example 7 Divide $x^3 + 2y^3 + 5x^2y + 7xy^2$ by $x + 2y$.

Solution Since we have two variables here, we arrange the terms in descending powers of one variable (say x) and then apply the procedure of long division to that variable.

$$\dfrac{x^3}{x} = x^2 \qquad \dfrac{3x^2y}{x} = 3xy \qquad \dfrac{xy^2}{x} = y^2$$

$$
\begin{array}{r}
x^2 + 3xy + y^2 \leftarrow \text{quotient} \\
\text{divisor} \rightarrow x + 2y \overline{) x^3 + 5x^2y + 7xy^2 + 2y^3} \leftarrow \text{dividend} \\
\text{subtract} \rightarrow \underline{x^3 + 2x^2y} \\
3x^2y + 7xy^2 \\
\text{subtract} \rightarrow \underline{3x^2y + 6xy^2} \\
xy^2 + 2y^3 \\
\text{subtract} \rightarrow \underline{xy^2 + 2y^3} \\
0 \leftarrow \text{remainder}
\end{array}
$$

Therefore, the quotient is $x^2 + 3xy + y^2$ and the remainder is 0.
To check, we calculate

$$\text{divisor} \times \text{quotient} + \text{remainder} = (x + 2y)(x^2 + 3xy + y^2) + 0$$
$$= x^3 + 5x^2y + 7xy^2 + 2y^3$$
$$= \text{dividend}.$$

PROBLEM SET 2.6

In problems 1–14, divide as indicated.

1. $12x^4$ by $3x$
2. $20y^6$ by $5y^4$
3. $24m^5$ by $-8m^3$
4. $-42r^7s^2$ by $7r^2s^4$
5. $-15y^3z^2$ by $3yz^5$
6. $18uv^5w^2$ by $-6u^4v^2w^5$
7. $4x^3 + 10x^2$ by $2x$
8. $10p^4 - 15p^5$ by $5p^3$
9. $30t^5 - 42t^5 + 18t^7$ by $6t^2$
10. $75z^8 + 50z^4$ by $-5z^3$
11. $6m^4n^4 - 9m^3n + 15mn^3$ by $3m^2n^2$
12. $33x^4y - 44xy^3 + 22x^2y^2$ by $-11x^2y^3$
13. $28z^2 - 8zw + 12w^2$ by $-4zw^2$
14. $7p^5q - 14p^4q^2 - 21p^3q^3 + 28p^2q^4$ by $7p^2q^3$

In problems 15–40, divide by the long-division method.

15. $x^2 + 5x + 6$ by $x + 2$
16. $x^2 - x - 12$ by $x + 3$
17. $m^2 + 6 - 7m$ by $m - 1$
18. $z^2 - 20 + z$ by $z - 4$
19. $3y + 1 + 2y^2$ by $1 + 2y$
20. $6z^2 - 5z - 5$ by $3z - 4$
21. $x^3 + 3x^2 + 3x + 1$ by $x + 1$
22. $12y - 8 + y^3 - 6y^2$ by $y - 2$
23. $8 + 13x + 8x^2 + 3x^3$ by $2 + 3x$
24. $1 + 2t + 2t^3 + t^4$ by $1 + t$
25. $m^3 - 2m^2 - 2m - 8$ by $m^2 + m + 1$
26. $7u + u^2 + 6 + 6u^3$ by $3 - u + 2u^2$
27. $3x^4 + 3x^3 - 6x^2 + x - 4$ by $x^2 - x + 2$
28. $4p^4 - 7p^3 + 2p^2 + p - 3$ by $p^2 - 2p + 1$
29. $6x^4 + 13x - 11x^3 - 10 - x^2$ by $3x^2 - 5 - x$
30. $m^4 - 2m^2 + 1$ by $m^2 - 2m + 1$
31. $x^3 + y^3 - 3xy^2 - 3x^2y$ by $x + y$
32. $2w^4 + 7w^3z + 9w^2z^2 + 5wz^3 + z^4$ by $2w + z$
33. $2m^4 - 3m^3n - 16mn^3 + 24n^4$ by $2m - 3n$
34. $a^4 - 6a^2b^2 + 8b^4 - a^3b + 4ab^3$ by $a - 2b$
35. $y^5 + y^4 + 5y^2 - 4y + 16$ by $y^3 + y^2 - 2y + 3$
36. $x^5 - 2x^2y^3 - 2x^3y^2 + 3xy^4 - y^5 + 2x^4y$ by $x^2 - y^2 + xy$
37. $8u^3 - 27v^3$ by $2u - 3v$
38. $81x^4 - z^4$ by $3x - z$
39. $p^5 + q^5$ by $p + q$
40. $y^5 + 32z^5$ by $2z + y$

2.7 Factoring Integers

The product of 3 and 5 is 15. *Factoring* is the reverse process of multiplication. For example, we can factor 15 by writing it as the product of 3 and 5. The following diagram illustrates the relationship between 15 and $3 \cdot 5$.

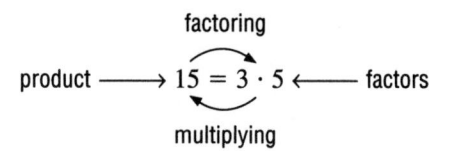

$$\text{product} \longrightarrow 15 \underset{\text{multiplying}}{\overset{\text{factoring}}{=}} 3 \cdot 5 \longleftarrow \text{factors}$$

The numbers 3 and 5 are called **factors,** or **divisors,** of 15. To factor the number 42, we can write

$$42 = 6 \cdot 7.$$

But 42 can be factored still further:

$$42 = 3 \cdot 2 \cdot 7.$$

The numbers 3, 2, and 7 are called *prime factors* because none of them can be factored any further. In general,

> A **prime number** is any positive integer greater than 1 whose only factors are itself and 1.

For example,

$$2, 3, 5, 7, 11, 13, 17, 19, 23, 29, \ldots$$

are prime numbers.

A positive integer that is not a prime number is called a **composite number.** If a number is composite, then it can be written as a product of prime numbers or prime factors. For example, 15 is a composite number, since it can be written as the product of the prime numbers 3 and 5.

Example 1 Factor each number into the product of prime numbers.

(a) 45 (b) 60

Solution (a) We begin by writing 45 as the product of two positive integers. One choice is

$$45 = 9 \cdot 5.$$

We then factor the composite number 9:

$$45 = 3 \cdot 3 \cdot 5$$
$$= 3^2 \cdot 5.$$

(b) We can write 60 as the product of 6 and 10.

$$60 = 6 \cdot 10$$
$$= 2 \cdot 3 \cdot 2 \cdot 5$$
$$= 2^2 \cdot 3 \cdot 5.$$

Here are some suggestions for finding the prime factors of a number.

1. If a number ends in an even digit, that is, if it ends with 0, 2, 4, 6, or 8, then it is divisible by 2.
2. If a number ends in 0 or 5, then it is divisible by 5.

3. A number is divisible by 3 if the sum of its digits is divisible by 3. For example, 621 is divisible by 3, since $6 + 2 + 1 = 9$, which is divisible by 3.

Example 2 Factor each number into the product of prime numbers.

(a) 42 (b) 55 (c) 423

Solution (a) Since 42 ends in an even digit, we can start with

$$42 = 2 \cdot 21$$
$$= 2 \cdot 3 \cdot 7.$$

(b) Since 55 ends in 5, then 5 is a factor, so that

$$55 = 5 \cdot 11.$$

(c) Since the sum of the digits of 423 is $4 + 2 + 3 = 9$, and since 9 is divisible by 3, then 423 is divisible by 3, so we have:

$$423 = 3 \cdot 141.$$

Now, the sum of the digits of 141 is $1 + 4 + 1 = 6$, which is also divisible by 3, so we have

$$423 = 3 \cdot 3 \cdot 47 = 3^2 \cdot 47.$$

The Greatest Common Factor

A number that is a factor of two or more numbers is called a *common factor* of these numbers. For instance, the factors of 36 are

1, 2, 3, 4, 6, 9, 12, 18 and 36,

and the factors of 60 are

1, 2, 3, 4, 5, 6, 10, 12, 15, 20, 30, and 60.

The common factors of 36 and 60 are

1, 2, 3, 4, 6, and 12.

The largest number that is a factor of two or more numbers is called the **greatest common factor (GCF)** of these numbers. In our example, the greatest common factor of 36 and 60 is 12. A procedure to find the greatest common factor is outlined in the following steps.

Step 1. Factor each number into a product of prime numbers, using prime numbers raised to a power (2^4 or 3^2, for example) for repeated factors.
Step 2. Determine the common prime factors. For each repeated factor. determine the highest common power (or exponent).
Step 3. The GCF is the product of these common prime factors, each raised to the highest common power.

Example 3 Find the greatest common factor for the given numbers.

(a) 30 and 42 (b) 60, 140, and 200

Solution (a) Step 1. Writing each number in prime factored form, we have

$$30 = 2 \cdot 3 \cdot 5 \quad \text{and} \quad 42 = 2 \cdot 3 \cdot 7$$

Step 2. The common prime factors are 2 and 3. (No powers are involved because none of the prime factors of either given number is repeated.

Step 3. The product of common prime factors 2 and 3 is 6. Therefore, the greatest common factor of 30 and 42 is 6.

(b) Step 1. Writing each number in prime factored form, we have

$$60 = 2^2 \cdot 3 \cdot 5$$
$$140 = 2^2 \cdot 5 \cdot 7$$
$$200 = 2^3 \cdot 5^2$$

Step 2. The common prime factors are 2 and 5. Raising each to its highest common power, we have 2^2 and 5.

Step 3. The product of 2^2 and 5 is 20. Therefore, the greatest common factor of 60, 140, and 200 is 20.

PROBLEM SET 2.7

In problems 1–40, write each number in prime factored form.

1. 18
2. 126
3. 51
4. 65
5. 63
6. 62
7. 80
8. 90
9. 84
10. 450
11. 200
12. 171
13. 221
14. 115
15. 58
16. 78
17. 132
18. 129
19. 1023
20. 243
21. 44
22. 52
23. 54
24. 96
25. 125
26. 140
27. 168
28. 135
29. 189
30. 19
31. 29
32. 101
33. 118
34. 216
35. 180
36. 225
37. 150
38. 102
39. 175
40. 5,005

In problems 41–50, find the greatest common factor (GCF) for the given numbers.

41. 14 and 21
42. 44 and 60
43. 60 and 126
44. 120 and 160
45. 12, 16, and 36
46. 26, 39, and 65
47. 10, 15, and 40
48. 8, 18, and 24
49. 30, 48, and 72
50. 12, 15, and 75

2.8 Common Factors and Factoring by Grouping

Now we use the procedure presented in Section 2.7 to find the greatest common factor of two or more monomials. For example, to find the greatest common factor (GCF) of the monomials $30x^2$ and $42x^3$, we proceed as follows.

First, we write each expression in factored form:

$$30x^2 = 2 \cdot 3 \cdot 5 \cdot x^2 \quad \text{and} \quad 42x^3 = 2 \cdot 3 \cdot 7 \cdot x^3.$$

Note that the largest number that divides each coefficient is 6 (the product of the common factors) and the highest power of x that is a factor of x^2 and x^3 is x^2. Therefore, $6x^2$ is the greatest common factor of $30x^2$ and $42x^3$.

The *greatest common factor* of two or more monomials is a monomial with the largest numerical coefficient and the highest powers of the variables that are common factors of the given monomials.

For example, the greatest common factor of $15x^3y^2$, $6x^2y^3$, and $9xy^4$ is obtained as follows:

$$15x^3y^2 = 3 \cdot 5 \cdot x^3 \cdot y^2$$
$$6x^2y^3 = 3 \cdot 2 \cdot x^2 \cdot y^3$$
$$9xy^4 = 3^2 \cdot x \cdot y^4.$$

2.8 COMMON FACTORS AND FACTORING BY GROUPING

The largest number that is a factor of each coefficient is 3. The highest powers of the variables that are common factors are x and y^2. Therefore, $3xy^2$ is the greatest common factor of the monomials.

A polynomial is **factorable** if it can be expressed as the product of two or more polynomials. If a polynomial has no factors other than itself and 1, or its negative and -1, the polynomial is called **prime. To factor** a polynomial means to express the polynomial as the product of prime polynomials. In this case, we say that the polynomial is **factored completely.** Note that throughout this book, unless otherwise stated, we only factor polynomials with integer coefficients, and we only look for factors that have integer coefficients.

The distributive property

$$P(Q + R) = PQ + PR$$

provides a bridge between products and factors. If we write the distributive property in the reverse order,

$$PQ + PR = P(Q + R),$$

we obtain a general principle of **common factoring.** Looking at the left side of this equation, we see that P is a common factor of both PQ and PR. The factors on the right side of the equation are P and $Q + R$. The parentheses show that $Q + R$ is a single factor.

Suppose, as an example, we wish to factor the expression

$$5x + 30,$$

We notice first that $5x + 30$ can be written as $5 \cdot x + 5 \cdot 6$. Next, we apply the distributive property:

$$5x + 30 = 5 \cdot x + 5 \cdot 6$$
$$= 5(x + 6).$$

Thus, $5(x + 6)$ is the factored form of $5x + 30$.

Notice that in this type of factoring. we look for numbers or variables that are **common factors** of the terms in the original expression. In the previous example, the number 5 is the common factor of the terms, and we *factored it out.*

Example 1 Factor $2x + 6$.

Solution We observe that the greatest common factor of $2x$ and 6 is 2, because

$$2x = 2 \cdot x \quad \text{and} \quad 6 = 2 \cdot 3.$$

We rewrite this expression as

$$2x + 6 = 2 \cdot x + 2 \cdot 3.$$

Then we apply the distributive property to factor out 2, so that

$$2 \cdot x + 2 \cdot 3 = 2(x + 3).$$

Therefore, the factored form is

$$2x + 6 = 2(x + 3).$$

Alternate Solution Once we place 2 outside the parentheses, we can obtain the expression inside by dividing each term in the polynomial $2x + 6$ by the greatest common factor 2. That is,

$$\frac{2x}{2} = x \qquad \frac{6}{2} = 3$$
$$2x + 6 = 2(x + 3).$$

146 CHAPTER 2 POLYNOMIALS AND FACTORING

The technique of common factoring can be summarized as follows:

Procedure for Factoring Polynomials by Common Factoring

> Step 1. We determine the greatest common factor of each of the terms of the polynomial.
> Step 2. We use the distributive property to place the greatest common factor determined in step 1 outside the parentheses. The factor inside the parentheses can be obtained by simply observing by what to multiply the greatest common factor in order to obtain each term of the original polynomial, or alternately, by dividing each term of the original polynomial by the greatest common factor.

Example 2 Factor $5y^3 + 5y$.

Solution Step 1. We see that the terms

$$5y^3 \quad \text{and} \quad 5y$$

are already written in factored form, and we observe that the greatest common factor is $5y$.

Step 2. We place the greatest common factor, $5y$, outside the parentheses:

$$5y^3 + 5y = 5y(\quad).$$

To obtain the expression inside the parentheses, we ask: What must $5y$ be multiplied by to give $5y^3$? The answer is y^2. Therefore, we write

$$5y^3 + 5y = 5y(y^2 + \underline{\quad}).$$

Then we ask: What must $5y$ be multiplied by to give $5y$? The answer is 1. Thus, the factored form is

$$5y^3 + 5y = 5y(y^2 + 1).$$

Example 3 Factor $6u^4 + 12u^2$.

Solution Step 1. We write the factored form of each term:

$$6u^4 = 2 \cdot 3 \cdot u^4$$
$$12u^2 = 2^2 \cdot 3 \cdot u^2.$$

We observe that the greatest common factor is $6u^2$, since the common prime factors are 2, 3, and u^2, and $2 \cdot 3 \cdot u^2 = 6u^2$.

Step 2. We place the greatest common factor, $6u^2$, outside the parentheses, so that

$$6u^4 + 12u^2 = 6u^2(\quad).$$

To obtain the expression inside the parentheses, we ask: By what must $6u^2$ be multiplied to give $6u^4$? The answer is u^2. Therefore. we write

$$6u^4 + 12u^2 = 6u^2(u^2 + \underline{\quad}).$$

Then we ask: By what must $6u^2$ be multiplied to give $12u^2$? The answer is 2. Thus. the factored form is

$$6u^4 + 12u^2 = 6u^2(u^2 + 2).$$

We can shorten step 1 by noting that the greatest common factor usually consists of a numerical part and a variable part. To determine the numerical part, take the largest number that is a factor of each coefficient. The variable part is determined by taking the product of the highest powers of each variable common to each term. Thus, the greatest common factor of

$$15x^5 + 10x^3y \text{ is } 5x^3,$$

and the greatest common factor of

$$12u^3v^2 - 18u^2v^4 \text{ is } 6u^2v^2.$$

Example 4 Factor $18h^5 + 12h^4 - 21h^3$.

Solution Step 1. By inspection. the greatest common numerical factor is 3. Because the highest power of the variable h that is a factor of each term is h^3, the greatest common factor is $3h^3$.

Step 2. We place the common factor $3h^3$ outside the parentheses:

$$18h^5 + 12h^4 - 21h^3 = 3h^3(\qquad).$$

The expression inside the parentheses is obtained by dividing each term of the original expression by $3h^3$:

$$\boxed{\dfrac{18h^5}{3h^3} = 6h^2} \quad \boxed{\dfrac{12h^4}{3h^3} = 4h} \quad \boxed{\dfrac{-21h^3}{3h^3} = -7}$$

$$18h^5 + 12h^4 - 21h^3 = 3h^3(6h^2 + 4h - 7).$$

Example 5 Factor $2r^4s^5 + 4r^3s^4 - 8r^2s^3$.

Solution Step 1. The greatest common numerical factor is 2. We observe that r and s appear in each term, and that the highest powers of r and s common to each term are r^2 and s^3. Therefore, the greatest common factor is $2r^2s^3$.

Step 2. We place the common factor $2r^2s^3$ outside the parentheses, and then obtain the expression within the parentheses by dividing each term of the original expression by this common factor:

$$\boxed{\dfrac{2r^2s^5}{2r^2s^3} = r^2s^2} \quad \boxed{\dfrac{4r^3s^4}{2r^2s^3} = 2rs} \quad \boxed{\dfrac{-8r^2s^3}{2r^2s^3} = -4}$$

$$2r^4s^5 + 4r^3s^4 - 8r^2s^3 = 2r^2s^3(r^2s^2 + 2rs - 4).$$

Example 6 Factor $4a^7b^5c^2 + 10a^3b^8 - 16a^4b^6 + 18a^5b^7$.

Solution Step 1. We note that the greatest common numerical factor is 2. We see that the variables a and b, but not c, appear in each term. The highest power of a that is a factor of each term is a^3, and the highest power of b that is a factor of each term is b^5. Thus, the greatest common factor is $2a^3b^5$.

Step 2. We place the greatest common factor $2a^3b^5$ outside the parentheses, and we divide each term of the original expression by $2a^3b^5$:

$$\boxed{\dfrac{4a^7b^5c^2}{2a^3b^5}=2a^4c^2} \quad \boxed{\dfrac{10a^3b^8}{2a^3b^5}=5b^3} \quad \boxed{\dfrac{-16a^4b^6}{2a^3b^5}=-8ab} \quad \boxed{\dfrac{18a^5b^7}{2a^3b^5}=9a^2b^2}$$

$$4a^7b^5c^2 + 10a^3b^8 - 16a^4b^6 + 18a^5b^7 = 2a^3b^5(2a^4c^2 + 5b^3 - 8ab + 9a^2b^2).$$

Factoring by Grouping

Certain polynomials, which do not contain factors common to every term, are still factorable. To factor these polynomials, first we group those terms that do have a common factor. For example, consider the polynomial $3xm + 3ym - 2x - 2y$. Note that two of the terms contain the factor x and the other two terms contain the factor y. We can group these terms as

$$3xm + 3ym - 2x - 2y = (3xm - 2x) + (3ym - 2y).$$

Next, using the common factor x in the first group and the common factor y in the second group, we can write

$$(3xm - 2x) + (3ym - 2y) = x(3m - 2) + y(3m - 2)$$

We now see that $3m - 2$ is a common factor of the two grouped terms. Thus we have

$$x(3m - 2) + y(3m - 2) = (x + y)(3m - 2).$$

Notice that we could have obtained the same answer by recognizing different common factors. Look again at $3xm + 3ym - 2x - 2y$. We could have grouped the terms as follows:

$$\begin{aligned}3xm + 3ym - 2x - 2y &= (3xm + 3ym) + (-2x - 2y) \\ &= 3m(x + y) - 2(x + y) \\ &= (3m - 2)(x + y).\end{aligned}$$

In Examples 7 and 8, factor each expression by grouping.

Example 7 $4y + xy + 4y^2 + x$

Solution We try to find groupings that will allow us to factor out a common binomial factor:

$$\begin{aligned}4y + xy + 4y^2 + x &= (4y + 4y^2) + (xy + x) &&\text{(We grouped the terms.)} \\ &= 4y(1 + y) + x(y + 1) &&\text{(We factored each grouping.)} \\ &= 4y(1 + y) + x(1 + y) &&\text{(We used the commutative property.)} \\ &= (4y + x)(1 + y). &&\text{(We factored out the common factor } 1 + y.)\end{aligned}$$

Example 8 $3a + 3b - ma - mb$

Solution
$$\begin{aligned}3a + 3b - ma - mb &= (3a + 3b) + (-ma - mb) &&\text{(We grouped the terms.)} \\ &= 3(a + b) - m(a + b) &&\text{(We factored each grouping.)} \\ &= (a + b)(3 - m). &&\text{(We factored out the common factor } a + b.)\end{aligned}$$

Solving Equations by Common Factoring

Recall that Property 7, from Chapter 1 (page 20) states: If a and b are real numbers, and if $a \cdot b = 0$, then $a = 0$ or $b = 0$ or both a and $b = 0$.

This property, along with common factoring, provides us with a technique for solving some equations whose left side contains common factors. For example, let us solve the equation

$$x^2 + 5x = 0 \quad \text{for } x.$$

First, we factor $x^2 + 5x$ as

$$x^2 + 5x = x \cdot x + 5 \cdot x = x(x + 5).$$

Thus,

$$x(x + 5) = 0.$$

The equation has the form

$$a \cdot b = 0,$$

where $a = x$ and $b = x + 5$. This means that either $x = 0$ or $x + 5 = 0$ or both x and $x + 5 = 0$. We find the solutions for the original equation by solving these two equations.

$$x = 0 \quad | \quad x + 5 = 0$$
$$\quad \quad \quad | \quad x = -5.$$

Therefore, the solutions are -5 and 0.

Example 9 Solve $2x^2 - 6x = 0$.

Solution We factor the left side of the equation:

$$2x^2 - 6x = 2x(x - 3) = 0.$$

We set each factor equal to zero:

$$2x = 0 \quad | \quad x - 3 = 0$$
$$x = 0 \quad | \quad x = 3.$$

Therefore, the solutions are 0 and 3.

PROBLEM SET 2.8

In problems 1–6, find the GCF of each set of monomials.

1. $6x^2, 8x^3$
2. $6x^2y, 15xy^2$
3. $4x^3y^2, 12x^2y^5$
4. $60xy^2, 72x^3y^3$
5. $8x^2y, 10xy^2, 15x^2y^2$
6. $6xy^5, 9x^2y^3, 10xy^2$

In problems 7–56, factor each polynomial.

7. $5x + 10$
8. $7t + 21$
9. $3r - 6$
10. $3c - 21$
11. $5f - 5$
12. $8w^3 + 4$
13. $12u^2 + 24$
14. $4y^4 + 12$
15. $3x^2 - x$
16. $6y^2 + 3y$
17. $21z^2 - 7z$
18. $4t^2 + 2t$
19. $64a^2 + 96$
20. $100r^2 + 125r$
21. $3u^3 + 6u^2$
22. $9r^2 + 3r^4$
23. $10m^2 - 5m^3$
24. $3c^4 - 4c^3$
25. $4x^8 + 9x^3$
26. $27x^4 - 18x^2$
27. $ax^2 + ay^2$
28. $a^2u^2 + a^2v^2$
29. $\pi r^2 h + 2\pi rh$
30. $6lw + 3wh$
31. $6rs^2 + 30r^2s$
32. $a^2b + 3ab^2$
33. $5a^2b^2 + 7ab^3$
34. $9m^2n - 18mn^2$

35. $6p^2q + 24pq^2$
36. $5abc + 20abc^2$
37. $2t^2 + 4t + 6$
38. $3y^2 - 6y - 3$
39. $w^3 + 3w^2 + 5w$
40. $2a^3 - 8a^2 - 6a$
41. $r^5 + 2r^4 + 3r^3$
42. $y^6 + 7y^5 - 11y^3$
43. $3x^3 - 6x^5 + 15x^2$
44. $5c^3 - 10c^4 + 3c^5$
45. $16A^3 + 12A^2 - 8A^4$
46. $-30x^6 + 24x^5 - 42x^4$
47. $4x^2y + 12xy - 7xy^2$
48. $121x^2y^3 + 55x^3y^2 + 77xy^4$
49. $-35r^3t - 25r^2t^2 - 15rt^3$
50. $24a^2b^3 - 36ab^2 + 18ab$
51. $5x^3y^6 - 10x^2y^2 - 25xy^2$
52. $-2xyz^2 - 6x^2yz + 10xy^2c$
53. $-4c^2d^4 - 8c^3d^3 + 2c^3d^4$
54. $12abc^2 - 8a^2b^2c + 4a^2b^2c^2$
55. $8a^2b^2c^2 - 4ab^2 - 3ab^3$
56. $3c^2d^3f^4 - 6c^3d^2 + 12cd^3$

In problems 57–78, factor each expression by grouping.

57. $am + bm + 4a + 4b$
58. $ax + ay - 3x - 3y$
59. $at - b + t - ab$
60. $cu - 3c - bu + 3b$
61. $uv + 8u + 2v + 16$
62. $ax + bx + 2a + 2b$
63. $6x + 3y - 2xz - yz$
64. $xz + x + yz + y$
65. $cx + cy + x + y$
66. $cd + 7c - d - 7$
67. $ab - 5a - 15 + 3b$
68. $mx - 4pn + pnx - 4m$
69. $7s - 2t - 14 + st$
70. $ab - xa + a^2 - xb$
71. $3ac - ad - 2bd + 6bc$
72. $3uw - 5v + uw - 15u$
73. $a^2x^2 - 5x^2 + 4a^2 - 20$
74. $ma^2 + mb^2 - ka^2 - kb^2$
75. $bc + c - b - 1$
76. $a^2m^2 - 3a^2 + 7m^2 - 21$
77. $5xy - xz - 2wz + 10wy$
78. $xz^2 + 5z^2 + x + 5 + xz + 5z$

In problems 79–88, solve each equation.

79. $y^2 + 3y = 0$
80. $t^2 + 7t = 0$
81. $2x^2 - 14x = 0$
82. $3x^2 - 15x = 0$
83. $-7u^2 + 28u = 0$
84. $-4y^2 + 16y = 0$
85. $8x^2 - 24x = 0$
86. $9t^2 - 18t = 0$
87. $-4t^2 + 12t = 0$
88. $-5u^2 - 25u = 0$

89. A bullet is fired vertically upward and is h feet above the ground t seconds after being fired, where $h = 256t - 16t^2$. Write the expression on the right side in factored form, then find t when $h = 0$.

90. The total surface area S of a right circular cylinder with radius r and height h is given by the formula $S = 2\pi r^2 + 2\pi rh$. Write the expression on the right side in factored form, then find r when $S = 0$.

2.9 Factoring the Difference of Two Squares and Perfect-Square Trinomials

In Chapter 2 (page 135), we used the special product

$$(a + b)(a - b) = a^2 - b^2.$$

If we reverse this equation, we get:

Factoring the Difference of Two Squares

$$a^2 - b^2 = (a + b)(a - b).$$

Thus. the factors of $a^2 - b^2$ are $a + b$ and $a - b$ and the left side of the equation is the **difference of two squares.**

Warning: The sum of two squares $a^2 + b^2$ is a prime polynomial and is *not* factorable.

Example 1 Factor $x^2 - 4$.

Solution We write the expression as

$$x^2 - 4 = x^2 - 2^2.$$

This shows us that we have the difference of two squares, with $a = x$ and $b = 2$. We substitute x for a and 2 for b in the previous equation:

$$x^2 - 4 = (x + 2)(x - 2).$$

Example 2 Factor $9 - y^2$.

Solution Because

$$9 - y^2 = 3^2 - y^2,$$

we have the difference of two squares. Therefore,

$$9 - y^2 = (3 + y)(3 - y).$$

Example 3 Factor $25c^2 - 1$.

Solution The first term is the square of $5c$ and the last term is the square of 1. Again, we have the difference of two squares:

$$25c^2 - 1 = (5c)^2 - 1^2.$$

Thus,

$$25c^2 - 1 = (5c + 1)(5c - 1).$$

Example 4 Factor $4t^2 - 9s^2$.

Solution Because

$$4t^2 - 9s^2 = (2t)^2 - (3s)^2,$$

we can write

$$4t^2 - 9s^2 = (2t + 3s)(2t - 3s).$$

Example 5 Factor $u^4 - 16$.

Solution Because u^4 is $(u^2)^2$ and 16 is 4^2, we have another difference of two squares:

$$u^4 - 16 = (u^2)^2 - 4^2,$$

If we let $a = u^2$ and $b = 4$, we have

$$u^4 - 16 = (u^2 + 4)(u^2 - 4).$$

Notice, though, that the factor $u^2 - 4$ is also the difference of two squares. Therefore, the factor $u^2 - 4$ can itself be factored. (The factor $u^2 + 4$ is the *sum* of two squares and cannot be factored again.) The complete factorization is

$$u^4 - 16 = (u^2 + 4)(u + 2)(u - 2).$$

Example 6 Factor $3a^2c - 12cb^2$.

Solution This is not the difference of two squares because $3a^2c$ and $12cb^2$ are not perfect squares. However, $3c$ is a common factor of both terms. Thus, we can factor the expression as follows:

$$3a^2c - 12cb^2 = 3c(a^2 - 4b^2).$$

Now $a^2 - 4b^2$ is the difference of two squares, so that the complete factorization is

$$3a^2c - 12cb^2 = 3c(a + 2b)(a - 2b).$$

Factoring Perfect-Square Trinomials

Any trinomial of the form

$$a^2 + 2ab + b^2 \quad \text{or} \quad a^2 - 2ab + b^2$$

is called a *perfect-square trinomial* and can be factored by using Special Products 1 and 2 (on page 134) written in reverse order. That is,

> 1. $a^2 + 2ab + b^2 = (a + b)^2$
> 2. $a^2 - 2ab + b^2 = (a - b)^2.$

Example 7 Factor $x^2 + 6x + 9$.

Solution We write the expression as

$$x^2 + 6x + 9 = x^2 + 2(3)x + 3^2.$$

This is in the form of Special Product 1 above, with $a = x$ and $b = 3$, so that

$$x^2 + 6x + 9 = x^2 + 2(3)x + 3^2 = (x + 3)^2.$$

Example 8 Factor $16z^2 - 8z + 1$.

Solution We write the expression as

$$16z^2 - 8z + 1 = (4z)^2 - 2(4z)(1) + 1^2.$$

This is in the form of Special Product 2, with $a = 4z$ and $b = 1$. Therefore, we can write

$$16z^2 - 8z + 1 = (4z)^2 - 2(4z)(1) + 1^2 = (4z - 1)^2.$$

Solving Equations by Factoring: The Difference of Two Squares

Here we see how factoring the difference of two squares can be used to solve certain kinds of equations.

Example 9 Solve $x^2 - 16 = 0$.

Solution We factor the left side of the equation:

$$(x - 4)(x + 4) = 0.$$

We set each factor equal to zero and then solve for x:

$x - 4 = 0$	$x + 4 = 0$
$x = 4$	$x = -4.$

Therefore, the solutions are -4 and 4.

Example 10 Solve $25y^2 - 36 = 0$.

Solution We factor the left side of the equation:
$$(5y - 6)(5y + 6) = 0.$$
We set each factor equal to zero, and we have

$$5y - 6 = 0 \quad | \quad 5y + 6 = 0$$
$$5y = 6 \quad | \quad 5y = -6$$
$$y = \frac{6}{5} \quad | \quad y = -\frac{6}{5}.$$

Therefore, the solutions are $-\frac{6}{5}$ and $\frac{6}{5}$.

PROBLEM SET 2.9

In problems 1–36, factor each expression.

1. $x^2 - 9$
2. $y^2 - 16$
3. $4 - t^2$
4. $36 - u^2$
5. $4c^2 - 1$
6. $1 - 9w^2$
7. $1 - 100r^2$
8. $1 - 81y^2$
9. $25 - 81c^2$
10. $4a^2 - b^2$
11. $9f^2 - 25$
12. $25 - 49h^2$
13. $4k^2 - 9$
14. $4t^2 - 25u^2$
15. $100 - 121w^2$
16. $400 - 9u^2$
17. $a^2b^2 - 1$
18. $u^2v^2 - 9$
19. $9x^2y^2 - 16$
20. $a^2b^2c^2 - 100$
21. $x^2y^2 - 625$
22. $16t^2 - 225u^2$
23. $9m^2 - 49n^2$
24. $16a^2b^2 - 25c^2$
25. $9x^2y^2z^2 - 16w^2$
26. $25a^2 - 49c^2e^2f^2$
27. $81a^2b^2c^2 - 25u^2v^2$
28. $64c^2 - 49y^2z^2w^2$
29. $b^2 - 100c^2d^2f^2$
30. $9x^2y^2z^2 - 64c^2d^2$
31. $c^4 - d^4$
32. $16v^4 - 1$
33. $1 - 81t^4$
34. $81 - m^4n^4$
35. $16a^4 - 81b^4$
36. $16 - a^4b^4c^4$

In problems 37–48, use common factoring and the difference of two squares to factor each expression.

37. $5x^3 - 5xy^2$
38. $28u^4v - 7u^2v^3$
39. $27ab^3 - 108a^3b$
40. $3c^2 - 75d^4$
41. $3a^2 - 75$
42. $2ab^3 - 8a^3b$
43. $50u^2v - 72v$
44. $36t^3 - 49ts^2$
45. $8c^2d - 18d$
46. $9a^3 - b^2ac^2$
47. $3r^3 - 48rs^2$
48. $49c^3 - 25cd^2$

In problems 49–58, use Special Products 1 and 2 to factor each perfect-square trinomial.

49. $x^2 + 2x + 1$
50. $t^2 - 4t - 4$
51. $y^2 - 10y + 25$
52. $4x^2 + 20x + 25$
53. $4u^2 - 12uw + 9v^2$
54. $16a^2 + 40a + 25$
55. $64t^2 + 48t + 9$
56. $25x^2 - 110x + 121$
57. $81t^2 + 126t + 49$
58. $9 - 12z + 4z^2$

In problems 59–68, solve each equation.

59. $x^2 - 25 = 0$
60. $x^2 - 9 = 0$
61. $9u^2 - 1 = 0$
62. $t^2 - 81 = 0$
63. $u^2 - 121 = 0$
64. $4y^2 - 1 = 0$
65. $49t^2 - 25 = 0$
66. $9 - 16y^2 = 0$
67. $16x^2 - 121 = 0$
68. $3y^2 - 108 = 0$

69. Find the value of c so that the polynomial $x^2 + 6x + c$ is expressed in the factored form $(x + 3)^2$.

70. Find the value of b so that the polynomial $x^2 + bx + 25$ is expressed in the factored form $(x - 5)^2$.

71. Equation $(x + 1)^2 - x^2 = 15$ expresses the fact that the difference between the squares of two consecutive integers x and $x + 1$ is 15. Find the integers.

72. Equation $(x + 2)^2 - x^2 = 32$ expresses the fact that the difference between the squares of two odd consecutive integers x and $x + 2$ is 32. Find the integers.

2.10 Factoring Trinomials of the Form $x^2 + bx + c$

In this section and the following one, we factor trinomials of the form $ax^2 + bx + c$, where a, b, and c are integers and $a \neq 0$.

Here we only consider trinomials in which $a = 1$. Recall from Section 3.5 that the product of two binomials can be a trinomial; for instance,

$$(x + 3)(x + 2) = x \cdot x + 2x + 3x + 3 \cdot 2$$
$$= x^2 + (2 + 3)x + 6$$
$$= x^2 + 5x + 6.$$

Notice that the first term in the trinomial is the *product* of the first terms in each binomial. The last term in the trinomial is the *product* of the last terms in each binomial. The middle term in the trinomial results from *adding the product of the outside terms and the product of the inside terms* of the binomials.

To factor the trinomial $x^2 + 5x + 6$, we assume that the factors are the binomials $x + p$ and $x + q$, in which p and q are integers, such that

$$x^2 + 5x + 6 = (x + p)(x + q)$$
$$= x^2 + px + qx + pq$$
$$= x^2 + (p + q)x + pq.$$

That is,

$$x^2 + 5x + 6 = x^2 + (p + q)x + pq.$$

We compare the coefficients of like terms on each side of this equation, and we see that

$$pq = 6 \quad \text{and} \quad p + q = 5.$$

Therefore, our goal is to find two numbers p and q whose product is 6 and whose sum is 5. Possible choices for these numbers are shown in the following table:

Product	Sum
6 = 6(1)	6 + 1 = 7
6 = 2(3)	2 + 3 = 5 (correct)

The numbers 2 and 3 are the correct choices. Therefore, $x + 2$ and $x + 3$ are the factors of the trinomial, and

$$x^2 + 5x + 6 = (x + 2)(x + 3).$$

This process can be shortened as follows:

To find two binomial factors with integer coefficients, we first write

$$2 + 5x + 6 = (x + \underline{})(x + \underline{}).$$

Then we fill in the blanks with a pair of numbers whose product is 6 and whose sum is 5. The obvious choice for these numbers is 2 and 3. Therefore,

$$x^2 + 5x + 6 = (x + 2)(x + 3).$$

2.10 FACTORING TRINOMIALS OF THE FORM $x^2 + bx + c$

Example 1 Factor $x^2 + 8x + 7$.

Solution We need to find two integers whose product is 7 and whose sum is 8:

$$x^2 + 8x + 7 = (x + \underline{})(x + \underline{}).$$

Product	Sum
$7 = 7(1)$	$7 + 1 = 8$

Therefore, the correct factorization is

$$x^2 + 8x + 7 = (x + 7)(x + 1).$$

Example 2 Factor $z^2 - 8z + 12$.

Solution We need two numbers whose product is 12 and whose sum is -8. We factor 12 in all possible ways, and we try to find those factors whose sum is -8. Since the *product* of the numbers is 12, they have the *same sign*. Because their sum is negative (-8), both numbers must be negative. Therefore, we write

$$z^2 - 8 + 12 = (z - \underline{})(z - \underline{}).$$

The following table gives all possible combinations:

Product	Sum
$12 = -1(-12)$	$(-1) + (-12) = -13$
$12 = -3(-4)$	$(-3) + (-4) = -7$
$12 = -2(-6)$	$(-2) + (-6) = -8$ (correct)

Thus, the correct factoring is

$$z^2 - 8z + 12 = (z - 6)(z - 2).$$

Example 3 Factor $t^2 + 2t - 15$.

Solution We need two numbers whose product is -15 and whose sum is 2. Since the product is negative, the numbers must have *opposite* signs. Thus, we write

$$t^2 + 2t - 15 = (t - \underline{})(t + \underline{}).$$

The following table gives all possible combinations:

Product	Sum
$-15 = 1(-15)$	$1 + (-15) = -14$
$-15 = -1(15)$	$(-1) + 15 = 14$
$-15 = 3(-5)$	$3 + (-5) = -2$
$-15 = -3(5)$	$(-3) + 5 = 2$ (correct)

Thus, the correct factoring is

$$t^2 + 2t - 15 = (t - 3)(t + 5).$$

With enough practice, you will usually be able to find the correct factors by checking possible combinations mentally, without actually listing all the possibilities. However, it will be helpful to keep in mind the rules we have been using to determine the signs of the terms in the binomials. The following rules summarize the relationship between the signs of the terms of the trinomial and the signs of the binomial factors:

Rule 1. If the sign of the last term of a factorable trinomial is positive, the signs of the second terms in both binomial factors are the same as the sign of the middle term of the trinomial.

Rule 2. If the sign of the last term of a factorable trinomial is negative, the signs of the second terms of the two binomial factors differ.

These rules apply to any factorable trinomial that takes the form $ax^2 + bx + c$, where a is a positive integer.

Example 4 Factor $r^2 - 7rs - 18s^2$.

Solution We need two numbers whose product is -18 and whose sum is -7. The product is negative; therefore, by Rule 2 we know that the numbers have *opposite signs*. We write

$$r^2 - 7rs - 18s^2 = (r - \underline{\quad} s)(r + \underline{\quad} s).$$

The following table gives all possible combinations:

Product	Sum
$-18 = -1(18)$	$(-1) + 18 = 17$
$-18 = -2(9)$	$(-2) + 9 = 7$
$-18 = -3(6)$	$(-3) + 6 = 3$
$-18 = 3(-6)$	$3 + (-6) = -3$
$-18 = 1(-18)$	$1 + (-18) = -17$
$-18 = 2(-9)$	$2 + (-9) = -7$ (correct)

Therefore, the correct factoring is

$$r^2 - 7rs - 18s^2 = (r - 9s)(r + 2s).$$

Test for Factorability

Sometimes, when we try to factor trinomials of the form $ax^2 + bx + c$ or $ax^2 + bxy + cy^2$, none of the combinations tested will work. It would be useful to know before we start looking for factors whether a trinomial is factorable. Fortunately, there is a *test for factorability*.

If the expression $b^2 - 4ac$ is the *square of an integer*, then the trinomial can be factored: otherwise, it is not factorable.

Example 5 Test each trinomial to see whether it is factorable or prime. If it is factorable, factor it.
(a) $x^2 - x - 6$ (b) $m^2 + 5mn + 8n^2$

Solution (a) Here $a = 1$, $b = -1$, and $c = -6$. We have

$$b^2 - 4ac = (-1)^2 - 4(1)(-6) = 1 + 24 = 25 = 5^2.$$

Therefore, the trinomial can be factored. The correct factorization is

$$x^2 - x - 6 = (x - 3)(x + 2).$$

(b) Here $a = 1$, $b = 5$, and $c = 8$. We have

$$b^2 - 4ac = 5^2 - 4(1)(8) = 25 - 32 = -7.$$

Because -7 is not the square of an integer, $m^2 + 5mn + 8n^2$ is not factorable.

Solving Equations by Factoring a Trinomial

Property 7 (page 20) continues to play an important role in solving equations containing factorable trinomials.

Example 6 Solve $x^2 + 10x + 16 = 0$.

Solution Factoring the trinomial, we have

$$x^2 + 10x + 16 = (x + 8)(x + 2).$$

So,

$$(x + 8)(x + 2) = 0.$$

Next, set each factor equal to zero and solve for x:

$$x + 8 = 0 \quad | \quad x + 2 = 0$$
$$x = -8 \quad | \quad x = -2.$$

The solutions are -8 and -2.

PROBLEM SET 2.10

In problems 1–42, factor each trinomial.

1. $x^2 + 3x + 2$
2. $y^2 + 4y + 3$
3. $r^2 + 6r + 5$
4. $z^2 + 7z + 6$
5. $t^2 - 10t + 21$
6. $u^2 - 9u + 20$
7. $m^2 - 7m + 10$
8. $n^2 - 12n + 35$
9. $x^2 + 2x - 3$
10. $y^2 - y - 20$
11. $u^2 - 4u - 12$
12. $z^2 + 6z - 27$
13. $a^2 - a - 30$
14. $b^2 + 2b - 35$
15. $c^2 + 12c - 45$
16. $v^2 - 4v - 96$
17. $w^2 + 2w - 48$
18. $z^2 - 6z - 16$
19. $a^2 - 19a + 48$
20. $t^2 + 24t - 81$
21. $m^2 - 7m + 12$
22. $u^2 - 4u - 5$
23. $x^2 + x - 110$
24. $y^2 - y - 132$
25. $w^2 - 4w + 4$
26. $z^2 - 2z + 1$
27. $x^2 + 8x + 16$
28. $u^2 + 14u + 49$
29. $y^2 + 10y + 25$
30. $z^2 + 16z + 64$
31. $y^2 + 2yz - 63z^2$
32. $u^2 + 14uv - 15v^2$
33. $a^2 + 7ab - 30b^2$
34. $m^2 + 13mn + 42n^2$
35. $x^2 - 20xy - 44y^2$
36. $r^2 - 9rs - 36s^2$
37. $w^2 - 15wu + 36u^2$
38. $a^2 + 11ab - 42b^2$
39. $x^2 + 18xy + 81y^2$
40. $u^2 - 7uv - 30v^2$
41. $a^2 + 13ab + 40b^2$
42. $c^2 - cd - 6d^2$

In problems 43–50, use the test for factorability to determine whether or not each trinomial is factorable. If it is factorable, factor it.

43. $y^2 - 7y + 12$
44. $t^2 + 5t + 6$
45. $u^2 - 4u - 15$
46. $x^2 - 3x - 2$
47. $m^2 - 7m + 6$
48. $c^2 - 5cd + 6d^2$
49. $x^2 - 2xy + 24y^2$
50. $u^2 + 9uv + 14v^2$

In problems 51–60, solve each equation.

51. $x^2 + 7x + 12 = 0$
52. $t^2 + 9t + 20 = 0$
53. $p^2 - 4p - 12 = 0$
54. $u^2 - 3u - 40 = 0$
55. $x^2 + 9x - 22 = 0$
56. $t^2 - 11t + 24 = 0$
57. $y^2 + 5y - 66 = 0$
58. $x^2 + 4x - 45 = 0$
59. $m^2 + 2m - 35 = 0$
60. $x^2 - x - 56 = 0$

61. A company manufactures x units for a total cost of C dollars, where C is given by the equation

$$C = 2{,}000 + 700x - 100x^2.$$

Find x if the total cost is $1,200.

62. Under perfect competition a firm can sell x units for a total cost of C dollars, where C is given by the equation

$$C = x^2 + 20x + 700.$$

Find x if the total cost is $2,200.

2.11 Factoring Trinomials of the Form $ax^2 + bx + c$, $a \neq 1$

So far, we have factored only trinomials of the form $ax^2 + bx + c$, where $a = 1$. Now we turn our attention to factoring trinomials in which $a \neq 1$. For example, we factor

$$6x^2 - 5x - 6 \quad \text{or} \quad 3x^2 - 4xy - 15y^2$$

into the product of two binomials with integer coefficients.

By now it should be easy for you to obtain a product such as

$$(2x + 7)(3x + 5) = 6x^2 + 31x + 35.$$

By reversing the process, we can write

$$6x^2 + 31x + 35 = (2x + 7)(3x + 5).$$

The relationship between the coefficients of the trinomial and the coefficients of the factors is illustrated below. The coefficients 6 and 35 are obtained as follows:

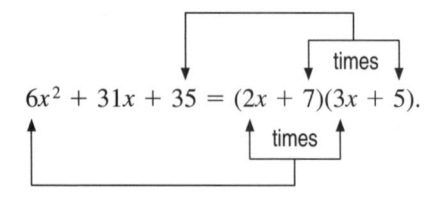

$$6x^2 + 31x + 35 = (2x + 7)(3x + 5).$$

The coefficient of the middle term in the trinomial, 31, is obtained as follows:

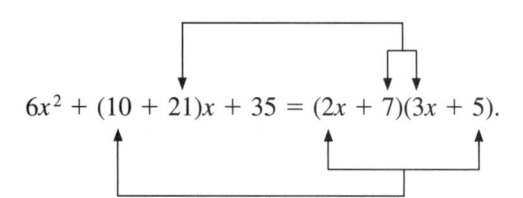

$$6x^2 + (10 + 21)x + 35 = (2x + 7)(3x + 5).$$

These diagrams illustrate how a trinomial of the form $ax^2 + bx + c$ is factored. We begin by writing

$$ax^2 + bx + c = (\underline{} x + \underline{})(\underline{} x + \underline{}).$$

Then we fill in the blanks with numbers so that:

1. The product of the first terms is ax^2.
2. The product of the last terms is c.
3. The sum of the product of the outside terms and the product of the inside terms is bx.

We try all possible choices of factors of a and c until we find a combination that gives us the desired middle term. For example, to factor $3x^2 - 7x - 6$, we look at the first term, $3x^2$ and we factor it in all possible ways:

$$3x^2 = 3x \cdot x$$

The sign of the last term is negative. Therefore the signs of the second terms of the binomials are different (one is positive and one is negative).

Because the last term, -6, has several pairs of factors, we list the following possible combinations:

2.11 FACTORING TRINOMIALS OF THE FORM $ax^2 + bx + c$, $a \neq 1$ 159

Possible Combination of Binomial Factors	First Term	Middle Term	Last Term
$(3x - 6)(x + 1) =$	$3x^2$	$-3x$	-6
$(3x - 1)(x + 6) =$	$3x^2$	$+17x$	-6
$(3x - 3)(x + 2) =$	$3x^2$	$+3x$	-6
$(3x - 2)(x + 3) =$	$3x^2$	$+7x$	-6
$(3x + 6)(x - 1) =$	$3x^2$	$+3x$	-6
$(3x + 1)(x - 6) =$	$3x^2$	$-17x$	-6
$(3x + 3)(x - 2) =$	$3x^2$	$-3x$	-6
$(3x + 2)(x - 3) =$	$3x^2$	$-7x$	-6

The combination $(3x + 2)(x - 3)$ gives the correct middle term, $-7x$. Therefore, the correct factorization is

$$3x^2 - 7x - 6 = (3x + 2)(x - 3).$$

Example 1 Factor $2x^2 + 11x + 12$.

Solution We begin by factoring the first term, $2x^2$, as

$$2x^2 = 2x \cdot x.$$

Because the signs of the middle and last terms in the trinomial are positive, the signs of the second terms in both binomials are also positive. So the trinomial can be expressed in the form

$$2x^2 + 11x + 12 = (2x + \underline{\qquad})(x + \underline{\qquad}).$$

Knowing that the factors of the last term, 12, must be positive, we can list all the possibilities:

$$12 = 12(1), \quad 12 = 6(2), \quad 12 = 4(3).$$

We try different possible combinations of the factors of 12, and we see that $(2x + 3)(x + 4)$ produces the correct middle term of the trinomial. This term is the sum of the product of the outside terms and the product of the inside terms of the binomials:

$(2x + 3)(x + 4)$ $11x$ (the correct middle term.)

$2x(4) + 3(x)$

Therefore, $2x + 3$ and $x + 4$ are the factors of the trinomial, and

$$2x^2 + 11x + 12 = (2x + 3)(x + 4).$$

Example 2 Factor $3m^2 - 10m + 8$.

Solution We factor the first term, $3m^2$, in all possible ways:

$$3m^2 = 3m \cdot m.$$

Next, we factor the last term, 8, in all possible ways:

$$8 = 8(1) \quad \text{and} \quad 8 = 4(2).$$

Because the sign of the last term is positive, the signs in the binomials must either both be positive or both be negative. They are both negative because the middle term is negative. Therefore, the trinomial can be expressed in the form

$$3m^2 - 10m + 8 = (3m - \underline{\qquad})(m - \underline{\qquad}).$$

We try different possible combinations of the factors of 8 to find a sum of the product of the outside terms and the product of the inside terms that gives the middle term of the trinomial. We find that

$(3m - 4)(m - 2) \qquad -10m \qquad$ (the correct middle term)

$$3m(-2) + m(-4)$$

Thus, $3m - 4$ and $m - 2$ are the factors of the trinomial, and

$$3m^2 - 10m + 8 = (3m - 4)(m - 2).$$

Example 3 Factor $6r^2 + 7r - 3$.

Solution We factor the first term, $6r^2$, in all possible ways:

$$6r^2 = (2r)(3r)$$
$$6r^2 = (6r)(r).$$

Next, we factor the number 3 in all possible ways as

$$3 = 1(3).$$

Now, because the sign of the last term is negative, the signs of the second terms of the binomials are different (one is positive and one is negative). So we have either

$$6r^2 + 7r - 3 = (\underline{\quad\quad} r + 3)(\underline{\quad\quad} r - 1)$$

or

$$6r^2 + 7r - 3 = (\underline{\quad\quad} r - 3)(\underline{\quad\quad} r + 1).$$

We try different possible combinations of the factors of 6 to find a sum of the product of the outside terms and the product of the inside terms that gives the middle term of the trinomial. We find that

$(2r + 3)(3r - 1) \qquad 7r \qquad$ (the correct middle term)

$$2r(-1) + 3r(3)$$

Thus, $2r + 3$ and $3r - 1$ are the factors of the trinomial, and

$$6r^2 + 7r - 3 = (2r + 3)(3r - 1).$$

Example 4 Factor $7x^2 - 9xy - 10y^2$.

Solution We factor the first term, $7x^2$, in all possible ways:

$$7x^2 = (7x)(x).$$

Next, we factor the expression $10y^2$ in all possible ways:

$$10y^2 = (2y)(5y) \quad \text{or} \quad 10y^2 = (10y)(y).$$

Because the sign of the last term is negative, the signs of the second terms of the binomials are opposite. So we have:

$$7x^2 - 9xy - 10y^2 = (7x - \underline{})(x + \underline{})$$

or

$$7x^2 - 9xy - 10y^2 = (7x + \underline{})(x - \underline{}).$$

We try different possible combinations of the factors of $10y^2$ to find one that produces the middle term. We find that

$(7x + 5y)(x - 2y)$ $-9xy$ (the correct middle term)

$7x(-2y) + x(5y)$

Therefore, $(7x + 5y)$ and $(x - 2y)$ are the factors of the trinomial, and

$$7x^2 - 9xy - 10y^2 = (7x + 5y)(x - 2y).$$

To factor a trinomial in which each term has a common factor, we begin by factoring out the greatest common factor (see Section 2.7). Then we examine the trinomial inside the parentheses to determine whether it can be factored into a pair of binomials. If this is possible, we complete the factorization.

Example 5 Factor $xy^2 - 3xy - 10x$.

Solution First we notice that x is a common factor of all three terms, so that

$$xy^2 - 3xy - 10x = x(y^2 - 3y - 10).$$

Now we try to factor the trinomial $y^2 - 3y - 10$. If it is factorable, we need to find two numbers whose product is -10 and whose sum is -3. The numbers are -5 and 2:

$$y^2 - 3y - 10 = (y - 5)(y + 2).$$

Therefore, the complete factorization is

$$xy^2 - 3xy - 10x = x(y - 5)(y + 2).$$

Example 6 Factor $60a^3b + 25a^2b^2 - 15ab^3$.

Solution First, we notice that $5ab$ is the greatest common factor, so that

$$60a^3b + 25a^2b^2 - 15ab^3 = 5ab(12a^2 + 5ab - 3b^2).$$

Now we attempt to factor the trinomial $12a^2 + 5ab - 3b^2$. We factor the first term, $12a^2$, in all possible ways:

$$12a^2 = 12a(a), \quad 12a^2 = 6a(2a), \quad 12a^2 = 3a(4a).$$

Next, we factor the expression $3b^2$ in all possible ways:

$$3b^2 = b(3b).$$

Because the sign of the last term is negative, the signs of the second terms of the binomials are different. We try different possible combinations to see if one will give the correct middle term. We find that

$(4a + 3b)(3a - b)$ $5ab$ (the correct middle term)

$4a(-b) + 3a(3b)$

Thus,
$$12a^2 + 5ab - 3b^2 = (4a + 3b)(3a - b).$$

Therefore, the original trinomial is factored completely as
$$60a^3b + 25a^2b^2 - 15ab^3 = 5ab(4a + 3b)(3a - b).$$

Solving Equations $ax^2 + bx + c = 0$ by Factoring

The factoring of a trinomial $ax^2 + bx + c$ enables us to solve some equations of the form
$$ax^2 + bx + c = 0, \quad a \neq 0.$$

Example 7 Solve $3x^2 - 4x - 4 = 0$.

Solution We factor the left side of the equation:
$$(x - 2)(3x + 2) = 0.$$

Now we set each factor equal to 0 and solve for x:

$$x - 2 = 0 \qquad \qquad 3x + 2 = 0$$
$$x = 2 \qquad \qquad 3x = -2$$
$$\qquad \qquad \qquad x = -\frac{2}{3}.$$

Therefore, the solutions are $-\frac{2}{3}$ and 2.

PROBLEM SET 2.11

In problems 1–44, factor each trinomial that can be factored. For those trinomials that you cannot factor, apply the test for factorability from Section 2.10 on page 156 to make sure they are prime.

1. $2x^2 + 5x + 3$
2. $3y^2 + 4y + 1$
3. $3t^2 + 7t + 2$
4. $5m^2 - 14m + 8$
5. $2t^2 - 7t + 6$
6. $4n^2 - 9n + 2$
7. $5z^2 + 3z - 2$
8. $3u^2 - 13u + 12$
9. $6c^2 + 7c + 2$
10. $4w^2 - w - 3$
11. $4z^2 + 7z - 2$
12. $4x^2 + 4x - 3$
13. $2y^2 - 11y + 15$
14. $12t^2 - 13t - 14$
15. $6r^2 - r - 12$
16. $30v^2 + 13v - 3$
17. $42c^2 - 5c - 2$
18. $2y^2 - 71 - 30$
19. $24f^2 - 34f + 5$
20. $63x^2 - 69x - 20$
21. $9y^2 - 27y + 20$
22. $4x^2 + x - 3$
23. $6a^2 + 11a + 2$
24. $2x^2 + 15x + 18$
25. $2u^2 + 7u + 1$
26. $3b^2 - b - 4$
27. $14x^2 + 29x - 15$
28. $14g^2 + 11g - 15$
29. $20h^2 + 9h - 20$
30. $6k^2 - 13k + 6$
31. $15x^2 - 11xy - 14y^2$
32. $12u^2 + uv - 6v^2$
33. $20a^2 - 9ab - 20b^2$
34. $32s^2 + 60st + 27t^2$
35. $6y^2 - 43yz + 72z^2$
36. $40c^2 + 11cd - 63d^2$
37. $7x^2 + 31xy + 12y^2$
38. $36y^2 + 40yz - 21z^2$
39. $12f^2 + 29fg + 16g^2$
40. $20u^2 + uv - 30v^2$
41. $32a^2 - 4ab - 21b^2$
42. $4a^2 - 9ab + 40b^2$
43. $9c^2 + 11cd + 21d^2$
44. $2x^2 + 7xy + 12y^2$

In problems 45–56, factor each trinomial completely. Look first for common factors.

45. $ax^2 - 8ax + 15a$
46. $3br^2 - 36br + 105b$
47. $cd^2 + 2cd - 8c$
48. $6ab^2 - 51ab + 63a$
49. $6y^3 + 28y^2 - 10y$

50. $3m^2p + 13mnp + 4n^2p$
51. $10x^2z^2 - 5x^2z - 5x^2$
52. $6u^3v - 5u^2v^2 - 6uv^3$
53. $100yw^2 + 120ywz + 36yz^2$
54. $36x^4y + 69x^3y^2 + 30x^2y^3$
55. $20p^3q^4 + 5p^2q^5 - 70pq^6$
56. $27y^2z^3 - 36yz^4 + 12z^5$

In problems 57–66, solve each equation.

57. $10x^2 + 23x - 5 = 0$
58. $6x^2 + x - 40 = 0$
59. $6t^2 - 11t + 4 = 0$
60. $16y^2 - 2y - 5 = 0$
61. $12u^2 - 16u + 5 = 0$
62. $2x^2 + 3x + 1 = 0$
63. $2m^2 + 5m - 3 = 0$
64. $12x^2 - 23x + 5 = 0$
65. $4x^2 - 16x - 9 = 0$
66. $12x^2 + 7x - 12 = 0$

REVIEW PROBLEM SET

In problems 1–6, find the value of each expression.

1. 4^3
2. 5^4
3. $(-3)^{-3}$
4. $(-6)^{-2}$
5. $-(-2)^0$
6. $-(-1)^0$

In problems 7–12, write each expression in equivalent exponential form.

7. $5 \cdot u \cdot u \cdot u \cdot v \cdot v$
8. $-9 \cdot x \cdot x \cdot y \cdot y \cdot y \cdot y$
9. $-2 \cdot t \cdot t \cdot t \cdot t \cdot s \cdot s \cdot s$
10. $11 \cdot p \cdot p \cdot p \cdot p \cdot q \cdot q \cdot q \cdot q \cdot q$
11. $4 \cdot x \cdot x \cdot y - 8 \cdot x \cdot y \cdot y \cdot y$
12. $-3 \cdot w \cdot w \cdot w \cdot z + 6 \cdot w \cdot w \cdot z \cdot z \cdot z \cdot z$

In problems 13–18, determine whether the given expression is a polynomial.

13. $7t^2 - 3t + 5$
14. $-8 + 2u - 3u^5$
15. $8z^3 + 7z - \dfrac{9}{z} + 2$
16. $11x^3 - 3x + 7x^2 + 1$
17. $\tfrac{2}{3}y - \tfrac{1}{5}y^3 + 10 - 6y^2$
18. $3t^3 - \dfrac{1}{t^2} + 3 + \dfrac{2}{t}$

In problems 19–26, identify the polynomial as a monomial, binomial, or trinomial if applicable. Find the degree of the polynomial and list the numerical coefficients.

19. $3y - 7$
20. -11
21. $4x^2 - 5x + 13$
22. $7 - 8t^3$
23. $-8u^5$
24. $2p^3 - 6p^2 + 4p + 3$
25. $7 + 9z^2 - z + 8z^3$
26. $4x^3 - 5x^2y^4 + 7y^5$

In problems 27–32, evaluate each polynomial for the given value of the variable.

27. $13 - 4t, t = -2$
28. $-5x + 17, x = 3$
29. $3y^2 - 5y + 7, y = 3$
30. $2z^3 - 3z^2 + 5z - 6, z = -3$
31. $2u^2 - 3uv + v^2, u = 1$ and $v = -2$
32. $4x^3y + 5xy^2 - 2y^3, x = 2$ and $y = -1$

In problems 33–38, evaluate each expression for the given values.

33. Prt; when $P = \$2{,}000$, $r = 12\%$, and $t = 4$ years.
34. Prt; when $P = \$12{,}000$, $r = 14.5\%$, and $t = 3\tfrac{1}{2}$ years.
35. $\tfrac{5}{9}(F - 32)$; when $F = 68°$.
36. $\tfrac{9}{5}C + 32$; when $C = 35°$.
37. $P(1 + rt)$; when $P = \$25{,}000$, $r = 10\%$, and $t = 5$ years.
38. $\ell w h$; when $\ell = 12$ centimeters, $w = 8$ centimeters, and $h = 11$ centimeters.

In problems 39–50, perform the addition or subtraction.

39. $8w^2 + 11w^2$
40. $-7x^3 + 13x^3$
41. $2x^3 - 5x^3 + 7x^3$
42. $-8y^2 + 12y^2 + (-9y^2)$
43. $-13t + (-6t) - 3t$
44. $17x^2y - 13xy^2 - 9x^2y$
45. $(3y^3 - 4y + 7) + (y^3 - 5y^2 + 7y - 3)$
46. $(7w^2 + 6w - 3) - (2 + 5w^2 - 8w)$
47. $(10p - 4p^3 + 15p^2 + 11) + (5p^3 - 8 - 3p - 12p^2)$
48. $(2m^3n - 3m^2n + 7mn^2 + mn^3) + (5mn^3 - 4m^3n + m^2n - 2mn^2)$
49. $(4x^2y - 5xy + 13xy^3) - (-9xy - x^2y + 6xy^3)$
50. $(9r^3 - 11r^2s + 12s^3) - (4r^3 - 7rs^2 + 6s^3 - 3)$

In problems 51–56, perform the indicated operations.

51. $(3y^2 + 5y - 1) + (5y^2 - 2y + 2) + (-3y^2 + y - 7)$
52. $(x^3 + 2x - 7) + (x^2 - 3x + 6) + (3x^3 - 2x^2 + 5)$
53. $(8z^2 - 11z + 7) + (2z^2 + 6z - 3) - (5z^2 + 3z + 13)$
54. $(4m^2n - 3mn^2 + 6mn) + (m^2n + 2mn^2 - 3mn) - (-5mn^2 + mn - 2m^2n)$
55. $(7u^2v - 3uv + 6uv^2) - (5uv^2 - 5uv + u^2v) - (-2uv + 8u^2v - 3uv^2)$
56. $(4p^4 + 3p^2 + 13) - (2p^4 - p^2 + 10) - (p^3 - 5p + 3)$

In problems 57–60, rewrite each number in scientific notation.

57. 5,892
58. 0.0351
59. 0.000792
60. 173,000

In problems 61–64, rewrite each number in expanded form.

61. 3.72×10^4
62. 8.95×10^{-3}
63. 1.23×10^{-2}
64. 7.66×10^5

In problems 65 and 66, simplify the expression and write the answer in scientific notation. Round off each answer to four decimal places.

65. $\dfrac{(3.21 \times 10^{-8}) \cdot (2.4 \times 10^5)}{1.78 \times 10^{-2}}$

66. $\dfrac{(4.31 \times 10^5) \cdot (5.21 \times 10^{-2})}{(1.38 \times 10^4) \cdot (3.4 \times 10^{-7})}$

In problems 67–94, use the properties of exponents to simplify the given expressions.

67. $(8^3)^5$
68. $[(-7)^6]^4$
69. $9^4 \cdot 9^3$
70. $5^2 \cdot 5^4 \cdot 5$
71. $(xy)^5$
72. $\left(\dfrac{-u}{v}\right)^7$
73. $(-u)^3(-u)^4$
74. $(-5x)^3$
75. $\left(\dfrac{a}{b}\right)^6$
76. $\dfrac{m^3}{m^7}$
77. $\dfrac{z^7}{z^4}$
78. $(-2pq)^4$
79. $(x^3)^4$
80. $(-x)^4(-x)^5(-x)^6$
81. $(-t)^3(-t)^2(-t)$
82. $(3u^2vw^3)^4$
83. $(2a^4)^3$
84. $(-7t^3)^2$
85. $\dfrac{(-y)^2}{(-y)^5}$
86. $\dfrac{(-x)^4}{(-x)^9}$
87. $(3x^3y)^2(-x^2y^4)^3$
88. $(-m^3n^5)^4(-3mn^4)^3$
89. $\dfrac{(2p^4q^2)^5}{(2p^2q^3)^4}$
90. $\dfrac{(z^4)^3(z^2)^5}{5(z^8)^2}$
91. $\left(\dfrac{x^4}{x^7}\right)^{-2}$
92. $(y^{-2})^3$
93. $\left(\dfrac{x^{-2}}{y^{-3}}\right)^{-2}$
94. $(x^{-3}y^{-4})^{-4}$

In problems 95–104, find the products and simplify.

95. $(7x^3y^2z)(-3xz^2)$
96. $(9y^3z^4)(5y^2z)$
97. $3t^2(2t^3 + 5t - 4)$
98. $5mn^2(2m^3n - 3m^2n + mn)$
99. $(v + 2)(v^2 - v + 3)$
100. $(y - 5)(2y^2 + 3y + 1)$
101. $(2x + y)(3x^2 - xy + y^2)$
102. $(3y - z)(y^2 - 2yz + z^2)$
103. $(m^2 + 2m + 2)(m^2 - 3m + 4)$
104. $(2x^2 - xy - y^2)(3x^2 + 2xy + y^2)$

In problems 105–114, find the products of the binomials by the FOIL method.

105. $(x + 6)(x + 7)$
106. $(y + 12)(y + 15)$
107. $(t - 1)(t - 9)$
108. $(z - 2)(z - 6)$
109. $(2y + 7)(y - 8)$
110. $(3x - 4)(7x + 6)$
111. $(x + 4y)(3x - 5y)$
112. $(2w + 7z)(3w - 5z)$
113. $(5u - 3v)(6u - 5v)$
114. $(9x - 8y)(7x + 6y)$

In problems 115–124, use the special products to perform each multiplication.

115. $(z + 4)^2$
116. $(7x + 1)^2$
117. $(2x - 7)(2x + 7)$
118. $(11w - 10z)(11w + 10z)$
119. $(5t - 2)^2$
120. $(3pq - 7)^2$
121. $(4x - y)^2$
122. $(2r^2 - 1)(2r^2 + 1)$
123. $(10 + 3v)(10 - 3v)$
124. $(\tfrac{1}{2}y - 4y)$

In problems 125–138, divide as indicated.

125. $40x^2y^3$ by $8xy$
126. $50m^4n^3p^2$ by $-25m^2p$
127. $24u^3v^5 - 30u^5v^2$ by $6u^2v$
128. $15x^8y^7z^5 - 35x^6y^5z^4$ by $5x^4yz^3$
129. $16t^7 - 12t^3 + 4t$ by $4t^3$
130. $7z^3 - 14y^2z + 21y^4$ by $7y^2z^4$
131. $x^2 + 8x - 9$ by $x - 1$
132. $4y - 21 + y^2$ by $y + 7$
133. $2m^2 - 5m + 12 + m^3$ by $4 + m$
134. $6z^3 + 3z + 2 - 7z^2$ by $1 + 3z$
135. $x^4 + 6x^3 - 3x^2 + 4x + 3$ by $x^2 - x + 1$
136. $2y^4 - y^3 - 6y^2 + 7y - 2$ by $2y^2 - 3y + 1$
137. $x^5 + y^5 + xy^4 + 4x^3y^2 + x^2y^3 + 4x^4y$ by $y^2 + x^2 + 2xy$
138. $u^6 - v^6$ by $u^2 - v^2$

CHAPTER 2 TEST

1. Find the value of each expression.
   (a) -3^4
   (b) 7^0
   (c) 4^{-2}
   (d) $(-2)^2$

2. Write each expression in equivalent exponential form.
   (a) $3 \cdot x \cdot x \cdot x \cdot y \cdot y \cdot y \cdot y$
   (b) $2 \cdot x \cdot y \cdot y + 3 \cdot x \cdot x \cdot x \cdot y$

3. Identify the polynomial as a monomial, binomial, or trinomial. Find the degree of the polynomial and list the numerical coefficients.
   (a) $3x^2 + 5$
   (b) $-7x^3y^2$
   (c) $2x^3 - 4x^2 + 7x$

4. Evaluate $2x^2 - 3xy + y^3$ for $x = -1$ and $y = -2$.

5. Evaluate the expressions for the given values.
   (a) $\frac{5}{9}(F - 32)$, when $F = 59°$
   (b) Prt, when $P = \$160$, $r = 6\%$, and $t = 3$ years.

6. Perform the indicated operations.
   (a) $(3x^2 - 2x + 7) + (x^2 + 5x - 11)$
   (b) $(4y^2 + 5y - 1) - (-2y^2 + 3y + 2)$
   (c) $(5m^2n + 2mn^2) + (6mn - 2m^2n) - (-mn^2 + 2mn + m^2n)$

7. Rewrite each number in scientific notation.
   (a) 438.5
   (b) 0.000729

8. Rewrite each number is expanded form.
   (a) 2.78×10^4
   (b) 1.98×10^{-3}

9. Simplify and write the answer in scientific notation. Round off the answer to four decimal places.
   $$\frac{(1.7 \times 10^{-6})(2.3 \times 10^3)}{(3.1 \times 10^5)}$$

10. Find each product and simplify.
    (a) $(-3x^2y)(4xy^4)$
    (b) $-2t^2(3t^4 - 4t^3 + 2t)$
    (c) $(2x - y)(x^2 - 3xy + y^2)$

11. Find each product by the FOIL method.
    (a) $(x + 3)(x + 8)$
    (b) $(x - 5)(x + 2)$
    (c) $(3x + 5y)(2x - 4y)$

12. Use special products to perform each multiplication.
    (a) $(x - 7)(x + 7)$
    (b) $(3z + 2)^2$
    (c) $(u - 4v)^2$

13. Divide as indicated.
    (a) $24u^3v^2w^4$ by $-6uvw^2$
    (b) $16x^4y - 8x^3y^2$ by $4x^2y$
    (c) $x^3 + 5x^2 + 2x - 8$ by $x + 2$

14. Use the properties of exponents to simplify the given expressions. Write the answers with positive exponents.
    (a) $(x^3)^4$
    (b) $(-y)^3 \cdot (-y^5)$
    (c) $\dfrac{4x^4}{2x^7}$
    (d) $\left(\dfrac{2x}{-y}\right)^3$
    (e) $(2x^3y^2)^{-2}$
    (f) $\dfrac{(x^{-3}y^2)^4}{(x^2y^3)^{-2}}$

15. The length of a side of a square is $3x + 4$ units. Find an expression for its area in terms of x.

Chapter 3

LINEAR EQUATIONS AND INEQUALITIES IN ONE VARIABLE

3.1 Linear Equations in One Variable
3.2 Literal Equations and Formulas
3.3 Applied Problems
3.4 Linear Inequalities

3.1 Linear Equations in One Variable

An **equation** is a statement that two mathematical expressions are equal. Substituting a particular number for the variable produces an equation that is either true or false. If a true statement results from such a substitution, we say the number **satisfies** the equation and this number is called a **solution** or a **root** of the equation. For instance, 3 is a solution of the equation $4x + 7 = 19$, because the substitution 3 for x makes the equation a true statement. That is,

$$4(3) + 7 = 12 + 7 = 19$$

But -4 is not a solution, because

$$4(-4) + 7 = -16 + 7 = -9 \neq 19$$

Two more equations are said to be **equivalent** if they have *exactly* the same solutions. For example, the following equations:

$$4x + 7 = 19$$
$$4x = 12$$
$$x = 3$$

are equivalent because they all have the same solution, namely, 3.

We can change an equation $P = Q$ into an equivalent one by performing any of the following operations.

Chapter 3, except section 3.4, from *Intermediate Algebra,* 5th edition by M.A. Munem and C. West. Copyright © 2000 by Kendall/Hunt Publishing Company. Reprinted by permission.

OBJECTIVES

1. Solve Linear Equations
2. Solve Linear Equations Involving Fractions
3. Classify Identities and Inconsistent Equations
4. Use Technology Exploration

Property	Let $P = Q$, then	Illustration Let $x = 3$
1. Addition (Subtraction) Property Add (subtract) the same quantity on both sides of the equation.	(a) $P + R = Q + R$ (b) $P - R = Q - R$	$x + 7 = 3 + 7$ $x - 7 = 3 - 7$
2. Multiplication (Division) Property Multiply (divide) both sides of the equation by the same nonzero quantity.	(a) $PR = QR$, $R \neq 0$ (b) $\dfrac{P}{R} = \dfrac{Q}{R}$, $R \neq 0$	$5x = (5)(3)$ $\dfrac{x}{6} = \dfrac{3}{6}$
3. Symmetric Property Interchange the two sides of the equation.	$Q = P$	$3 = x$

To **solve** an equation means to find all of its solutions. The usual method for solving an equation is to write a sequence of equations, starting with the given one, in which each equation is equivalent to the previous one. The last equation should express the solution directly.

Solving Linear Equations

Each of the following equations,

$$5x - 3 = 7, \qquad 5 + 4x = 33,$$
$$4 + 12m = 5m - 10, \qquad 5 + 8(x + 2) = 23 - 2(2x - 5)$$

and

$$\frac{x}{7} - \frac{x-1}{2} = \frac{11}{4}$$

contain one variable, and each variable has degree one. They are examples of **linear** or **first degree equations.** In this section, we consider linear equations in one variable involving a polynomial of degree one. For instance, we solve the linear equation $5x - 3 = 7$ as follows:

$$\begin{aligned} 5x - 3 &= 7 & &\text{Start with the given equation} \\ 5x - 3 + 3 &= 7 + 3 & &\text{Use the Addition Property} \\ \frac{5x}{5} &= \frac{10}{5} & &\text{Use the Division Property} \\ x &= 2 & &\text{An equivalent equation with an obvious solution} \end{aligned}$$

Check:

If $x = 2$, then
$$5(2) - 3 = 10 - 3$$
$$= 7.$$

The solution is 2 because it makes the equation true.

It should be noted that this linear equation has only *one* solution; that is, no other value of x will make the original equation true.

In general, we may solve linear equations in one variable by using the following strategy.

STRATEGY FOR SOLVING LINEAR EQUATIONS

1. **COMBINE**
   (a) Multiply each side of the equation by the least common denominator (LCD) if fractions exist in the equation.
   (b) Remove all parentheses or grouping symbols on each side of the equation.
   (c) Combine like terms on each side of the equation.

2. **REARRANGE**
   Apply the addition (or subtraction) property to collect all terms containing the variable for which we are solving on one side of the equation and all other terms on the other side.

3. **ISOLATE**
   Divide each side of the equation by the coefficient of the variable.

4. **CHECK**
   Substitute the solution into the original equation to verify that the equation is true.

Example 1 Solving a Linear Equation
Solve each equation.

(a) $5 + 4x = 33$ (b) $4t + 7 = 3t + 5$

Solution (a)
$$5 + 4x = 33 \quad \text{Given}$$
$$5 + 4x - 5 = 33 - 5 \quad \text{Rearrange}$$
$$4x = 28$$
$$\frac{4x}{4} = \frac{28}{4} \quad \text{Isolate}$$
$$x = 7$$

Check: If $x = 7$, then $5 + 4(7) = 5 + 28 = 33$.
So that 7 is the solution and the only solution since this is a linear equation.

(b)
$$4t + 7 = 3t + 5 \quad \text{Given}$$
$$4t + 7 - 7 = 3t + 5 - 7 \quad \text{Subtract}$$
$$4t = 3t - 2 \quad \text{Rearrange}$$
$$4t - 3t = -2 \quad \text{Subtract}$$
$$t = -2$$

Check: If $t = -2$, then

Left Side	Right Side
$4(-2) + 7 =$	$3(-2) + 5 =$
$-8 + 7 = -1$	$-6 + 5 = -1$

Note that the "sides" of an equation are separated by the "=" sign.

Example 2 Using the Distributive Property to Solve an Equation
Solve the equation

$$5 + 8(x + 2) = 23 - 2(2x - 5)$$

Solution We begin by using the distributive property and combining like terms.

$$5 + 8(x + 2) = 23 - 2(2x - 5) \quad \text{Given}$$
$$5 + 8x + 16 = 23 - 4x + 10 \quad \text{Apply the distributive property}$$
$$8x + 21 = -4x + 33 \quad \text{Combine like terms}$$
$$8x + 4x + 21 - 21 = -4x + 4x + 33 - 21 \quad \text{Add } 4x - 21 \text{ to each side}$$
$$12x = 12 \quad \text{Combine like terms}$$
$$x = 1 \quad \text{Divide each side by 12}$$

Check: Let $x = 1$, then

Left side	Right side
$5 + 8(1 + 2) =$	$23 - 2(2(1) - 5) =$
$5 + 8(3) =$	$23 - 2(-3) =$
$5 + 24 = 29$	$23 + 6 = 29$

Example 3 Solving an Equation Involving Decimals
Solve the equation
$$1.3(2 - 5.2y) = 0.2 - (1.76y - 3.4)$$

Solution Here we can use a calculator for some computations.

$1.3(2 - 5.2y) = 0.2 - (1.76y - 3.4)$ Given
$2.6 - 6.76y = 0.2 - 1.76y + 3.4$ Apply the distributive property
$-6.76y + 1.76y = 0.2 + 3.4 - 2.6$ Add $1.76y$ and -2.6 to each side
$-5y = 1.00$ Combine like terms
$y = -0.2$ Divide each side by -5

If an equation involves decimals, the solution should be expressed in decimals.

A check will show that -0.2 is the solution.
We recommend that the reader check the answer.

Solving Linear Equations Involving Fractions

We indicated in step 1 of our strategy that when an equation contains fractions, we multiply each side of that equation by the least common denominator (LCD) to clear the fractions. This process produces an equivalent equation with integer coefficients which is easier to solve.

Example 4 Solving Equations Involving Fractions
Solve each equation.

(a) $\dfrac{2}{3}x + \dfrac{1}{2} = \dfrac{5}{6}$

(b) $\dfrac{y - 3}{4} - \dfrac{y - 2}{3} = \dfrac{y - 11}{6}$

Solution To find an equivalent equation without fractions, multiply each side of the equation by the LCD.

(a)
$\dfrac{2}{3}x + \dfrac{1}{2} = \dfrac{5}{6}$ Given

$6\left(\dfrac{2}{3}x + \dfrac{1}{2}\right) = 6\left(\dfrac{5}{6}\right)$ Multiply each side by 6, the LCD

$6\left(\dfrac{2}{3}x\right) + 6\left(\dfrac{1}{2}\right) = 6\left(\dfrac{5}{6}\right)$ Apply the distributive property

$4x + 3 = 5$ Multiply
$4x = 2$ Subtract 3 from each side
$x = \dfrac{2}{4} = \dfrac{1}{2}$ Divide each side by 4

Therefore, the solution is $\frac{1}{2}$. A check will show the answer is correct.

(b)

$$\frac{y-3}{4} - \frac{y-2}{3} = \frac{y-11}{6} \quad \text{Given}$$

$$12\left(\frac{y-3}{4} - \frac{y-2}{3}\right) = 12\left(\frac{y-11}{6}\right) \quad \text{Multiply each side by 12, the LCD}$$

$$12\left(\frac{y-3}{4}\right) - 12\left(\frac{y-2}{3}\right) = 12\left(\frac{y-11}{6}\right) \quad \text{Apply the distributive property}$$

$$3(y-3) - 4(y-2) = 2(y-11) \quad \text{Multiply}$$

$$3y - 9 - 4y + 8 = 2y - 22 \quad \text{Apply the distributive property}$$

$$3y - 4y - 2y = -22 + 9 - 8 \quad \text{Add } -2y, 9 \text{ and } -8 \text{ to each side}$$

$$-3y = -21 \quad \text{Combine like terms}$$

$$y = \frac{-21}{-3} = 7 \quad \text{Divide each side by } -3$$

Thus, 7 is the solution. A check will show the answer is correct.

Classifying Identities and Inconsistent Equations

Each of the equations considered in Examples 1–4 were true for only one value of the variable. These linear equations are called **conditional equations.** There are two other possibilities; identities and inconsistent equations. An equation that is true for all real numbers is called an **identity.** For instance,

$$7x + 2 = 10x + 2 - 3x$$

is an example of an *identity.*

An equation that has no solution is called an **inconsistent equation.** The equation

$$4(5x + 1) = 26x + 5 - 6x$$

is *inconsistent* as can be seen by simplifying each side.

$$20x + 4 = 20x + 5$$

There are no values of x which would make this equation true.

Example 5 Classifying Equations
Determine which of the given equations is inconsistent and which is an identity.
(a) $3(4x + 5) = 4 + 12x$ (b) $2(x + 5) = 5x + 10 - 3x$

Solution (a) $12x + 15 = 4 + 12x$ Apply the distributive property
$15 = 4$ Subtract $12x$ from each side

Since $15 = 4$ is a false statement, the equation has no solution and it is called *inconsistent.*

(b) $2x + 10 = 5x + 10 - 3x$ Apply the distributive property
$2x + 10 = 2x + 10$ Combine like terms

Notice that any number we substitute for x in the original equation leads to a true statement. Therefore, the solution consists of all real numbers and the equation is called an *identity.*

Remember that 0 can be a solution to an equation. This is not the same as "no solution."

Using Technology Exploration

We can use a grapher to solve all the linear equations we have considered so far. This is accomplished by first writing an equivalent equation for each linear equation with all the variables and the constants on the right side and zero on the left. That is, by writing each equation of the form $0 = mx + b$, and then graphing the line $y = mx + b$.

G Example 6 Using a Grapher to Solve an Equation

Use a grapher to solve the equation

$$5 + 4x = 33.$$

Solution We begin by writing an equivalent equation and setting it equal to zero.

$$0 = 4x - 28$$

Then we graph the equation

$$y = 4x - 28$$

with a grapher (Figure 1). The solution of the equation $0 = 4x - 28$ is the x intercept of the graph since $y = 0$ in the original equation. It appears from the graph in Figure 1 that there is an x intercept at $x = 7$. We can TRACE and ZOOM to the x intercept of the graph. Thus, 7 is the solution to the original equation $5 + 4x = 33$. This method can be used with any conditional equation.

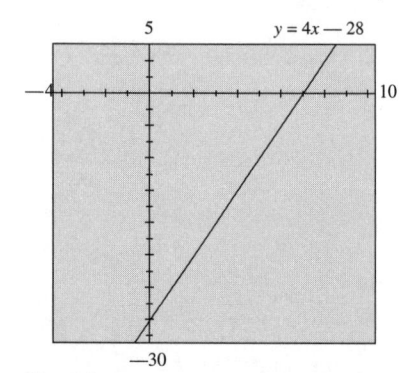

Figure 1

PROBLEM SET 3.1

Mastering the Concepts

In Problems 1–50, solve each equation and check the results.

1. (a) $2x - 5 = 11$
   (b) $2x + 7 = 7$
2. (a) $3w - 8 = 13$
   (b) $4 - 3w = 16$
3. (a) $19 - 15x = -131$
   (b) $-21 = 15x - 131$
4. (a) $104 - 12t = 152$
   (b) $12t + 104 = 144$
5. (a) $18 - x = 5 + 3x$
   (b) $12 - 3m = 4m + 26$
6. (a) $2 - 5y = -3y + 6$
   (b) $2x - 14 = 7 - x$
7. (a) $2u + 1 = 5u + 18$
   (b) $2x - 3x = 7 - 15x$
8. (a) $7y - 15 = 3y - 10$
   (b) $7 + 8x - 12 = 3x - 8 + 5x$
9. $3y - 2y + 7 = 12 - 4y$
10. $2t - 9t + 3 = 6 - 5t$
11. $1 - 2(5 - 2y) = 26 - 3y$
12. $7t - 3(9 - 5t) = -5t$
13. $8(5x - 1) + 36 = -3(x - 5)$
14. $2(1 + 2y) = 3(2y - 4)$
15. $2(x + 5) - (3 - x) = 15$
16. $7(x - 3) = 4(x + 5) - 47$
17. $6(c - 10) + 3(2c - 7) = -45$
18. $5(x - 3) - (x - 1) = -14$
19. $3(x - 2) + 5(x + 1) = -45$
20. $4(x + 4) - 5(2 - x) = 3(6x - 2)$
21. $2[3x - (x - 3)] = 4(x - 3)$
22. $16 - 2[4 - 3(1 - x)] = 14 - 5x$
23. $\dfrac{t}{6} - \dfrac{t}{7} = 5$
24. $\dfrac{y}{2} - \dfrac{y}{3} = 4$
25. $\dfrac{4x}{3} - 1 = \dfrac{5x}{6}$
26. $\dfrac{x}{4} + \dfrac{2x}{3} = \dfrac{35}{12}$
27. $\dfrac{5x - 15}{7} - \dfrac{x}{3} = \dfrac{2}{3}$
28. $\dfrac{3y}{5} - \dfrac{13}{15} = \dfrac{y + 2}{12}$
29. $\dfrac{t + 5}{7} + \dfrac{t - 3}{4} = \dfrac{5}{14}$
30. $\dfrac{w - 2}{3} + \dfrac{w + 1}{4} = -3$
31. $\dfrac{4x + 1}{10} = \dfrac{5x + 2}{4} - \dfrac{5}{4}$
32. $\dfrac{4x - 1}{10} - \dfrac{5x + 2}{4} = -3$
33. $\dfrac{2y + 3}{3} + \dfrac{3y - 4}{6} = \dfrac{y - 2}{9}$
34. $\dfrac{x - 2}{2} + \dfrac{x + 3}{4} = \dfrac{6x + 7}{6}$
35. $\dfrac{x}{7} - \dfrac{x - 1}{2} = \dfrac{11}{4}$
36. $2z - \dfrac{9z + 1}{4} = \dfrac{2}{3}$
37. $\dfrac{2t - 1}{4} + \dfrac{3t + 4}{8} = \dfrac{1 - 4t}{12}$
38. $\dfrac{y - 2}{4} - \dfrac{y + 5}{6} = \dfrac{5y - 12}{9}$
39. $\dfrac{y + 9}{4} - \dfrac{6y - 9}{14} = 2$
40. $\dfrac{8t + 10}{5} - \dfrac{6t + 1}{4} = \dfrac{3}{20}$
41. $\dfrac{3x - 2}{3} + \dfrac{x - 3}{2} = \dfrac{5}{6}$
42. $\dfrac{3u - 6}{4} - \dfrac{u + 6}{6} + \dfrac{2u}{3} = 5$
43. $0.7x + 3 = 0.5x + 2$
44. $1.5y + 4 = 1.2y - 2$
45. $x - 0.1x - 4.5 = 0.75$

46. $0.6x - 2.5 = 0.03x + 8.9$
47. $0.02(y - 100) = 62 - 0.06y$
48. $0.05(t - 100) = 0.2t - 35$
49. $0.5x - 0.1x - 0.2x = 0.05$
50. $0.75x - 0.5x = 0.625x + 0.75$

In Problems 51–53, determine whether each equation is an identity, a conditional equation or an inconsistent equation.

51. (a) $6(x + 2) = 6x + 12$ (b) $4(x + 1) - 4 = 4x$
52. (a) $5(x + 3) = 5x + 3$ (b) $5(x + 1) - 5 = 0$
53. (a) $4(w - 2) + 2 = 3w - 7 + w$
    (b) $16 - 2[4 - 3(1 - x)] = 14 - 6x$

Applying the Concepts

54. **Work Rate Problem:** Two workers can dig trenches for a sprinkler system in t hours, where t satisfies the equation
$$\frac{t}{15} + \frac{t}{10} = 1$$
Find the required time t

55. **Coin Problem:** A vending machine accepts nickels, dimes and quarters. When the coin box is emptied, the total value of the coins is found to be $24.15. Suppose that the box contains n nickels, $n + 5$ dimes and $n/2$ quarters, where n satisfies the equation
$$0.05n + 0.10(n + 5) + 0.25(n/2) = 24.15$$
Solve the equation for n, and then find the number of coins of each kind in the box.

56. **Temperature Conversion:** A rule of thumb for converting a temperature from degrees Celsius, C, to degrees Fahrenheit, F, is given by the equation
$$F = 2C + 30 \quad 0° \le C \le 100°$$
(a) If a bank thermometer registers 17° in the Celsius scale, what is the corresponding temperature in the Fahrenheit scale?
(b) If a thermometer registers 70° in the Fahrenheit scale, what is the corresponding temperature in the Celsius scale?

57. **Car Rental:** A car rental company charges for its midsize models $75 a week plus $0.15 for each mile driven for that week. The total charges C (in dollars) for a week is given by the equation
$$C = 75 + 0.15n$$
where n is the number of miles driven for a week.
(a) What is the total charge if a midsize rented car is driven 400 miles in a week?
(b) If the total charge for a week is $165, how many miles are driven for that week?

58. **Telephone Charges:** The cost of a long distance telephone call between two cities during business hours is $0.28 for the first minute and $0.15 for each additional minute or a fraction of a minute. The total cost C (in dollars) of long distance calls is given by the equation
$$C = 0.28 + 0.15(t - 1)$$
where t is the time for a call in minutes and $t > 1$.

(a) What is the total charge for a phone call that lasted 41 minutes?
(b) What is the length of a long distance phone call if the total cost is $3.28?

59. **Real Estate:** A real estate agent gets a commission of 7% of the sale price of a house. The net price N (in dollars) of a house after the commission is paid is given by the equation
$$N = p - 0.07p$$
where p (in dollars) is the sale price of the house.
(a) What is the net price of a house if it is sold for $140,000?
(b) If a couple wants to sell their house and have $148,800 after they pay the commission, what should be the selling price of the house?

Developing and Extending the Concepts

In Problems 60 and 61, solve each equation as follows:
(a) Multiply each side of the equation by a power of 10 that will eliminate the decimals.
(b) Without changing to integer form.
(c) Explain if you prefer method (a) or (b) and why.

60. $3y - 3(1.9y - 4.1) - 8 = 2(1.6y)$
61. $0.2(2x + 6.1) + 0.4(2.1x + 3) = 0.56$

G In Problems 62–65, use a grapher to solve each equation.
(a) Solve the equation first.
(b) Graph the given equation in y. The x intercept is the solution to the original equation.
(c) Simplify the right side of the given equation in y and graph this new expression.
(d) What do you see and why? If no mistake is made in simplifying, the given equation in y and your simplified equation in y have the same graphs.

62. $5x + 2(x + 1) = 23$
graph $y = 5x + 2(x + 1) - 23$
xMin $= -8$, xMax $= 8$, xScl $= 1$
yMin $= -30$, yMax $= 30$, yScl $= 1$

63. $7x - (3x - 4) = 12$
graph $y = 7x - (3x - 4) - 12$
xMin $= -8$, xMax $= 8$, xScl $= 1$
yMin $= -10$, yMax $= 10$, yScl $= 1$

64. $2(2x + 5) - 3x = 3$
graph $y = 2(2x + 5) - 3x - 3$
xMin $= -8$, xMax $= 8$, xScl $= 1$
yMin $= -10$, yMax $= 10$, yScl $= 1$

65. $5x - 4(x - 2) = 10$
graph $y = 5x - 4(x - 2) - 10$
xMin $= -8$, xMax $= 8$, xScl $= 1$
yMin $= -10$, yMax $= 10$, yScl $= 1$

66. Is the equation that expresses the associative property of addition
$$a + (b + c) = (a + b) + c$$
a conditional equation, an inconsistent equation, or an identity?

67. If the equation $a = b$ is true, what can you say about the equation $a + c = b + c$?

68. From geometry, two angles that are on opposite sides of the intersection of two lines are called *vertical angles*. It can be shown that vertical angles have the same measure.

Figure 2
(a) Use Figure 2a to find x.
(b) Use Figure 2b to find y.

69. From geometry, the sum of the interior angles of a triangle is 180°. Use Figure 3 with the given measurements to find t.

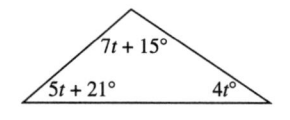

Figure 3

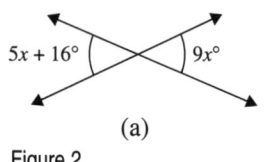

OBJECTIVES

1. Solve Literal Equations
2. Solve Formulas
3. Use Formulas to Solve Applied Problems
4. Use Technology Exploration

3.2 Literal Equations and Formulas

Thus far we have worked with equations that contain one variable, represented by a letter. To analyze a relationship between more than one variable or letter, we often use **literal equations.** In these equations, some of these letters represent variables, others represent constants. In mathematics, common examples of literal equations that express relationships between two or more quantities are called **formulas.** For example,

$$I = Prt, \quad F = \frac{9}{5}C + 32 \quad \text{and} \quad ax + by = c$$

are formulas for the simple interest, the conversion from Celsius to Fahrenheit, and the general form of a linear equation. These formulas describe real-life situations in mathematical terms.

Solving Literal Equations

Because literal equations and formulas are types of equations, we may use the equation solving strategy illustrated in section 2.1 to solve them. That is, we bring all terms containing the variable for which we wish to solve to one side of the equation, and all other terms to the opposite side. Then, we divide each side by the coefficient of that variable. As always, we must be careful not to divide by zero.

Example 1 Solving a Literal Equation
Solve each equation.
(a) $4x - a = x + 8a$ for x
(b) $3x + 2y - 6 = 0$ for y

Solution (a)
$4x - a = x + 8a$	Given
$4x - a - x = 8a$	Subtract x from each side
$4x - x = 8a + a$	Add a to each side
$3x = 9a$	Combine like terms
$x = 3a$	Divide by 3

Check: If $x = 3a$, then

Left side	Right side
$4(3a) - a =$	$(3a) + 8a =$
$12a - a = 11a$	$11a$

(b) $\quad 3x + 2y - 6 = 0 \quad$ Given
$\quad\quad\quad 2y - 6 = -3x \quad$ Subtract $3x$
$\quad\quad\quad\quad\;\; 2y = -3x + 6 \quad$ Add 6
$\quad\quad\quad\quad\;\;\; y = \dfrac{-3x + 6}{2} \quad$ Divide by 2
$\quad\quad\quad\quad\;\;\; y = \dfrac{-3x}{2} + 3$

Check: Let $y = \dfrac{-3x}{2} + 3$, then

Left side	Right side
$3x + 2\left(\dfrac{-3x}{2} + 3\right) - 6 =$	0
$3x - 3x + 6 - 6 = 0$	0

Example 2   Solving a Literal Equation Involving Fractions

Solve the equation $\dfrac{y - 3x}{b} = \dfrac{2x}{b} + y$ for y, $b \neq 0$

Solution

$\dfrac{y - 3x}{b} = \dfrac{2x}{b} + y \quad$ Given

$b\left(\dfrac{y - 3x}{b}\right) = b\left(\dfrac{2x}{b} + y\right) \quad$ Multiply each side by b, the LCD

$y - 3x = b\left(\dfrac{2x}{b}\right) + by \quad$ Apply the distributive property

$y - 3x = 2x + by \quad$ Multiply
$y - by - 3x = 2x \quad$ Subtact by
$y - by = 2x + 3x \quad$ Add $3x$
$y(1 - b) = 5x \quad$ Combine like terms and factor y

$y = \dfrac{5x}{1 - b} \quad$ Divide each side by $1 - b$, $b \neq 1$

A check will show the answer is correct.

Solving Formulas

Many applications from different sciences require the use of *formulas* for their solutions. For instance,

$$A = lw, \quad d = rt, \quad \text{and} \quad A = P(1 + rt)$$

are formulas for the area of a rectangle, the distance in terms of rate and time, and the accumulated amount of money at simple interest rate. To solve a formula for a specific variable or letter, we follow the same strategies as we do for solving equations.

Example 3 Solving a Formula

Solve the formula $A = lw$ for w.

Solution
$$A = lw \qquad \text{Given}$$
$$lw = A \qquad \text{Interchange the two sides}$$
$$w = \frac{A}{l} \qquad \text{Divide each side by } l$$

Example 4 Solving a Formula

The formula $A = P(1 + rt)$ gives the accumulated sum of money in dollars, when a principal of P dollars is invested at an interest rate r for t years. Solve the formula for t.

Solution
$$A = P(1 + rt) \qquad \text{Given}$$
$$P(1 + rt) = A \qquad \text{Interchange the two sides}$$
$$P + Prt = A \qquad \text{Apply the distributive property}$$
$$Prt = A - P \qquad \text{Subtract } P \text{ from each side}$$
$$t = \frac{A - P}{Pr} \qquad \text{Divide each side by } Pr$$

Using Formulas to Solve Applied Problems

The process of describing a real-world situation in mathematical terms is called **mathematical modeling.** This process may involve constructing an equation or collecting data. Once a model is established, it can be used to analyze the situation and make predictions. Some mathematical models are very accurate because they can be described by formulas. Others are too complicated to be precisely modeled. For instance, if a volume is being sought in a real-world situation, we write a *formula* for the volume and replace quantities such as length, width and height by letters or values given in the situation. These quantities should be represented by the same units of measure such as feet, meters, or miles. Recall from section R5 that *dimensional analysis* is needed to convert from one system of measurement into another.

Example 5 Modeling by Using a Formula for Volume

Each of the public tennis courts in a city measure 40 feet wide and 88 feet long. When these asphalt courts need to be resurfaced, sixteen 5 yard loads of asphalt are used for each court.

(a) Solve the volume formula $V = lwh$, for h (its depth), where l is the length of the court and w is the width.

(b) Determine the depth of asphalt for each court in inches.

Solution (a) First we solve the formula $V = lwh$ for h.
$$V = lwh \qquad \text{Given}$$
$$lwh = V \qquad \text{Interchange the sides}$$
$$h = \frac{V}{lw} \qquad \text{Divide each side by } lw$$

(b) Volume $= 16 \; \cancel{\text{loads}} \cdot \dfrac{5 \; \cancel{\text{yd}^3}}{1 \; \cancel{\text{load}}} \cdot \dfrac{27 \; \text{ft}^3}{1 \; \cancel{\text{yd}^3}} = 2160 \; \text{ft}^3$

With length $= 88$ ft and width $= 40$ ft, we have
$$h = \frac{V}{lw} = \frac{2160 \; \text{ft}^3}{40 \; \text{ft} \cdot 88 \; \text{ft}} = \frac{2160 \; \text{ft}^3}{3520 \; \text{ft}^2} = 0.61 \; \text{ft}$$

Therefore, the depth of the asphalt would be 0.61 foot rounded to two decimal places or $0.61 \; \text{foot} \cdot \frac{12 \; \text{inches}}{1 \; \text{foot}} = 7.32$ inches, rounded to two decimal places.

Example 6 Modeling by Using a Formula from Chemistry

The formula $F = \frac{9}{5}C + 32$ expresses temperature F in degrees Fahrenheit in terms of the temperature C in degrees Celsius.

(a) Solve the formula for C.

(b) Complete the following table then use the table to predict when the temperature is the same in both scales.

F°	−40	32	212
C°			

Solution (a)

$$F = \frac{9}{5}C + 32 \quad \text{Given}$$

$$F - 32 = \frac{9}{5}C \quad \text{Subtract 32}$$

$$\frac{5}{9}(F - 32) = C \quad \text{Multiply by } \frac{5}{9}$$

$$C = \frac{5}{9}(F - 32) \quad \text{Interchange}$$

(b) If we replace F by −40, 32, and 212 in the above result, we obtain the corresponding values for C. Notice as the table shows, the temperature reading is the same, at −40 on both scales.

F°1	−40	32	212
C°	−40	0	100

Using Technology Exploration

Graphers can be used to graph straight lines. Quite often people get the impression that if they have a grapher, they don't need to use algebra. This is not the case. For example, to graph the equation

$$3x + 2y - 6 = 0$$

in Example 1b by a grapher, it is always required (by the grapher) to solve this equation for y in terms of x. Thus it provides a formula for calculating the y coordinate for a given x coordinate as the next example illustrates.

Example 7 Using a Grapher

Use a grapher to graph the equation

$$3x + 2y - 6 = 0$$

Solution In Example 1b, we solved for y in terms of x and obtained

$$y = -\frac{3}{2}x + 3$$

The graph of this equation is shown in the viewing window in Figure 1. Notice that the x intercept of the graph is 2.

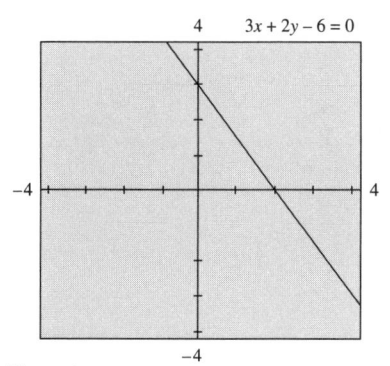

Figure 1

PROBLEM SET 3.2

Mastering the Concepts
In Problems 1–14, solve each equation for the indicated variable and check the results.

1. $6x + 7c = 37c$ for x
2. $4u - 19a = 5u$ for u
3. $ad + b = c$ for d
4. $ad - b = c$ for d
5. $12z - 4b = 6z - 7b$ for z
6. $13f + 6g = 8f - 9g$ for g
7. $4x - 3a - (10x + 7a) = 0$ for x
8. $27t - 4b - (15t - 6b) = 0$ for t
9. $5(4r - 3c) - 2(7r - 9c) = 0$ for r
10. $3(a - 2b) + 4(b + a) = 5$ for a
11. $5(mu - 2d) - 3(mu - 4d) = 9d$ for u
12. $9t + 7h - (11h - 13t) = 6t$ for t
13. $8(y - 2b) - 3(5y + 11b) = 0$ for y
14. $a(y - a) = -ab + 2b(y - b)$ for y

In Problems 15–24, solve each equation for y.

15. $\dfrac{x}{3} + \dfrac{y}{4} = 1$
16. $\dfrac{x}{7} - \dfrac{y}{4} = 1$
17. $\dfrac{x}{9} + \dfrac{y}{11} = 1$
18. $-\dfrac{x}{9} + \dfrac{y}{5} = 1$
19. $\dfrac{3x}{7} + \dfrac{2y}{9} = 1$
20. $\dfrac{7x}{11} - \dfrac{3y}{2} = 1$
21. $x = 0.01y - 0.03$
22. $8.7x = 3.9y + 9.6$
23. $\dfrac{x - 2y}{5} = \dfrac{3(y + 2)}{4} + 3$
24. $\dfrac{x + 3y}{5} = \dfrac{x - y}{2} + \dfrac{1}{2}$

In Problems 25–36, solve each formula for the indicated variable.

25. $A = P + Prt$, for P (Business)
26. $IR + Ir = E$, for r (Physics)
27. $\dfrac{T}{C} = \dfrac{x}{y}$ for C (Ecology)
28. $S(1 - r) = a$, for r (Mathematics)
29. $P = c - 30\%c$, for c (Business)
30. $BS = F + BV$, for B (Business)
31. $y = mx + b$, for x (Geometry)
32. $a = x - \dfrac{y^1}{y^2}$ for y_1 (Calculus)
33. $C = 2\pi r$, for r (Geometry)
34. $V = \pi r^2 h$, for h (Geometry)
35. $P = 2l + 2w$, for l (Geometry)
36. $A = 2\pi r^2 + 2\pi rh$, for h (Geometry)

In Problems 37 and 38, solve each formula for the indicated variable.

37. $V = \dfrac{Bh}{3}$ (volume of a pyramid)
    (a) For base B (b) For height h

38. $V = \dfrac{\pi r^2 h}{3}$ (volume of a cone)
    (a) For height h (b) For radius r

In Problems 39–42, solve each formula for the indicated variable.

39. $v = -32t + v_0$ (velocity in free-fall)
    (a) For initial velocity v_0 (b) For time t
40. $A = \left(\dfrac{b_1 + b_2}{2}\right)h$ (area of a trapezoid)
    (a) for b_1 (b) For height h
41. $1 = \dfrac{x}{a} + \dfrac{y}{b}$ (intercept form of a line)
    (a) For a, the x intercept (b) For b, the y intercept
42. $y - y_1 = m(x - x_1)$ (point-slope form of a line)
    (a) For slope m (b) For coordinate x_1

Applying the Concepts

43. **Circumference:** A runner jogged 5 miles in 35 laps around a circular track in a field house. What is the greatest distance measured straight across the track in feet? (Use $C = \pi d$, $\pi \approx 3.14$, and 5280 feet = 1 mile, round to one decimal place.)

44. **Area:** Suppose we wish to carpet the room shown in Figure 2. If the carpet, pad and labor total $18 per square yard, what is the cost?

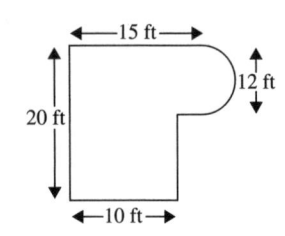

Figure 2

45. **Area:** You want to make a circular tablecloth with a ruffle along the bottom. The tablecloth is to hang 30 inches over the edge of a round table top (Figure 3).
    (a) Find the length of the ruffle. (Use $C = 2\pi r$)
    (b) If the ruffle costs $2.39 per yard, how much will it cost?

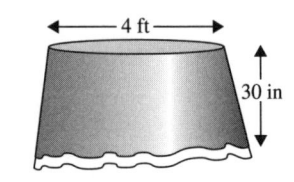

Figure 3

46. **Volume of a Teepee:** A tent is in the shape of a cone. Its volume is given by the formula
$$V = \dfrac{\pi r^2 h}{3}$$
where r is the radius of the circular base and h is the height of the tent (Figure 4).
    (a) Solve the formula for h.
    (b) What is the height of the tent if its volume is 1000 cubic meters and its radius is 8.7 meters.

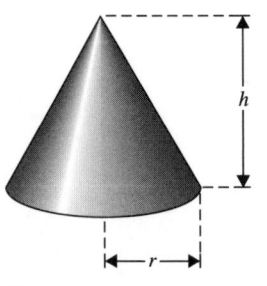

Figure 4

3.2 LITERAL EQUATIONS AND FORMULAS 179

47. Hot Tub Volume: A hot tub is in the shape of a right circular cylinder (Figure 5). It measures 6 feet in diameter and 4 feet in depth.
  (a) Find the volume V of the tub using $V = \pi r^2 h$ where r is the radius and h the depth.
  (b) Solve the formula for h.
  (c) Find the depth of a hot tub if its surface area is 266.9 square feet and its radius is 5 feet. (Use $S = \pi r^2 + 2\pi rh$)

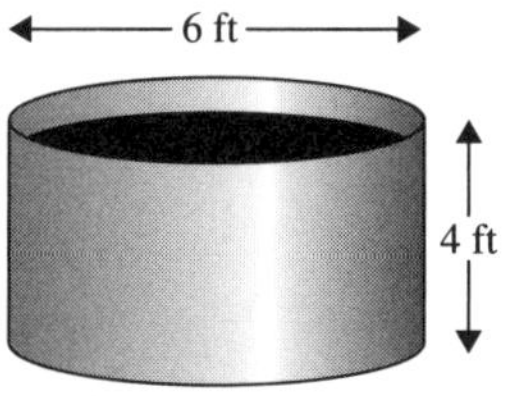

Figure 5

48. Depth of Aquarium: Suppose you have two fish aquariums. One measures 4 feet long and 18 inches wide with a depth of 2 feet (Figure 6). If you pour all the water into the other aquarium which is 7 feet long and 1.5 feet wide, what is the depth of water in the second aquarium (Use $V = lwh$)?

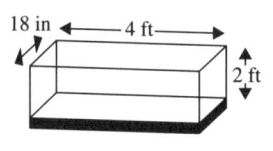

Figure 6

49. ▣ **Tax Credit:** A certain state offers to give a tax credit for installing solar energy panels. The credit is modeled by the equation
$$T = 0.15(C - 1000)$$
where T is the tax credit and C is the cost of the panels and $C > 1000$.
  (a) Solve this equation for C in terms of T.
  (b) Use part (a) to determine the cost of panels if a tax credit of $252 is desired.
  (c) Use a grapher to graph $T = 0.15(C - 1000)$ in the viewing window: $x\text{Min} = 1000$, $x\text{Max} = 2520$, $x\text{Scl} = 100$, $y\text{Min} = 0$, $y\text{Max} = 300$, $y\text{Scl} = 50$.
  (d) Use the grapher to predict the tax credit for panels costing $1760 ($C = 1760$).

50. Depreciation: For tax purposes, companies use the depreciation formula
$$D = \frac{C - S}{n}$$
where D (in dollars) is the yearly depreciation of an item, C is its original cost, S is its salvage value and n is its useful life.
  (a) Solve the formula for s.
  (b) What is the salvage value of a copy machine if its original cost was $20,000 and its yearly depreciation is $2000 over a life of 10 years?

51. Volume: The volume V of a tank in the shape of a right circular cylinder of height h feet with hemispheres at each end (Figure 7) is given by the formula
$$V = \pi r^2 \left(h + \frac{4r}{3}\right).$$
  (a) Solve the formula for h.
  (b) The external liquid hydrogen tank of

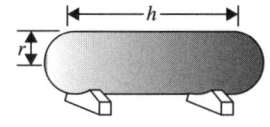

Figure 7

the space shuttle Endeavor is fabricated in this shape. Find the height h of the tank in meters if its total volume is 1174.86 cubic meters and the radius of each hemisphere is 4.2 meters. Round off the answer to two decimal places.

52. Storage: A cistern is to be fabricated in the shape of a right circular cylinder with a themisphere at the bottom (Figure 8). The volume V of the cistern is given by the formula
$$V = \pi r^2 h + \frac{2}{3}\pi r^2$$
  (a) Solve this formula for h.
  (b) Find h if r is 3 feet and V is 810 cubic feet.

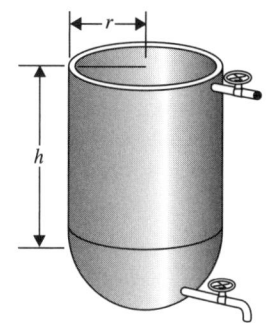

Figure 8

53. Postal Charges: In 1999 the U.S. Postal Service used the following charges to mail a first class item domestically: 33 cents for the first ounce and 22 cents for each additional ounce.
  (a) Write a formula for the total cost C (in dollars) for mailing a first class item whose weight is w (in ounces).
  (b) Solve the formula in part (a) for w in terms of C.
  (c) Use part (b) to determine the weight of a first class item if it costs $2.53 to mail it in the U.S.

▣ In Problems 54–56,
  (a) Solve each equation for y in terms of x.
  (b) Use a grapher to graph each equation in the given viewing window:
    $x\text{Min} = -10$, $x\text{Max} = 10$, $x\text{Scl} = 1$,
    $y\text{Min} = -10$, $y\text{Max} = 10$, $y\text{Scl} = 1$.
  (c) Use the ZOOM and TRACE features to find the x intercepts each equation.

54. (a) $3x - 2y - 6 = 0$ (b) $2x - 3y + 6 = 0$
55. (a) $y + 7 + 2(x - 5) = 0$ (b) $y - 8 - 4(x - 2) = 0$
56. (a) $\dfrac{x}{3} + \dfrac{y}{4} = 1$ (b) $\dfrac{y}{2} - \dfrac{3x}{5} = 1$

Developing and Extending the Concepts

In Problems 57–60, solve each equation for the indicated variable.

57. $(ax + 7)(ax - 3) = ax(ax + 1)$, for x
58. $A = 2lw + 2lh + 2wh$ for l
59. $\dfrac{a^2 y + 8}{b} = \dfrac{a^2 y - 10}{3b}$ for y
60. $\dfrac{b - x}{a} = \dfrac{2a + b}{4} - \dfrac{3x}{5}$ for x

61. Polygons: The number of diagonals d of a polygon is one half the product of the number of sides n and the number of sides minus 3.
  (a) Write the formula for finding the number of diagonals d.
  (b) Find the number of diagonals for a six sided polygon.

62. Polygons: The measure m (in degrees) of each interior angle of a regular polygon having n sides is the product of 180 and two less than the number of sides divided by the number of sides.
 (a) Write the formula to find the measure of each interior angle m.
 (b) Solve the formula in part (a) for the number of sides n.

63. Chain Length: The approximate length L of a chain joining two sprockets (toothed wheels) of a bicycle of radii r and R (Figure 9) is given by the formula

$$L = 2d + \frac{1}{2}(2\pi r) + \frac{1}{2}(2\pi R)$$

where d is the distance between the center of the two sprockets. Consider a bicycle having a chain length of 64 inches and a distance between the center of the sprockets of 24 inches.

 (a) Find r, rounded to one decimal place, if $R = 3.5$ inches.
 (b) Find R, rounded to one decimal place, if $r = 1.5$ inches.

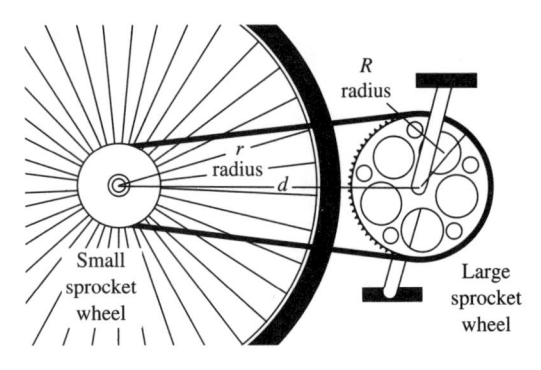

Figure 9

OBJECTIVES

1. Introduce a Problem Solving Format
2. Solve a Variety of Applied Problems
3. Use Technology Exploration

3.3 Applied Problems

In this section, we introduce a strategy for solving applied problems. These techniques were described in (the Hungarian mathematician) George Polya's book, "How to Solve It." The method focused on the problem solving process rather than just obtaining the correct answer.

Introducing a Problem Solving Format

One reason algebra is so important is that it can be used to solve problems in many disciplines, including business, engineering, economics, social sciences, and the physical sciences. Often the most challenging step in solving a word problem or application is translating the situation described in the problem into algebraic form. The complete result, including clearly stated assumptions and the appropriate formula, is a mathematical model. Polya developed the following three- step process for solving applied problems.

PROBLEM SOLVING PROCESS

Step 1. Make a plan
Step 2. Write the solution
Step 3. Look back

Step 1. Make a Plan

The following list is very useful in applying this plan.

(a) Identify the *given* (What information or assumptions are known).
(b) Identify the *goal* (What questions need to be answered).
(c) Draw a diagram whenever possible.
(d) Arrange and organize the given information in the form of a table or a chart whenever possible.
(e) Select one of the unknown quantities and label it with a symbol or a variable.
(f) Write algebraic relationships among the quantities involved, and then combine these relationships into a single equation.

Step 2. Write the Solution

(a) Solve the resulting equation in part (f) of step 1.
(b) Justify each step of the solution.

Step 3. Look Back

Interpret the solution in terms of the original problem to determine whether or not the answer is reasonable.

Example 1 Solving a Number Problem
Find three consecutive odd integers whose sum is 63.

Solution Make a Plan
The *given* is that there are three odd integers and they are consecutive. This means that these integers follow one another on the number line; that is, they are in order. Since they are consecutive odd integers they differ by 2. The *goal* is to find the three integers.

$$\text{Let } x = \text{first odd integer, then}$$
$$x + 2 = \text{second consecutive odd integer, and}$$
$$(x + 2) + 2 = x + 4 = \text{third consecutive odd integer.}$$

The following diagram is a visual description of the *given*.

First odd integer	+	Second odd integer	+	Third odd integer	+	Sum of the integers
x	+	$x + 2$	+	$x + 4$	=	63

Translating the problem into symbols, we have the equation

$$x + (x + 2) + (x + 4) = 63$$

Write the Solution

Now we solve the equation

$x + (x + 2) + (x + 4) = 63$	Given
$3x + 6 = 63$	Remove parentheses
$3x = 57$	Subtract 6 from each side
$x = 19$	Divide by 3
$x + 2 = 19 + 2 = 21$	
$x + 4 = 19 + 4 = 23$	

Therefore, the three integers are 19, 21, and 23.

Look Back
These three integers, 19, 21, and 23, are consecutive and odd. Also, their sum is 63. So our answers, 19, 21, and 23, are reasonable.

As you try to work applied problems, you will notice the frequent occurrence of certain words in the statements of these problems. Table 1 presents some common English phrases used in word problems together with their symbolic translations, where x and y represent variables.

Table 1

English Phrases	Algebraic Expressions
The sum of x and y	$x + y$
The difference of x and y	$x - y$
The product of x and y	xy
The quotient of x and y	x/y
The square of the difference of x and y	$(x - y)^2$
Greater than x by y	$x + y$
y less than x	$x - y$
Twice the sum of x and y	$2(x + y)$
The sum of twice x and y	$2x + y$
The sum of two consecutive integers	$x + (x + 1)$

Solving a Variety of Applied Problems

Our equation-solving skills enable us to tackle a wide range of applications.

Example 9   Solving a Geometry Problem

The length of a rectangular soccer field is 20 meters more than twice the width. The field's perimeter is 310 meters. What are the dimensions of the field?

Solution   Make a Plan

The key words in this problem are *rectangular* and *perimeter*. The two unknowns are the length, l, and width, w, of the rectangular field. The perimeter P of a rectangle is found using the formula $P = 2l + 2w$. Here we make a sketch to help us visualize the situation involved (Figure 1). Let w = the width of the field in meters, $l = 2w + 20$ = length of the field in meters. The following diagram is a visual description of the *given*.

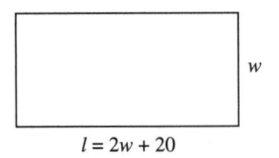

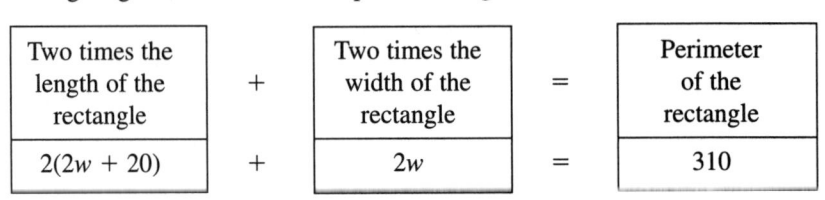

$l = 2w + 20$

Figure 1

Two times the length of the rectangle	+	Two times the width of the rectangle	=	Perimeter of the rectangle
$2(2w + 20)$	+	$2w$	=	310

Translating the problem into symbols, we obtain the equation

$$2(2w + 20) + 2w = 310$$

Write the Solution

$2(2w + 20) + 2w = 310$	Given
$4w + 40 + 2w = 310$	Apply the distributive property
$6w + 40 = 310$	Combine like terms
$6w = 270$	Subtract 40
$w = 45$	Divide by 6
$2w + 20 = 110$	

The width of the soccer field is 45 meters and the length is 110 meters.

Look Back

We can check this result in the words of the problem. The perimeter is

$$P = 2l + 2w$$
$$= 2(110) + 2(45)$$
$$= 220 + 90$$
$$= 310.$$

Also, the length of 110 meters is 20 meters more than twice the width of 45 meters.

Example 3 **Cutting a Length of Wire**

A wire is 150 centimeters long and is to be cut into three pieces. The second piece must be 6 centimeters shorter than three times the first, and the third piece must be 2 centimeters longer than two-thirds of the first. How long should the first piece be?

Solution **Make a Plan**

The *given* is a length of wire which is 150 centimeters and is being cut into three pieces. Our *goal* is to determine the length of the first piece. We assign a variable to the first piece.

Let x = the length of the first piece in centimeters.
$3x - 6$ = the length of the second piece in centimeters.
$\frac{2}{3}x + 2$ = the length of the third piece in centimeters.

The following diagram is a visual description of the *given*.

length of the first piece	+	Length of the second piece	+	length of the third piece	=	Overall length
x	+	$3x - 6$	+	$\frac{2}{3}x + 2$	=	150

Translating the problem into symbols, we have the equation

$$x + (3x - 6) + \left(\frac{2}{3}x + 2\right) = 150$$

Write the Solution

$$x + 3x - 6 + \frac{2}{3}x + 2 = 150 \qquad \text{Remove parentheses}$$

$$\frac{14}{3}x - 4 = 150 \qquad \text{Combine like terms}$$

$$\frac{14}{3}x = 154 \qquad \text{Add 4 to each side}$$

$$\left(\frac{3}{14}\right)\left(\frac{14}{3}x\right) = \left(\frac{3}{14}\right)(154) \qquad \text{Multiply both sides by } \frac{3}{14}$$

$$x = 33$$

The length of the first piece is 33 centimeters.

Look Back

To check the results, we observe that the lengths of the three pieces are as follows:

The length of the first piece is 33 centimeters.
The length of second piece is $3(33) - 6 = 99 - 6 = 93$ centimeters.
The length of third piece is $\frac{2}{3}(33) + 2 = 22 + 2 = 24$ centimeters.
The sum of the length of the three pieces is $33 + 93 + 24 = 150$ centimeters.

In many applied problems, rates, discounts, increases and decreases are often written as percents (%) which means *by the hundred*. Also, in most applications percents are expressed in decimals.

Example 4 **Solving an Investment Problem**

A financial planner advised some customers to invest part of their money in a safe money market fund that paid $4\frac{1}{4}\%$ simple interest and the rest in a risky investment that earned $13\frac{3}{4}\%$ simple interest. If the customers invested $24,000 that earned a total interest of $1960 the first year, how much money did they invest in each fund?

Solution **Make a Plan**

The formula for simple interest I is $I = Prt$ where P is the principal, r is the rate, and t is the time in years. The *given* is a total principal of $24,000, two rates of interest earning a total of $1960, and time is one year.

Let x = amount (in dollars) invested at $13\frac{3}{4}\%$
$24,000 - x$ = amount (in dollars) invested at $4\frac{1}{4}\%$

We organize the information in the following table.

	Principal (dollars)	rate (% per year)	time (year)	interest (dollars)
Risky Investment	x	0.1375	1	$0.1375x$
Safe Investment	$24,000 - x$	0.0425	1	$0.0425(24,000 - x)$

We know that the sum of interest earned by both investments is a total of $1960 for the year. The following diagram is a visual description of this *given*.

Interest from risky investment	+	Interest from safe investment	=	Total interest earned
$0.1375x$	+	$0.0425(24,000 - x)$	=	1960

Translating the problem into symbols, we have the equation

$$0.1375x + 0.0425(24,000 - x) = 1960$$

Write the Solution

$$0.1375x + 0.0425(24,000 - x) = 1960 \quad \text{Given}$$
$$0.1375x + 1020 - 0.0425x = 1960 \quad \text{Distributive property}$$
$$0.095x = 940$$
$$x = 9894.74$$
$$24,000 - x = 24,000 - 9894.74 = 14,105.26$$

Thus, $9894.74 was invested in the risky investment and $14,105.26 was invested in the safe investment.

Look Back

We check this solution by calculating the earnings from each investment.

$$\$9894.74 \text{ at } 13\tfrac{3}{4}\% \text{ yields } \$9894.74(0.1375) = \$1360.53$$
$$\$14{,}105.26 \text{ at } 4\tfrac{1}{4}\% \text{ yields } \$14{,}105.26(0.0425) = \$599.47$$

The total amount of interest is $1960. The total amount invested is

$$\$14{,}105.26 + \$9894.74 = \$24{,}000.$$

Example 5 Solving a Mixture Problem

A cereal processing plant introduces a recipe for a new breakfast cereal that combines rice crispies and wheat flakes. It is determined that a cup of rice crispies contains 2 grams of protein and a cup of wheat flakes contains 4.5 grams of protein. How much of each kind of cereal is included in one cup of the mixture that provides 3 grams of protein?

Solution Make a Plan

We are combining into a single blend two kinds of cereal that have different amounts of protein. We identify the unknowns by using the fact that the total amount of protein in one cup is 3 grams.

$$\text{Let } x = \text{ the amount of rice crispies in one cup.}$$
$$1 - x = \text{ the amount of wheat flakes in one cup.}$$

The table below summarizes the *given* information.

	Grams of protein in one cup of each cereal	Amount of cereal in one cup	Grams of protein
rice crispies	2	x	$2x$
wheat flakes	4.5	$1 - x$	$4.5(1 - x)$

The following diagram is a visual description of the *given*.

Grams of protein in the portion of rice crispies in one cup	+	Grams of protein in the portion of wheat flakes in one cup	=	Total grams of protein in one cup
$2x$	+	$4.5(1 - x)$	=	3

The equation that describes this model is:

$$2x + 4.5(1 - x) = 3$$

Write the Solution

$$\begin{aligned}
2x + 4.5(1 - x) &= 3 &&\text{Given} \\
2x + 4.5 - 4.5x &= 3 &&\text{Apply the distributive property} \\
-2.5x &= -1.5 \\
x &= \frac{-1.5}{-2.5} = \frac{3}{5} = 0.6
\end{aligned}$$

and $\quad 1 - x = 1 - 0.6 = 0.4$

The final mixture must contain 0.6 cup of rice crispies and 0.4 cup of wheat flakes.

Look Back

The total amount of protein in one cup is:

$$2(0.6) + 4.5(0.4) = 3$$

These numbers check in the original problem.

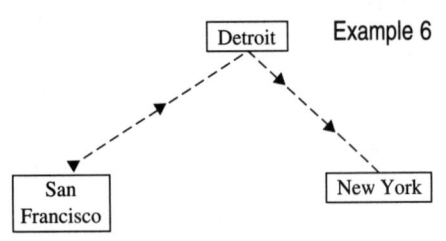

Example 6 **Solving a Rate Problem**

On a recent trip across the country, a jetliner flew from San Francisco to Detroit at an average speed of 500 miles per hour. It then continued to New York at an average speed of 550 miles per hour. If the entire trip covered 3075 miles and the total flying time was 6 hours, what was the distance of each leg of the trip?

Solution **Make a Plan**

Motion problems are based on the distance formula $d = rt$ where d is the distance, r is the average rate, and t is time. This trip consists of two legs: San Francisco to Detroit and Detroit to New York.

Let t = time (in hours) traveled on the first leg,
and $6 - t$ = time (in hours) traveled on the second leg

For each leg of the trip, we write expressions to represent rate, time and distance. This information is summarized in the following table.

	rate r (miles per hour)	time t (hours)	distance $d = rt$ (miles)
first leg of the trip	500	t	$500t$
second leg of the trip	550	$6 - t$	$550(6 - t)$

The following diagram is a visual description of the *given*.

Distance from San Francisco to Detroit	+	Distance from Detroit to New York	=	Total Distance
$500t$	+	$550(6 - t)$	=	3075

Since the total distance traveled is 3075 miles, we obtain the equation

$$500t + 550(6 - t) = 3075$$

Write the Solution

$500t + 550(6 - t) = 3075$ Given
$500t + 3300 - 550t = 3075$ Apply the distributive property
$-50t = -225$
$t = 4.5$

Substituting 4.5 for t, yields

$$500t = 500(4.5) = 2250 \text{ mi.}$$
$$550(6 - t) = 550(6 - 4.5) = 825 \text{ mi.}$$

Therefore, the first leg was 2250 miles long; the second was 825 miles.

Look Back

As a check, notice that the time spent traveling from San Francisco to Detroit is 4.5 hours. To find this distance, substitute the values of r and t in the equation $d = rt$, so that we have

$$d = 500(4.5) = 2250 \text{ miles}$$

The time spent traveling from Detroit to New York is $6 - 4.5 = 1.5$ hrs. Thus, the distance is

$$d = 550(1.5) = 825$$

Total distance traveled is $2250 + 825 = 3075$ miles.

Examples 4, 5, and 6 are *analogous;* they are solved using the same mathematical techniques. An important problem solving strategy is recognizing when a technique used to solve one problem can be used to solve another seemingly different, but analogous problem.

Using Technology Exploration

We can use a grapher to solve applied problems involving linear equations in one variable. We suggest the following guidelines for solving these applied problems with a grapher:

1. Use step 1: "make a plan."
2. Obtain a linear equation from step 1, and then write the equation in the form: $0 =$ expression and graph $y =$ expression.
3. Set the proper viewing window on your grapher to include all the important parts of the graph.

Let us solve the equation in example 2 with a grapher.

G Example 7 Using a Grapher

Use a grapher to solve the equation $2(2w + 20) + 2w = 310$.

Solution To graph this equation, we first rewrite it in the simpler form,

$3w + 20 = 155$ Divide each term by 2
$3w - 135 = 0$ Zero form

then graph $y = 3w - 135$. The viewing window in Figure 2 shows a portion of the graph where the line crosses the w axis at 45, which represents the solution of the equation. This can be confirmed by using the ZOOM and TRACE keys repeatedly.

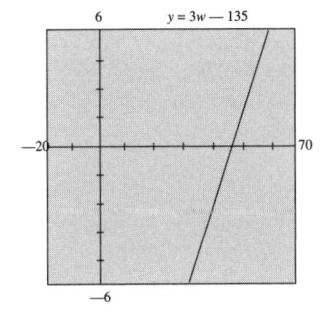

Figure 2

PROBLEM SET 3.3

Mastering the Concepts

1. **Number Problem:** Find three consecutive even integers whose sum is 60.
2. **Number Problem:** (a) Find a number such that three more than twice the number is 57.
   (b) Find a number such that twice the sum of three and the number is 57.
3. **Number Problem** Find three consecutive even integers such that twice the sum of the first and third is twelve more than twice the second.
4. **Age Problem:** A mother is five times as old as her son. If twice the age of the son is 3 years less than half of the age of the mother, how old are they?

5. **Number Problem:** Find three consecutive even integers such that twice the sum of the first and third is four more than three times the second.

6. **Home Address:** Four houses on one side of a street have addresses that are consecutive odd numbers. Find each address if the sum of these numbers is 23,672.

7. **Swimming Pool Dimensions:** The length of a rectangular swimming pool is 2 feet more than twice its width (Figure 3). The perimeter of the pool is 94 feet. What are the length and width of the pool?

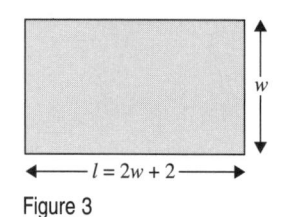

Figure 3: $l = 2w + 2$, w

8. **Rectangular Dimensions:** A carpet layer wishes to determine the cost of carpeting a rectangular room whose perimeter is 52 feet. Suppose that the length of the room is 4 feet more than its width. Find
   (a) the length and width of the room;
   (b) the cost of the carpet if it costs $18.95 per square yard.

9. **Plot Dimensions:** A rectangular plot of farmland is bounded on one side by a river and the other three sides by a single-strand fence 797 meters long. What is the length and the width of the plot if its length is 1 meter more than twice its width. No fence is used on the river side (Figure 4).

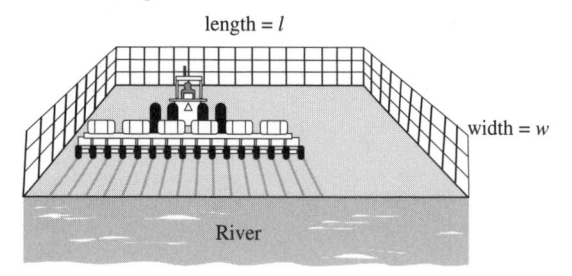

Figure 4

10. **Playground Dimensions:** A recreation department plans to build a rectangular playground enclosed by 1000 meters of fence. If the length of the playground is 4 meters less than three times the width, find the length and the width of the playground (Figure 5).

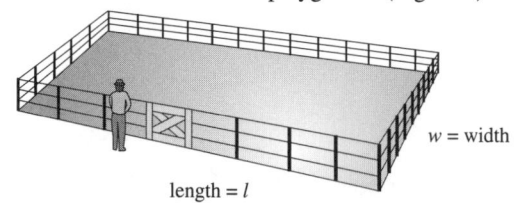

Figure 5

11. **Computer Monitor Dimensions:** The length of a rectangular computer monitor is 6 inches less than twice its width. The perimeter of the face of the monitor is 36 inches. Find the length and width of the monitor.

12. **Frame Dimensions:** The molding of a picture frame (the frame's perimeter) is 2.78 meters. The length of the finished frame is 0.1 meter more than twice the width (Figure 6). What are the dimensions of the frame?

13. **Plumbing:** A plumber wishes to cut a 26 foot length of pipe into two pieces. The longer piece needs to be 7 feet less than twice the length of the shorter piece. Where should the plumber cut the pipe?

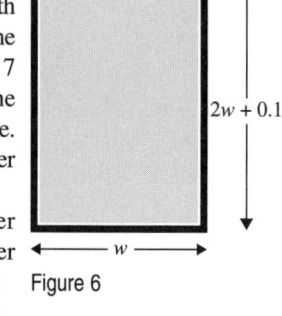

14. **Carpentry:** A carpenter needs two pieces of lumber which together are 21 feet long. The longer piece needs to be 1 foot longer than three times the length of the shorter piece. Find the length of each piece.

Figure 6: $2w + 0.1$, w

15. **Rope Cutting Problem:** A scout leader wishes to cut a rope for tents into three pieces whose lengths are consecutive odd integers. If the sum of the lengths of the first and the third piece is 58 feet, find the length of the middle piece.

16. **Cable Cutting Problem:** A cable installer has 105 meters of cable which is to be cut into three pieces. The second piece is 5 meters shorter than the first piece, and the length of the third piece is twice the length of the second. Find the length of each piece.

17. **Wire Cutting Problem:** A wire 180 centimeters long is to be cut into three pieces. The length of one piece is three times the length of the second. The length of the third piece is 12 centimeters shorter than twice the length of the second. Find the length of each piece.

18. **Fabric Cutting Problem:** A fabric store worker has a 57 yard bolt of cloth which is to be cut into three lengths. The length of the second piece is 5 yards shorter than twice the length of the first. The length of the third piece is 2 yards longer than twice the length of the first. Find the length of each piece.

19. **Investments:** Suppose an investor earned $2820 at the end of one year from a $32,000 investment in mutual funds. The investor bought two types of funds. One investment fund earned 9% annual interest, and the other earned 8.5% annual interest. How much was invested in each type of fund?

20. **Investments:** Suppose that $18,000 is invested in two types of bonds for one year. Part of the money is invested in a high-risk, high-growth fund that earned 24% annual interest. The rest is invested in a safe fund that earned 5% annual interest. If $2320 in interest is earned, how much was invested at each rate?

21. **Investments:** An inheritance is divided into two investments. One investment pays 7% annual interest whereas the second investment, which is twice as large as the first, pays 10% annual interest. If the combined annual income from both investments is $4050, how much money is the total inheritance?

22. **Assets of a Retirement Fund:** A portfolio manager purchased 10,000 shares in mutual stock funds and mutual bonds funds valued at $146,000 for a retirement fund. At the time of the purchase, the stock funds sold for $11 per share, while the bond funds sold for $17 per share. How many shares of each did the retirement fund purchase?

23. **Investments:** Suppose one investment in a money market fund earned 4.25% annual interest; the other investment in a stock fund earned 7.75% annual interest. Together the two investments totaled $3450. If a total of $230 for the year is earned, how much was invested in each fund?

24. **Auto Loan:** A customer purchased a car and financed $16,000. The customer borrowed part of the money from a bank charging 10% annual interest and the rest from a credit union at 8% annual interest. If the total interest for the year was $1390, how much was borrowed from the bank and how much was borrowed from the credit union?

25. **Juice Mixture:** A fruit juice company mixes pineapple juice that sells for $5.50 per gallon with 100 gallons of orange juice that sells for $3.00 per gallon. How much pineapple juice is used to make a pineapple-orange juice drink selling for $3.50 per gallon?

26. **Milk Mixture:** How many liters of whole milk containing 3.5% butterfat must be mixed with 3 liters of skim milk (containing no butter fat, 0% butterfat) to obtain a mixture containing 2% butterfat?

27. **Meat Mixture:** A meat market manager mixes hamburger having 30% fat content with hamburger that has 10% fat content in order to obtain 400 pounds of hamburger with 25% fat content. How much hamburger of each type should the manager use?

28. **Antifreeze Mixture:** A car radiator has a capacity of 8 quarts of coolant which is 20% antifreeze. A mechanic needs to drain part of the radiator fluid and replace it with 100% antifreeze to bring the coolant in the system to 60% antifreeze. How much of the original coolant should be drained and replaced by 100% antifreeze to achieve the desired mix?

29. **Gravel Driveway:** A farmer wishes to make a gravel driveway. The mixture needed is two parts pea gravel which costs $9.00 per ton and one part sand which costs $8.00 per ton. The gravel company gave a bid of $390. How much gravel and sand are they going to deliver?

30. **Seed Mixture:** A nursery has one kind of grass seed selling at 75 cents per pound and another kind of grass seed selling at $1.10 per pound. How many pounds of each kind should be mixed to produce 50 pounds of a mixture of seed that will sell for 90 cents per pound?

31. **Travel Distance:** A train travels through the mountains at an average speed of 35 miles per hour. It then continues on flat land at an average speed of 85 miles per hour. If the entire trip covers 1200 miles and each leg of the trip takes the same number of hours, how many miles is each leg?

32. **Travel Time:** Last weekend two college students were returning to school 200 miles away. During a blizzard, they were only able to average 25 miles per hour. For the rest of their trip, they averaged 55 miles per hour. If the entire trip took 4 hours, how long were they driving in the blizzard?

33. **Average Speed:** A cross country skier follows a trail to a nearby camp in 50 minutes. A snowmobile, averaging 16 miles per hour faster than the skier, starts from the same point on the trail and following the skier's track covers the same distance in 10 minutes. What was the average speed of the snowmobile?

34. **Travel Time:** On the first part of a 27.5 mile trip, the average speed was 48 miles per hour. Later due to traffic, the average speed was reduced to 35 miles per hour. If one spent five times as long on the first part as on the second part, what is the total time of the trip?

35. **Average Speed:** An Amtrak train leaves Las Vegas heading west towards Los Angeles. At the same time, 272 miles away a bus leaves Los Angeles traveling east along the same route towards Las Vegas. If the train averages 31 miles per hour faster than the bus and they pass each other in 2 hours, what are the average speeds of the train and the bus?

36. **Average Speed:** A helicopter travels 30 miles per hour faster than a speeding car. The car had a half hour head start, and both traveling the same route. If the helicopter overtakes the car in 1.5 hours, what is the speed of each?

Developing and Extending the Concepts

37. **Cash Register Change:** A fitness center usually needs three times as many $5 bills as $10 bills to transact its daily business. If the cash register contains $750 in fives and tens at the beginning of a business day, how many of each bill does the register contain?

38. **Coin Problem:** A vending machine accepts nickels, dimes and quarters. When the coins are emptied, the total value of the coins is found to be $24.15. Find the number of coins of each kind in the box, if there are twice as many nickels as quarters and 5 more dimes than nickels.

39. **Age Difference:** A man is 25 years older than his son. Fourteen years from now the man will be twice as old as his son will be. What are the present ages of the man and his son?

40. **Clothing Sale:** A clothing store has discounted the price of a sweater by 20%. If the sale price of the sweater is $49.95, what was the original price?

41. **Insurance Discount:** Some insurance companies reward students with safe driving records and good grades by discounting their annual premiums. If a student's annual premium is discounted by 15% to $874.65, what is the original annual premium?

42. **Computing Salaries:** An employee's new annual salary is $36,516, which includes a 5% pay raise and a 2.4% cost-of-living allowance. What was the employee's original salary?

43. **Tire Sale:** A tire was sold for $58.50 which included a 6% sales tax and an 11% federal tax. What was the price of the tire before the taxes?

44. **Consumer Cost:** A customer has $70 to spend in a grocery store. She spends $28.20 for canned goods. The remaining amount is needed to buy a total of 9 pounds of steak and chicken. If the cost per pound for steak and chicken is $6.95 and $2.80 respectively, how many pounds of each meat can she afford to buy?

45. **Tipping:** For groups of four or more diners, many restaurants automatically add a 15% gratuity as well as a tax of 8% to the original bill. If the bill (including gratuity and tax) for four people came to $193, how much was the tip?

46. **Credit Card Charge:** A customer carries an average daily balance on her credit card of $12.70 per day. The

customer is offered a card from Bank A with no annual fee and 18% finance charge. Bank B offers a card with 15% finance charge and $39 annual fee. Which card is more economical for the customer? What average daily balance would give the same total yearly charges?

47. **Aerobics:** A runner burns 3 times as many calories running as walking the same distance. Suppose a runner burns a total of 765 calories running 3 kilometers and walking 8 kilometers. At this rate, how many calories are burned by running 1 kilometer?

48. **Operating Budget:** Suppose 60% of the salary budget for a college goes to faculty salaries, 25% to administrative salaries, 10% to support staff salaries, and the remaining $1,820,000 to maintenance salaries. How much money goes to each group?

49. **Work Force:** When a new factory opened recently, twice as many men as women applied for work. Five percent of all the applicants are hired. If 3% of the men who applied were hired, what percentage of the women who applied were hired?

50. **Stock Prices:** Two different stocks were sold for $3000 each. On one stock the gain was 10% of its original value; on the other the loss was 10% of its original value. Determine the total gain or loss from both stocks.

51. **Automobile Maintenance:** The mileage on a car's odometer reads 6253.6. The next service is scheduled to take place at about 10,000 miles. If the car is driven 170 miles per week, in how many weeks will it need to be serviced?

52. **Bank Charges:** A bank charges its customers a general fee of $6 per month plus an additional fee of 10 cents per check written from a checking account. Over the past nine months, a customer was charged $62 in service fees. How many checks did the customer write during that period?

53. **Geology:** While some climbers began a hike up Mt. St. Helens, the temperature at sea level was 20°C. The temperature dropped 5°C with every 1000 meters of elevation gained. When the climbers reached the top of their ascent, the temperature was 7.5°C. How high had they climbed?

54. **Car Rental:** For one of its compact cars, a car rental company charges customers $33 per day. In addition, the company charges 25 cents per mile for each mile driven over 400 miles. A customer rented a compact car for four days and paid $186.50. How many miles did the customer drive?

55. **Car Racing:** After 125 miles of The Indianapolis 500 car race, Car A has a lead of 1650 feet over Car B. If Car A averages 185 miles per hour and Car B averages 192 miles per hour, how long will it take Car B overtake Car A?

56. **Energy Saving:** By installing a $50 thermostat that automatically reduces the temperature setting while at work and at night, a condominium owner hopes to cut the annual heating bill by 15%. If the monthly heating bill is $85 before the new thermostat, how long will it take to recover the cost of the thermostat?

57. **Factory Outlet Sales:** A retail factory outlet store sells running shoes for $45 a pair. At this price, the profit on each pair of shoes is 25% of the cost to the retailer. What is the retailer's profit on each pair of shoes?

58. **Metal Alloys:** An archaeologist discovers a crown weighing 800 grams in the tomb of an ancient Egyptian king. Evidence indicates that the crown is made of a mixture of gold and silver. Gold weighs 19.3 grams per cubic centimeter; silver weighs 10.5 grams per cubic centimeter. It is found that the crown weighs 16.2 grams per cubic centimeter. How many grams of gold does the crown contain?

59. **Geometry:** The sum of the measures of the interior angles of a triangle is 180°. Is it possible that the three angles can have the following measurements? If yes, find the measures of the angles. If no, explain why.
   (a) Three consecutive integers. (Hint: Let the first angle measure x degrees.)
   (b) Three consecutive even integers. (Hint: Let the first angle measure $2b$ degrees.)
   (c) Three consecutive odd integers. (Hint: Let the first angle measure $2c + 1$ degrees.)

60. **Norman Doors and Win-dows:** The Norman style of architecture is characterized by massive construction and rectangular windows and doors surmounted by semi-circular arches (Figure 7). If such a door has an area of 29.4 square feet and a width of 3.3 feet, find its height h rounded to one decimal place.

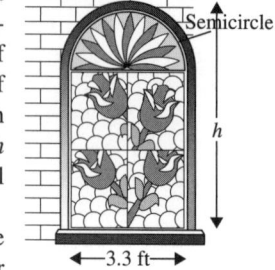

61. **Basketball Court:** The perimeter of a rectangular basketball court is 80 meters. Its length is 2 meters shorter than twice its width. Find the dimensions of the court.

Figure 7

62. **Fencing Cost:** The length of a rectangular yard is 10 meters less than 4 times its width. The total cost for a fence to enclose the entire yard is $2706. Find the dimensions of the yard if the fence costs $8.20 per meter.

63. **Package Design:** An open box is to be constructed by removing equal squares, each of length x inches, from the corners of a piece of cardboard and then folding up the sides (Figure 8). If the cardboard measures 12 inches by 16 inches, find the volume of the box if its length is twice its width.

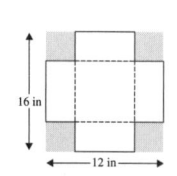

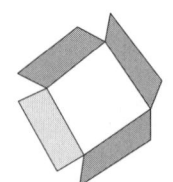

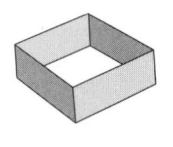

Figure 8

64. **Biology:** A biologist wishes to determine the volume of blood in the circulatory system of an animal. She injects 6 milliliters of a 9% solution of a biologically inert chemical and waits until it is thoroughly mixed with the animal's blood. Then she withdraws a small sample of blood and determines that 0.03% of the sample consists of the biologically inert chemical. Find the original volume of blood in the circulatory system.

OBJECTIVES

1. Introduce Interval Notation
2. Solve Linear Inequalities
3. Solve Compound Inequalities
4. Solve Applied Problems
5. Use Technology Exploration

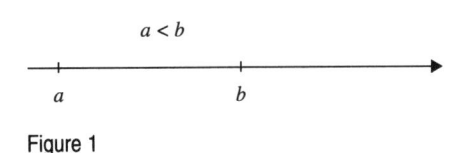

3.4 Linear Inequalities

So far we have used linear equations to model certain types of applied problems. Certain relationships among quantities cannot always be described by linear equations. For instance, suppose that a mathematics department wishes to purchase a number of graphing calculators, where the cost of each calculator is $60 plus 8% sales tax. Also suppose that the delivery charge for the entire shipment is $50. To determine the number of calculators to be purchased in order to allocate "at most" $3200 for the entire order, we need to solve a linear inequality. In this section, we study such *linear inequalities*, which contain one or more variable terms separated by one or more of the following symbols:

$>$ greater than $\qquad$ $<$ less than
$\geq$ greater than or equal to $\qquad$ $\leq$ less than or equal to

Recall that if point a lies to the left of point b on a number line (Figure 1), we say that b is *greater than* a (or equivalently, that a is *less than* b) and we write $b > a$ (or $a < b$). In other words, $b > a$ (or $a < b$) means that $b - a$ is positive. The symbols $<$, $>$, $\leq$ and $\geq$ are called **inequality signs** and the expressions on the left and right of these signs are called the **sides** or **members** of the inequality. The inequalities $b < a$ or $a > b$ are said to be **strict** because they do not allow the possibility of equality. However, the inequalities $b \leq a$ and $a \geq b$ which allow the possibilities of equality are said to be **nonstrict**. If the inequality $x \leq a$ is true when x is replaced by a or any number less than a, we say that these replacements **satisfy** the inequality. The set of all points on a real number line that satisfy an inequality is called the **graph** of the inequality.

Figure 1

Introducing Interval Notation

Certain sets of real numbers, called *intervals*, have an important role to play in the study of inequalities. The basic types of intervals are summarized in Table 1. (Note that a *parenthesis* is used to graph inequalities with symbols $<$

Table 1

Interval Type	Inequality Notation	Interval Notation	Graph
Closed interval: Numbers between a and b, inclusive	$a \leq x \leq b$	$[a,b]$	
Open interval: Numbers between a and b	$a < x < b$	(a,b)	
Numbers greater than a	$x > a$	(a, ∞)	
Numbers less than b	$x < b$	$(-\infty, b)$	
Half-open interval: Numbers greater than or equal to a and less than b	$a \leq x < b$	$[a,b)$	
Numbers greater than a and less than or equal to b	$a < x \leq b$	$(a,b]$	
Numbers greater than or equal to a	$x \geq a$	$[a, \infty)$	
Numbers less than or equal to b	$x \leq b$	$(-\infty, b]$	

or >, and a *bracket* is used to graph inequalities with symbols ≤ or ≥.) In Table 1, we let a and b be real numbers such that $a < b$. The special symbols ∞ and $-\infty$, called **positive infinity** and **negative infinity,** are used to indicate that the interval extends indefinitely to the right or to the left.

Example 1 Graphing Inequalities

Graph each inequality and express each using interval notation.

(a) $-3 \leq x \leq 5$ (b) $x \geq 4$ (c) $x < -3/2$

Solution (a) The graph of this inequality is the set of all real numbers on the number line that are greater than or equal to -3 and less than or equal to 5 (Figure 2a). The interval notation is $[-3, 5]$.

(b) The graph of $x \geq 4$ is the set of all real numbers on the number line that are greater than or equal to 4 (Figure 2b). The interval notation is $[4, \infty)$.

(c) The graph of $x < -3/2$ is the set of all real numbers on the number line that are less than $-3/2$ (Figure 2c). The interval notation is $(-\infty, -3/2)$.

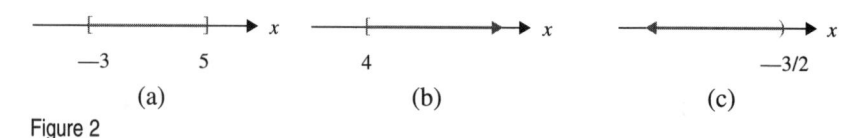

(a) number line with bracket at −3 and bracket at 5
(b) number line with bracket at 4 extending right
(c) number line with parenthesis at −3/2 extending left

Figure 2

Solving Linear Inequalities

Inequalities such as

$$x - 3 < 5,\ y + 2 \geq 7,\ \text{and}\ 2x + 3 < 11 - 2x$$

are examples of linear inequalities in one variable. We **solve** an inequality for the variable x by finding all values of x for which the inequality is true. Such values are called **solutions.** The set of all solutions of an inequality is called its **solution set.** Two inequalities are said to be **equivalent** if they have exactly the same solution set. To solve a linear inequality, we proceed in much the same way as in solving linear equations. Use the following **properties** of **inequalities** to complete a solution.

Properties of Inequalities	Equivalent Inequality	Illustration
1. Addition (Subtraction) Property Add (subtract) any real number to each side of an inequality.	If $a < b$, then $a + c < b + c$ $a - c < b - c$	Since $5 < 13$, then $5 + 7 < 13 + 7$ $5 - 7 < 13 - 7$
2. Multiplication (Division) Property Multiplying (dividing) each side of an inequality by a *positive* number, yields an equivalent inequality.	If $a < b$, then $ac < bc,\ c > 0$ $\dfrac{a}{c} < \dfrac{b}{c},\ c > 0$	$5 < 13$, then $5(7) < (13)(7)$ $\dfrac{5}{7} < \dfrac{13}{7}$
Multiplying (dividing) each side of an inequality by a *negative* number, yields an equivalent inequality with the order of the inequality reversed.	If $a < b$, then $ac > bc,\ c < 0$ $\dfrac{a}{c} > \dfrac{b}{c},\ c > 0$	$5 < 13$, then $5(-7) > (13)(-7)$ $\dfrac{5}{-7} < \dfrac{13}{-7}$

Each of these properties is true if the symbols $<$ and $>$ are replaced by $\leq$ and $\geq$, respectively. In addition a, b, and c can be either real numbers or algebraic expressions. Consider the effect of adding (subtracting) and multiplying (dividing) each side of the inequality $4 < 6$ by a positive number (see Table 2) or a negative number (see Table 3).

3.4 LINEAR INEQUALITIES

Table 2

	Add 2	Subtract 2	Multiply by 2	Divide by 2
Given	$4 < 6$	$4 < 6$	$4 < 6$	$4 < 6$
Result	$6 < 8$	$2 < 4$	$8 < 12$	$2 < 3$

Note that if $-3x > 0$, then $x < 0$, because the product of two negative numbers is positive. Thus, remember when you multiply or divide by a negative number, you must reverse the inequality symbol.

Table 3

	Add -2	Subtract -2	Multiply by -2	Divide by -2
Given	$4 < 6$	$4 < 6$	$4 < 6$	$4 < 6$
Result	$2 < 4$	$6 < 8$	$-8 > -12$	$-2 > -3$

Notice in Table 3, when we multiply or divide each side of the inequality $4 < 6$ by -2, the inequality symbol is reversed.

Example 2 Solving a Linear Inequality

Solve the inequality $2x + 3 < 11 - 2x$ and graph the solution.

Remember that $x < 2$ is equivalent to $2 > x$. Traditionally, we rewrite inequalities so that the variable is on the left side.

Solution

$2x + 3 < 11 - 2x$	Given
$4x + 3 < 11$	Add $2x$ to each side
$4x < 8$	Subtract 3 from each side
$x < 2$	Divide each side by 4

The solution set of the inequality consists of all real numbers less than 2. In interval notation, $(-\infty, 2)$ is the solution set. The graph of this solution set is in Figure 3.

To check, choose a value in the solution set; the inequality is true. Also choose a value not in the solution set; the inequality is false.

Check: If $x = 1$, then

Left side	Right side
$2(1) + 3 = 5$	$11 - 2(1) = 9$
$5 < 9$ true

Check: If $x = 3$, then

Left side	Right side
$2(3) + 3 = 9$	$11 - 2(3) = 5$
$9 < 5$ false

$x < 2$

Figure 3

Example 3 Solving a Linear Inequality

Solve the inequality $4(x - 2) \geq 3(x - 2) - 4$ and graph the solution.

Solution

$4(x - 2) \geq 3(x - 2) - 4$	Given
$4x - 8 \geq 3x - 6 - 4$	Apply the distributive property
$4x - 8 \geq 3x - 10$	Add like terms
$x - 8 \geq -10$	Add $-3x$ to each side
$x \geq -2$	Add 8 to each side

The solution set of the inequality consists of all real numbers greater than or equal to -2. In interval notation $[-2, \infty)$ is the solution set. The graph of this solution set is in Figure 4.

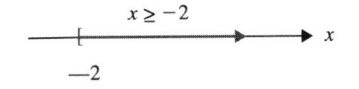

$x \geq -2$

Figure 4

Check: If $x = -1$, then

Left side	Right side
$4(-1 - 2) = -12$	$3(-1 - 2) - 4 = -13$
$-12 \geq -13$ true

Check: If $x = -3$, then

Left side	Right side
$4(-3 - 2) = -20$	$3(-3 - 2) - 4 = -19$
$-20 \geq -19$ false

Example 4 Solving a Linear Inequality

Solve the Inequality $\dfrac{2-x}{2} \le \dfrac{x+1}{3}$ and graph the Solution.

Solution

$\dfrac{2-x}{2} \le \dfrac{x+1}{3}$ Given

$6\left(\dfrac{2-x}{2}\right) \le 6\left(\dfrac{x+1}{3}\right)$ Multiply each side by 6, the LCD

$3(2-x) \le 2(x+1)$ Multiply

$6 - 3x \le 2x + 2$ Apply the distributive property

$6 - 5x \le 2$ Subtract $2x$ from each side

$-5x \le -4$ Subtract 6 from each side

$x \ge \dfrac{4}{5}$ Divide each side by -5 and reverse the inequality

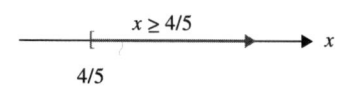

The solution set of the inequality consists of all real numbers greater than or equal to 4/5. In interval notation, [4/5,∞) is the solution set. The graph of this solution set is in Figure 5.

The check is left to the reader.

x ≥ 4/5

Figure 5

Solving Compound Inequalities

Two inequalities connected with the words "and" or "or" are called **compound inequalities.** Examples of these inequalities are:

$-1 \le x$ and $x \le 2$, $x \le -5$ or $x \ge 1$

The solution set of a compound inequality with the word "and" is the set of real numbers *common* to the solution sets of each inequality. For example, the solution set of

$-1 \le x$ and $x \le 2$

consists of all real numbers x that are greater than or equal to -1 *and* less than or equal to 2. That is, all real numbers between -1 and 2 including the end points. This is also written as

$-1 \le x \le 2.$

Graphing these inequalities provides a visual idea of the solutions. Figure 6a shows the graph of the inequalities $-1 \le x$ and $x \le 2$.

The solution set of a compound inequality containing the word "or" is the set of real numbers in the solution set of either or both of the two inequalities. For example, the solution set of

$x \le -5$ or $x \ge 1$

consists of all real numbers x that are less than or equal to -5, or all real numbers that are greater than or equal to 1 (Figure 6b).

Note that a compound inequality containing the word "or" may not be compressed as the "and" compound inequality is compressed.

$-1 \le x \le 2$ $x \le -5$ or $x \ge 1$

(a)   (b)

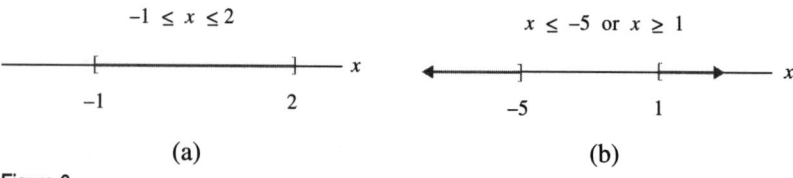

Figure 6

Example 5　Solving Compound Inequalities

Solve each compound inequality. Express the solution set in interval notation and graph.

(a) $-3 < 2x + 1 \leq 5$ 　　(b) $5x - 2 < -7$ or $5x - 2 > 8$

Solution　(a) Our goal is to isolate the variable x in the middle,

$-3 < 2x +$	$1 \leq 5$		Given
$-3 - 1 < 2x + 1 - 1 \leq 5 - 1$			Subtract 1 from all three parts
$-4 < 2x$	≤ 4		Combine like terms
$-2 < x$	≤ 2		Divide each part by 2

Therefore, the solution set consists of all real numbers greater than -2 and less than or equal to 2. In interval notation the solution set is $(-2, 2]$ and the graph is drawn in Figure 7a.

(b) We solve each inequality individually.

$5x - 2 < -7$	or	$5x - 2 > 8$	Add 2 to each side of both inequalities
$5x < -5$		$5x > 10$	Combine like terms
$x < -1$		$x > 2$	Divide each side of both inequalities by 5

The solution set consists of all real numbers less than -1 or greater than 2. In interval notation the solution set is $(-\infty, -1)$ or $(2, \infty)$ and the graph is drawn in Figure 7b.

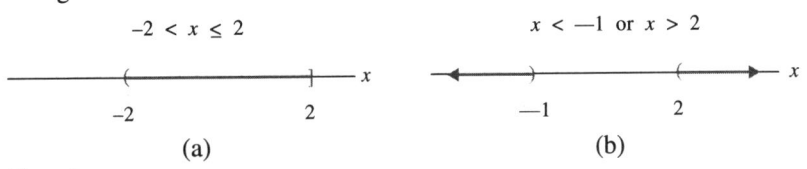

$-2 < x \leq 2$　　　　　　　　　　　　$x < -1$ or $x > 2$

(a)　　　　　　　　　　　　　　　　　(b)

Figure 7

A check is left to the reader.

Solving Applied Problems

The strategy for solving word problems in section 2.3 applies equally to word problems involving linear inequalities. Table 4 illustrates how inequality statements are translated into symbols. We use these symbols to solve the cost problem stated at the opening statement of this section.

Table 4

Word Statement	Algebraic Statement
x is at least 20	$x \geq 20$
x is at most 15	$x \leq 15$
x is no more than 20	$x \leq 20$
x is not less than 7	$x \geq 7$

Example 6　Solving a Cost Problem

A mathematics department wishes to purchase a number of graphing calculators. Each calculator costs $60, plus 8% sales tax. In addition, a $50 delivery charge for the entire shipment must be paid. If the department has allocated at most $3200 for the purchase, how many calculators can be ordered?

Solution　Make a Plan

The cost of each calculator is $60 + 0.08(\$60) = \$64.80$.

Let x = the number of graphing calculators, then
$\$64.80x$ = total cost of the graphing calculators.

To solve this problem we write an inequality that requires the cost of the calculators, $\$64.80x$, plus the delivery charge, $\$50$, to be less than or equal to $\$3200$. The following diagram visualizes the *given*.

Total cost of calculators		Delivery charge		total purchase price
	+		$\leq$	
$\$64.80x$	+	$\$50$	$\leq$	$\$3200$

Write the Solution

We solve the inequality as follows:

$64.80x + 50 \leq 3200$	Given
$64.80x \leq 3150$	Subtract 50 from each side
$x \leq 48.6$	Isolate by dividing by 64.80

Enough money was allocated to order 48 or fewer calculators.

Look Back

Notice that we do not round up to 49, since

$$\$64.80(49) + \$50 = \$3225.20.$$

Such a purchase would be over the budget amount of $3200.

Example 7 **Temperature Range**

The average daily range of temperature for a city during a late summer month varies from at least 72°F to no more than 98°F.

(a) Describe the temperature range for one day.
(b) If the average temperature decreases 2°F per day, describe the temperature range after one week.

Solution **Make a Plan**

Let T represent the temperature during a summer month. We write a compound inequality stating that the temperature is to be between 72°F and 98°F inclusive. The temperature drops 2°F each day for a week. We decrease the temperatures by 7(2°F).

Write the Solution

(a) The average range of temperature is $72° \leq T \leq 98°$.
(b) Since the temperature decreases 2°F per day for a week, the temperature range after one week is

$$72° - 7(2°) \leq T \leq 98° - 7(2°)$$
$$72° - 14° \leq T \leq 98° - 14°$$
$$58° \leq T \leq 84°$$

Look Back

By decreasing the temperature 2°F each day for seven days, we see that 72°F drops to 58°F and 98°F to 84°F.

Using Technology Exploration

We can use a grapher to solve linear inequalities. The techniques are similar to those used in solving linear equations.

1. Rewrite the inequality in the zero form. That is, in the form $Q < 0$ or $Q > 0$ where Q is a linear expression.
2. Graph $y = Q$.
3. Examine the graph to determine the solutions of the inequalities that correspond to all values of x for which the values of y are below ($Q < 0$) or above the x axis ($Q > 0$).

We solve example 2 graphically.

Example 8 Using a Grapher

Use a grapher to solve
$$2x + 3 < 11 - 2x$$

Solution
1. Rewrite the inequality in the zero form,
$$2x + 2x + 3 - 11 < 0$$
$$\text{or} \quad 4x - 8 < 0$$
2. Graph $y = 4x - 8$.
3. The solution set corresponds to all x values on the graph of the line for which the y values are below the x axis; that is, $y < 0$. The viewing window in Figure 8 shows the solution is $x < 2$ and the solution set is $(-\infty, 2)$.

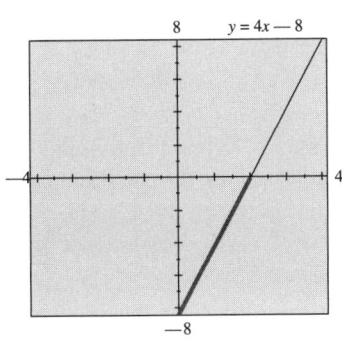

Figure 8

PROBLEM SET 3.4

Mastering the Concepts

In Problems 1–8, graph each inequality and express each using interval notation.

1. (a) $-1 < x \leq 1$
   (b) $-1 \leq x < 1$
2. (a) $-1 \leq x \leq 1$
   (b) $-1 < x < 1$
3. (a) $x \geq -1$
   (b) $x < -1$
4. (a) $x < 1$
   (b) $x \leq 1$
5. (a) $x \leq 2$ or $x \geq 4$
   (b) $x \geq 2$ or $x \leq 4$
6. (a) $x \geq 4$ or $x \leq 5$
   (b) $x \leq 4$ or $x \geq 5$
7. (a) $x \leq -10$ or $x \geq -10$
   (b) $x \leq -10$ and $x \geq -10$
8. (a) $x < 0$ or $x > 0$
   (b) $x < 0$ and $x > 0$

In Problems 9 and 10, state the property of inequalities that justifies each statement.

9. (a) If $x < 4$ then $x < x + 4$
   (b) If $x < 4$ then $2x < 8$
   (c) If $y \leq -2$ then $y + 2 \leq 0$
   (d) If $y \leq -2$ then $-y \geq 2$
10. (a) If $y \geq y - 1$ then $2y \geq 2y - 2$
    (b) If $y \geq y - 1$ then $-2y \leq 2 - 2y$
    (c) If $x - 1 \leq 4$ then $x \leq 5$
    (d) If $a \leq x$ then $a - 2 \leq x - 2$

In Problems 11–40, solve each inequality and express the solution using interval notation. Also graph the solution on a number line.

11. $5 + 3x \geq 8$
12. $-5x + 2 > 12$
13. $-3x + 4 < 14 + 2x$
14. $x + 6 \leq 4 - 3x$
15. $8 - 9x \leq -x$
16. $4 - x \geq 3x$
17. $2(3x - 2) < 7 + 4x - 1$
18. $6 - (7 - x) \geq 2(x - 3)$
19. $-(x - 1) > 2(x + 1/2)$
20. $4(3 - x) \geq 2(x - 1)$
21. $5x \geq -3(x - 2)$
22. $-3(x + 1) \geq -4(2x - 1)$
23. $-9(x - 3) - 8(4 - x) \geq -2x$
24. $-4(x - 1) \geq 2(x + 1) - 3$
25. $3(x - 5) + 10 < 2(x + 4)$
26. $9(x + 2) < -6(4 - x) + 18$
27. $3(x + 2) - 2 \geq -(x + 5) + x$
28. $4(x + 4) > -2(x - 3) + 1$
29. $\dfrac{1}{9}(3x - 2) < \dfrac{1}{3}(1 - 4x)$
30. $\dfrac{1}{6}(2x - 7) < \dfrac{1}{2}(x + 1)$
31. $\dfrac{x}{3} + 2 < \dfrac{x}{4} - 2x$
32. $\dfrac{3x}{2} \geq -6 - \dfrac{x}{2}$
33. $\dfrac{x}{2} - \dfrac{x}{3} \leq 4$
34. $\dfrac{x}{6} + 1 \geq \dfrac{x}{3}$
35. $\dfrac{4x - 2}{2} > \dfrac{3x + 6}{3}$
36. $\dfrac{x + 4}{2} < \dfrac{2x - 3}{3}$
37. $\dfrac{x + 4}{-2} \leq \dfrac{x - 3}{5}$
38. $\dfrac{4x + 17}{7} > x + 2$
39. $\dfrac{5x + 1}{3} > \dfrac{3x + 5}{4}$
40. $\dfrac{2x + 1}{-5} \leq \dfrac{3x - 3}{-3}$

In Problems 41–52, solve each compound inequality, express the solution in interval notation, and graph the solution on a number line.

41. $3 \leq x + 1 \leq 5$
42. $-2 \leq x - 3 \leq 3$
43. $-9 < 3x < 6$
44. $-4 \leq 2x \leq 8$
45. $-3 \leq 4x - 1 \leq 5$
46. $1 < 8 - 3x < 12$
47. $4x - 5 < -3$ or $4x - 5 > 3$
48. $5x - 4 > 1$ or $5x - 4 < -1$
49. $5 - 4x > 2$ or $5 - 4x < -2$
50. $4x - 7 \geq 3$ or $4x - 7 \leq -3$
51. $3x - 1 < -7$ or $3x - 1 > 7$
52. $4x + 3 < -5$ or $4x + 3 > 5$

Applying the Concepts

53. Sales Commission: A sales representative for a tennis club earns a base salary of $500 per week plus a commission of $15 for every new membership sold. How many memberships must be sold in a week in order to earn at least $725 for that week?

54. Weekly Income: A newspaper carrier earns $7 per week plus $0.18 for each newspaper delivered to a home. How many newspapers should be delivered in order for the carrier to earn more than $30 per week?

55. Telephone Charges: Suppose the cost for an international telephone call is $3.50 for the first minute and $1.85 for each additional minute (or part of a minute). If you do not want your total cost to exceed $30, how long can you talk?

56. Car Rental: A vacationer has two choices of car rental agencies. The first agency charges a flat fee of $44.95 per day with unlimited miles. The second charges $19.95 per day plus $0.19 per mile.

   (a) Assuming a seven day rental, what range of miles driven would make the flat fee rental more economical?

   (b) What range of miles driven per day would make the second rental more economical?

   (c) What range of miles driven in three days would make the flat fee rental more economical?

57. Test Averages: A student scored 65, 80 and 74 on the first three tests during the term. What does the student need to score on the fourth test to ensure an average score that is above 75?

58. Diet Program: A clinic advertises that a person can reduce their weight by at least 1.5 pounds per week by exercising and special dieting. At the beginning of a diet program, a person weighs 195 pounds. What is the maximum number of weeks before this person's weight will be reduced to 170 pounds?

59. Energy Consumption: The average American home uses at least 90 but no more than 120 kilowatt hours of electricity per month.

   (a) Use an inequality to express the average range in kilowatt hours per day, assuming that one month is 30 days.

   (b) Use an inequality to express the average range in kilowatt hours per week and per year.

60. Investment: Suppose you invest $7000 in a mutual fund that pays simple interest for one year. If you earn at most $924 at the end of 18 months, what is the maximum annual interest rate for this investment?

61. Geometry: The length of a rectangle is 17 centimeters less than three times the width. Find, to the nearest positive integer, the width of the rectangle if the perimeter is between 222 and 238 centimeters.

62. Travel Times: A traveler flying between New York and Paris can choose among several planes with different average speeds. Suppose the slowest plane averages 515 miles per hour for the trip, and the fastest averages mach 1.30 (1.30 times the speed of sound which is 1100 feet per second).

   (a) Find the range in travel times for the 3645 mile trip.

   (b) Assume the jet stream (high-altitude winds) blows west to east at a velocity of 50 miles per hour. Find the range in travel time of this trip, traveling both with and against the jet stream.

Developing and Extending the Concepts

In Problems 63 and 64, solve each inequality, write it in interval notation and graph the solution set.

63. $2x - 3 \leq 3x + 1 \leq 4x - 5$

64. $2x - 1 \leq 3x + 7 \leq x + 9$

G In Problems 65 and 66, use the given grapher display and viewing window for each equation to find the solution set in interval notation of the associated inequalities.

65. $y = (6 - x) - (4x + 1)$

   (a) $6 - x \leq 4x + 1$

   (b) $6 - x > 4x + 1$

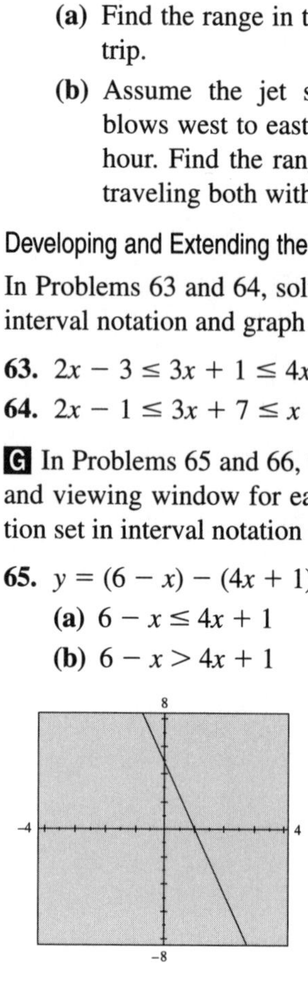

66. $y = (3x - 5) - (x + 1)$

   (a) $3x - 5 \leq x + 1$

   (b) $3x - 5 > x + 1$

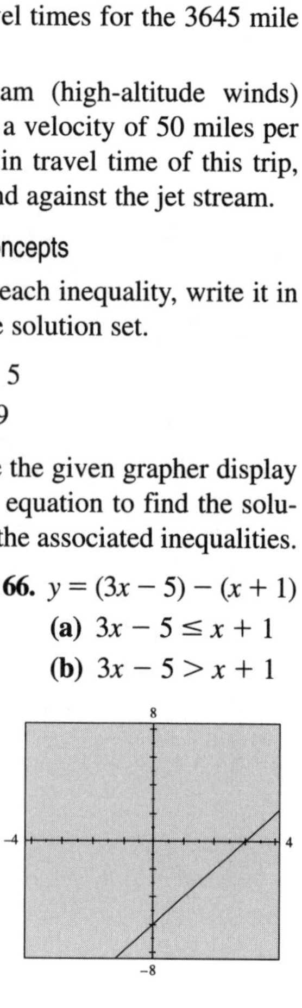

G In Problems 67 and 68, use a grapher to graph the associated equation for y in terms of x, and then use the graph of the equation to estimate the solution of the given inequality. Also write the solution in interval notation.

67. (a) $x + 7 \leq 4x - 8$; $y = (x + 7) - (4x - 8)$

   (b) $3x + 1 \geq 7x - 18$; $y = (3x + 1) - (7x - 18)$

68. (a) $\dfrac{2x}{9} - \dfrac{5}{6} > \dfrac{x}{12}$; $y = \left(\dfrac{2x}{9} - \dfrac{5}{6}\right) - \left(\dfrac{x}{12}\right)$

   (b) $\dfrac{x}{2} - \dfrac{3}{4} < \dfrac{7x}{4}$; $y = \left(\dfrac{x}{2} - \dfrac{3}{4}\right) - \left(\dfrac{7x}{4}\right)$

Chapter 4

QUADRATIC EQUATIONS

4.1 Solving Quadratic Equations
4.2 Solving Quadratic Equations by Using the Quadratic Formula
4.3 Applications of Quadratic Equations
4.4 Complex Solutions to Quadratic Equations (Optional)

4.1 Solving Quadratic Equations

The easiest way to solve a quadratic equation in standard form is to factor the polynomial expression on the left side of the equation. (If it is impossible to factor the polynomial, we can use other methods—discussed later in this chapter—to solve the equation.) In Chapter 4, we solved quadratic equations by *factoring* and applying the zero-product property:

$$\text{If } ab = 0, \text{ then } a = 0 \text{ or } b = 0.$$

As a review, let us examine a few examples of solving quadratic equations by factoring

In Examples 1–3, solve each equation by factoring.

Example 1

$x^2 - 5x + 6 = 0$

Solution

We factor the left side of the equation:

$$(x - 3)(x - 2) = 0.$$

Now we set each factor equal to zero and solve the resulting first-degree equations:

$$x - 3 = 0 \quad | \quad x - 2 = 0$$
$$x = 3 \quad | \quad x = 2.$$

Check For $x = 3$.
$$3^2 - 5(3) + 6 = 9 - 15 + 6 = 0.$$
For $x = 2$,
$$2^2 - 5(2) + 6 = 4 - 10 + 6 = 0.$$
Therefore, both 3 and 2 are solutions of the original equation.

Example 2 $\quad 6r^2 + 5r = 4$

Solution First, we rewrite the equation in standard form:
$$6r^2 + 5r - 4 = 0.$$
Factoring the left side of the equation, we have:
$$(2r - 1)(3r + 4) = 0.$$
Now we set each factor equal to zero and solve the resulting first-degree equations:

$$\begin{array}{c|c} 2r - 1 = 0 & 3r + 4 = 0 \\ 2r = 1 & 3r = -4 \\ r = \dfrac{1}{2} & r = -\dfrac{4}{3}. \end{array}$$

Substituting $\frac{1}{2}$ and $-\frac{4}{3}$ in the original equation confirms that both are solutions of the original equation.

Example 3 $\quad x^2 = 36$

Solution First, we write the equation in the form:
$$x^2 - 36 = 0.$$
Factoring the left side of the equation, we have:
$$(x - 6)(x + 6) = 0$$
$$\begin{array}{c|c} x - 6 = 0 & x + 6 = 0 \\ x = 6 & x = -6. \end{array}$$
Therefore, 6 and −6 are the solutions of the original equation.

Example 4 A rocket is fired vertically upward. Its height h (in feet) above the ground after t seconds is given by
$$h = 800t - 16t^2.$$
After how many seconds will the rocket be 6,400 feet above the ground?

Solution When the rocket is 6,400 feet above the ground, $h = 6{,}400$. Therefore,
$$6{,}400 = 800t - 16t^2.$$

We write the equation in standard form:
$$16t^2 - 800t + 6{,}400 = 0$$
$$t^2 - 50t + 400 = 0 \quad \text{(We divided by 16.)}$$
$$(t - 40)(t - 10) = 0$$

$$t - 40 = 0 \quad | \quad t - 10 = 0$$
$$t = 40 \quad | \quad t = 10.$$

Therefore, the rocket will be 6,400 feet above the ground 10 seconds after it is fired (as it ascends) and again at 40 seconds after it is fired (as it descends).

Solving Quadratic Equations by Roots Extraction

Let us consider the equation $x^2 = 36$ again. Instead of solving this equation by factoring, we could have determined the numbers whose squares are 36. That is, we could have determined the two square roots of 36. Thus, if
$$x^2 = 36,$$
then
$$x = \sqrt{36} \quad \text{or} \quad x = -\sqrt{36}$$
$$x = 6 \quad \text{or} \quad x = -6$$
since $6^2 = 36$ and $(-6)^2 = 36$.

In general, if p is a nonnegative real number and if
$$x^2 = p,$$
then
$$x = \sqrt{p} \quad \text{or} \quad x = -\sqrt{p}.$$

Therefore, the solutions are $\sqrt{p}$ or $-\sqrt{p}$. The two solutions can also be written as
$$x = \pm\sqrt{p}.$$

This method is called **roots extraction,** because solutions are obtained by extracting square roots. When applying this method, we must be sure to include both the positive and the negative square roots as solutions. For example, to solve $x^2 = 25$, we have:

$$x = \sqrt{25} \quad | \quad x = -\sqrt{25}$$
$$x = 5 \quad | \quad x = -5$$

since $5^2 = 25$ and $(-5)^2 = 25$.
Therefore, the solutions are -5 and 5.

In Examples 5–8, solve each equation by roots extraction.

Example 5 $\quad x^2 = 4$

Solution By extracting the square roots, we have:
$$x = \sqrt{4} \quad | \quad x = -\sqrt{4}$$
$$x = 2 \quad | \quad x = -2.$$

Check For $x = 2$, $2^2 = 4$. For $x = -2$, $(-2)^2 = 4$. Therefore, the solutions are -2 and 2.

Example 6 $y^2 - 11 = 0$

Solution We solve the equation for y^2:
$$y^2 = 11.$$

We extract the square roots:
$$y = \sqrt{11} \quad | \quad y = -\sqrt{11}.$$

The solutions are $\sqrt{11}$ and $-\sqrt{11}$.

Example 7 $4u^2 = 9$

Solution We solve for u^2:
$$u^2 = \frac{9}{4}.$$

We extract the square roots:
$$u = \sqrt{\frac{9}{4}} \quad | \quad u = -\sqrt{\frac{9}{4}}$$
$$u = \frac{3}{2} \quad | \quad u = -\frac{3}{2}.$$

The solutions are $\frac{3}{2}$ and $-\frac{3}{2}$.

Example 8 $25x^2 - 8 = 0$

Solution We solve for x^2:
$$25x^2 = 8$$
$$x^2 = \frac{8}{25}.$$

We extract the square roots:
$$x = \sqrt{\frac{8}{25}} \quad | \quad x = -\sqrt{\frac{8}{25}}$$
$$x = \frac{\sqrt{8}}{5} \quad | \quad x = -\frac{\sqrt{8}}{5}$$
$$x = \frac{2\sqrt{2}}{5} \quad | \quad x = -\frac{2\sqrt{2}}{5}.$$

Therefore, the solutions are
$$\frac{2\sqrt{2}}{5} \quad \text{and} \quad -\frac{2\sqrt{2}}{5}.$$

We can also use the method of roots extraction to solve equations of the form
$$(ax + b)^2 = p,$$
for any nonnegative real number p and any nonzero real number a.

In Examples 9 and 10, solve each equation by roots extraction.

Example 9 $(z - 1)^2 = 9$

Solution We extract the square roots and solve for z:

$z - 1 = \sqrt{9}$	$z - 1 = -\sqrt{9}$
$z - 1 = 3$	$z - 1 = -3$
$z = 4$	$z = -2.$

Check For $z = 4$, $(4 - 1)^2 = 3^2 = 9$. For $z = -2$, $(-2 - 1)^2 = (-3)^2 = 9$. Thus, -2 and 4 are the solutions.

Example 10 $(4y - 5)^2 - 6 = 0$

Solution Solving for $(4y - 5)^2$, we get

$$(4y - 5)^2 = 6.$$

By extracting the square roots and solving for y, we have

$4y - 5 = \sqrt{6}$	$4y - 5 = -\sqrt{6}$
$4y = 5 + \sqrt{6}$	$4y = 5 - \sqrt{6}$
$y = \dfrac{5 + \sqrt{6}}{4}$	$y = \dfrac{5 - \sqrt{6}}{4}.$

Therefore, the solutions are

$$\frac{5 + \sqrt{6}}{4} \quad \text{and} \quad \frac{5 - \sqrt{6}}{4}$$

Example 11 The square of three times a positive integer plus 2 is 64. Find the integer.

Solution Let x be the integer, then

$$(3x + 2)^2 = 64$$

so

$$3x + 2 = \pm 8$$

$3x + 2 = 8$	$3x + 2 = -8$
$3x = 8 - 2$	$3x = -8 - 2$
$3x = 6$	$3x = -10$
$x = 2$	$x = -\dfrac{10}{3}.$

We reject $-\frac{10}{3}$, since it is not a positive integer, so the integer is 2.

PROBLEM SET 4.1

In problems 1–14, solve each equation by factoring.

1. $z^2 - 4z - 21 = 0$
2. $t^2 - 7t - 18 = 0$
3. $x^2 + 11x + 30 = 0$
4. $v^2 - 12v + 32 = 0$
5. $u^2 - 7u - 44 = 0$
6. $m^2 + 13m + 30 = 0$
7. $3x^2 - x - 14 = 0$
8. $6y^2 - y - 2 = 0$
9. $21v^2 + 11v - 2 = 0$
10. $45x^2 - 34x + 5 = 0$
11. $15t^2 - 19t + 6 = 0$
12. $10w^2 + w - 2 = 0$
13. $18x^2 + 56x - 7 = 0$
14. $10z^2 - 31 - 14 = 0$

In problems 15–54, solve each equation by roots extraction.

15. $t^2 = 16$
16. $z^2 = 36$
17. $v^2 = 81$
18. $x^2 = 64$
19. $9x^2 = 25$
20. $4m^2 = 81$
21. $16z^2 = 9$
22. $25y^2 = 16$
23. $4u^2 = 7$
24. $81x^2 = 11$
25. $9y^2 = 50$
26. $16v^2 = 27$
27. $x^2 - 1 = 0$
28. $z^2 - 100 = 0$
29. $4z^2 - 121 = 0$
30. $9w^2 - 64 = 0$
31. $25x^2 - 18 = 0$
32. $100y^2 - 63 = 0$
33. $(z - 3)^2 = 4$
34. $(x + 7)^2 = 81$
35. $(y + 5)^2 = 16$
36. $(v - 9)^2 = 144$
37. $(t - 4)^2 = 7$
38. $(w + 6)^2 = 5$
39. $(x + 6)^2 - 3 = 0$
40. $(t - 2)^2 - 5 = 0$
41. $(2u - 3)^2 - 9 = 0$
42. $(5x - 7)^2 - 25 = 0$
43. $(3w + 1)^2 - 8 = 0$
44. $(6m + 11)^2 - 18 = 0$
45. $\left(x + \dfrac{3}{2}\right)^2 - 1 = 0$
46. $\left(y - \dfrac{3}{8}\right)^2 - 4 = 0$
47. $\left(t - \dfrac{5}{2}\right)^2 = \dfrac{9}{4}$
48. $\left(x + \dfrac{11}{2}\right)^2 = \dfrac{13}{4}$
49. $\left(y + \dfrac{5}{6}\right)^2 = \dfrac{37}{36}$
50. $\left(z - \dfrac{3}{4}\right)^2 = \dfrac{17}{16}$
51. $\left(x - \dfrac{1}{5}\right)^2 - \dfrac{6}{25}$
52. $\left(y + \dfrac{3}{8}\right)^2 = \dfrac{137}{64}$
53. $\left(v + \dfrac{5}{3}\right)^2 = \dfrac{8}{9}$
54. $\left(t - \dfrac{4}{7}\right)^2 = \dfrac{18}{49}$

55. If the diagonal of a rectangle is 50 centimeters and the width is 30 centimeters. find the length of the rectangle.

56. If the diagonal of a square garden is $8\sqrt{2}$ meters, how long is each side?

57. If the square of the sum of a positive number and 3 is equal to 25, find the number.

58. If the square of the difference of a positive number and 2 is 16, find the number.

59. A ball is thrown straight up from the ground. Its height h (in feet) above the ground after time t (in seconds) is given by

$$h = 64t - 16t^2.$$

After how many seconds will the ball be 48 feet above the ground?

60. A rocket is fired vertically upward from the ground. Its height h in feet above the ground t seconds after firing is given by

$$h = 480t - 16t^2.$$

After how many seconds will the rocket be 2,000 feet above the ground?

61. A projectile is fired straight up from the ground. Its height h in meters above the ground t seconds after firing is given by

$$h = 24.5t - 4.9t^2$$

After how many seconds will the projectile be 29.4 meters above the ground?

4.2 Solving Quadratic Equations by Using the Quadratic Formula

The method of completing the square can be used to solve any quadratic equation. However, it is often more efficient to use what is called the *quadratic formula*. To derive this formula we use the method of completing the square to solve, for x, the literal quadratic equation

$$ax^2 + bx + c = 0,$$

in which a, b, and c are constants and $a \ne 0$. We follow the procedure outlined in Section 4.1.

Step 1. We subtract c from each side of the equation:

$$ax^2 + bx = -c.$$

Step 2. We divide both sides by a:
$$x^2 + \frac{b}{a}x = -\frac{c}{a}.$$

Step 3. We complete the square of the left side by adding
$$\left[\frac{1}{2}\left(\frac{b}{a}\right)\right]^2 = \frac{b^2}{4a^2}$$
to both sides:
$$x^2 + \frac{b}{a}x + \frac{b^2}{4a^2} = \frac{b^2}{4a^2} - \frac{c}{a}.$$

Step 4. Since
$$x^2 + \frac{b}{a}x + \frac{b^2}{4a^2} = \left(x + \frac{b}{2a}\right)^2$$
and
$$\frac{b^2}{4a^2} - \frac{c}{a} = \frac{b^2}{4a^2} - \frac{4ac}{4a^2} = \frac{b^2 - 4ac}{4a^2},$$
we rewrite the equation in step 3 as
$$\left(x + \frac{b}{2a}\right)^2 = \frac{b^2 - 4ac}{4a^2}.$$

We solve the equation by roots extraction and by assuming that $b^2 - 4ac \geq 0$:

$$x + \frac{b}{2a} = \sqrt{\frac{b^2 - 4ac}{4a^2}} \qquad \bigg| \qquad x + \frac{b}{2a} = -\sqrt{\frac{b^2 - 4ac}{4a^2}}$$

$$x + \frac{b}{2a} = \frac{\sqrt{b^2 - 4ac}}{2a} \qquad \bigg| \qquad x + \frac{b}{2a} = -\frac{\sqrt{b^2 - 4ac}}{2a}$$

$$x = -\frac{b}{2a} + \frac{\sqrt{b^2 - 4ac}}{2a} \qquad \bigg| \qquad x = -\frac{b}{2a} - \frac{\sqrt{b^2 - 4ac}}{2a}$$

$$x = \frac{-b + \sqrt{b^2 - 4ac}}{2a} \qquad \bigg| \qquad x = \frac{-b - \sqrt{b^2 - 4ac}}{2a}$$

The two solutions $\dfrac{-b + \sqrt{b^2 - 4ac}}{2a}$ and $\dfrac{-b - \sqrt{b^2 - 4ac}}{2a}$ are usually written in the compact form shown in the following statement. The symbol $\pm$ means $+$ or $-$.

The Quadratic Formula

If $ax^2 + bx + c = 0$ and $a \neq 0$, then
$$x = \frac{-b \pm \sqrt{b^2 - 4ac}}{2a},$$
assuming that $b^2 - 4ac \geq 0$.

Recall that on page 173 we pointed out that if the expression $b^2 - 4ac$ is the square root of an integer, then the trinomial $ax^2 + bx + c$ can be factored. Now you can see how we obtain this test for factorability.

In Examples 1–3, use the quadratic formula to solve each equation.

Example 1 $3x^2 - 4x + 1 = 0$

Solution Here $a = 3$, $b = -4$, and $c = 1$. We substitute these values in the quadratic formula

$$x = \frac{-b \pm \sqrt{b^2 - 4ac}}{2a}$$

and obtain

$$x = \frac{-(-4) \pm \sqrt{(-4)^2 - 4(3)(1)}}{2(3)} = \frac{4 \pm \sqrt{4}}{6}.$$

Thus,

$$x = \frac{4 + 2}{6} = \frac{6}{6} = 1 \quad \text{or} \quad x = \frac{4 - 2}{6} = \frac{2}{6} = \frac{1}{3}.$$

The solutions are

$$1 \quad \text{and} \quad \frac{1}{3}.$$

Example 2 $2y^2 + 3y - 1 = 0$

Solution Here $a = 2$, $b = 3$, and $c = -1$. We substitute these values in the quadratic formula

$$y = \frac{-b \pm \sqrt{b^2 - 4ac}}{2a},$$

so that

$$y = \frac{-3 \pm \sqrt{3^2 - 4(2)(-1)}}{2(2)} = \frac{-3 \pm \sqrt{17}}{4}.$$

Therefore, the solutions are

$$\frac{-3 + \sqrt{17}}{4} \quad \text{and} \quad \frac{-3 - \sqrt{17}}{4}.$$

Example 3 $3w^2 - 7 = 10w$

Solution First, we write the equation in standard form:

$$3w^2 - 10w - 7 = 0.$$

We have $a = 3$, $b = -10$, and $c = -7$. Substituting these values in the quadratic formula

$$w = \frac{-b \pm \sqrt{b^2 - 4ac}}{2a},$$

we obtain

$$w = \frac{-(-10) \pm \sqrt{(-10)^2 - 4(3)(-7)}}{2(3)} = \frac{10 \pm \sqrt{100 + 84}}{6}$$

$$= \frac{10 \pm \sqrt{184}}{6} = \frac{10 \pm 2\sqrt{46}}{6} = \frac{2(5 \pm \sqrt{46})}{6} = \frac{5 \pm \sqrt{46}}{3}.$$

Therefore, the solutions are

$$\frac{5 + \sqrt{46}}{3} \quad \text{and} \quad \frac{5 - \sqrt{46}}{3}.$$

Check For $w = \dfrac{5 + \sqrt{46}}{3}$:

$$3\left(\frac{5 + \sqrt{46}}{3}\right)^2 - 7 \stackrel{?}{=} 10\left(\frac{5 + \sqrt{46}}{3}\right)$$

$$\cancel{3}\left(\frac{25 + 10\sqrt{46} + \sqrt{46}}{\cancel{9}_3}\right) - 7 \stackrel{?}{=} \frac{50 + 10\sqrt{46}}{3}$$

$$\frac{25 + 10\sqrt{46} + 46 - 21}{3} \stackrel{?}{=} \frac{50 + 10\sqrt{46}}{3}$$

$$\frac{50 + 10\sqrt{46}}{3} = \frac{50 + 10\sqrt{46}}{3}.$$

For $w = \dfrac{5 - \sqrt{46}}{3}$:

$$3\left(\frac{5 - \sqrt{46}}{3}\right)^2 - 7 \stackrel{?}{=} 10\left(\frac{5 - \sqrt{46}}{3}\right)$$

$$\cancel{3}\left(\frac{25 - 10\sqrt{46} + 46}{\cancel{9}_3}\right) - 7 \stackrel{?}{=} \frac{50 - 10\sqrt{46}}{3}$$

$$\frac{25 - 10\sqrt{46} + 46 - 21}{3} \stackrel{?}{=} \frac{50 - 10\sqrt{46}}{3}$$

$$\frac{50 - 10\sqrt{46}}{3} = \frac{50 - 10\sqrt{46}}{3}.$$

PROBLEM SET 4.2

In problems 1–40, solve each equation by using the quadratic formula.

1. $2x^2 - 5x + 1 = 0$
2. $3u^2 + 4u - 2 = 0$
3. $y^2 - 6y + 2 = 0$
4. $z^2 - 18z + 56 = 0$
5. $t^2 + 3t - 3 = 0$
6. $4x^2 + 16x + 15 = 0$
7. $6u^2 - 7u - 5 = 0$
8. $v^2 + 3v - 10 = 0$
9. $6x^2 - 5x - 6 = 0$
10. $y^2 - 4y - 2 = 0$
11. $3z^2 + 4z + 1 = 0$
12. $2m^2 + 5m + 3 = 0$
13. $2x^2 + 7x - 15 = 0$
14. $12t^2 - 23t + 10 = 0$
15. $15t^2 + 2t - 8 = 0$
16. $6x^2 + 37x + 6 = 0$
17. $3v^2 + 8v + 4 = 0$
18. $21z^2 - 43z + 20 = 0$
19. $10y^2 - 7y - 12 = 0$
20. $10v^2 - 29v + 10 = 0$
21. $6x^2 + 17x - 14 = 0$
22. $x^2 + 16x + 48 = 0$
23. $z^2 - 2z = 63$
24. $6t^2 + 6 = 13t$
25. $5 + m = 6m^2$
26. $4y^2 + 11y = 3$
27. $2x^2 + 1 = 3x$
28. $11 = 12x^2 + 29x$
29. $32u^2 = 4u + 21$
30. $9x^2 = 16x - 5$
31. $7z^2 - 3 = 12z$
32. $3t^2 - 8 = 2t$
33. $-3x^2 - 5x + 17 = 0$
34. $-6x^2 - 3x + 5 = 0$
35. $-2x^2 - x + 4 = 0$
36. $-15x^2 + 4z + 3 = 0$
37. $y^2 + 7y + 5 = 0$
38. $v^2 - 9v + 11 = 0$
39. $5x^2 - 12x + 2 = 0$
40. $12x^2 - 14x + 3 = 0$

C In problems 41–44, use the quadratic formula and a calculator to find the solutions of each quadratic equation approximated to three decimal places.

41. $3x^2 - 7x - 1 = 0$
42. $2x^2 - x - 5 = 0$
43. $2x^2 + 4x - 3 = 0$
44. $3x^2 - 9x + 3 = 0$

45. The equation
$$2x^3 - 7x^2 - 4x = 0$$
can be solved by first factoring out the common factor x, so that
$$x(2x^2 - 7x - 4) = 0.$$
Use the quadratic formula to find the two other solutions, then write the three solutions to this equation.

46. Solve the equation
$$4x^3 - 6x^2 + x = 0$$
by first factoring out the common factor x, then use the quadratic formula to find the other solutions to this equation.

C 47. Donna invested $300 in an account with interest rate r compounded annually. The amount of money, s, in the account after two years is given by the equation
$$s = 300r^2 + 600r + 300.$$
Find r if $s = \$343$.

C 48. What interest rate r compounded annually will make $1,000 grow to $1,214, if the accumulated amount of money s is given by the equation
$$s = 1{,}000r^2 + 2{.}000r + 1{,}000.$$

4.3 Applications of Quadratic Equations

In this section, we see how quadratic equations can be utilized in a wide variety of applications. The procedures we used in Chapter 3 to solve word problems involving linear equations are just as appropriate here for word problems that give rise to quadratic equations. When quadratic equations are involved. however, it is important to check both solutions in terms of the original problem: A number may satisfy an equation numerically and yet have no meaning in the context of the original problem.

Example 1 Find two numbers whose sum is 7 and whose product is 12.

Solution If we let
$$x = \text{the first number,}$$
then
$$7 - x = \text{the second number.}$$
The product of the two numbers is 12, so an equation is
$$x(7 - x) = 12$$
$$7x - x^2 = 12$$
$$x^2 - 7x + 12 = 0.$$
We solve this equation by factoring:
$$(x - 3)(x - 4) = 0$$

$x - 3 = 0$	$x - 4 = 0$
$x = 3$	$x = 4$
and	and
$7 - x = 4$	$7 - x = 3.$

Therefore, the two numbers are 3 and 4.

Check
$$3 + 4 = 7$$
$$3 \cdot 4 = 12.$$

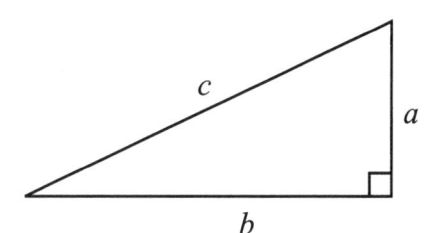

Recall from Section 7.7 (page 318) that the *Pythagorean theorem* expresses the following property for a right triangle (Figure 1):

$$c^2 = a^2 + b^2.$$

Figure 1

Example 2 In order to support a solar collector at the correct angle, the roof trusses of a house are designed as right triangles. Rafters form the right angle, and the base of the truss is the hypotenuse (Figure 2). If the rafter on the same side as the solar collector is 10 feet shorter than the other rafter and if the base of each truss is 50 feet long, how long is each of the rafters?

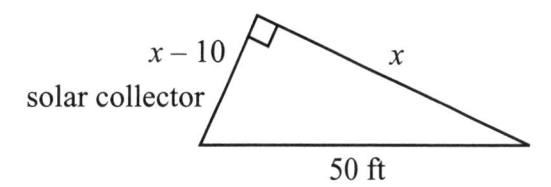

Figure 2

Solution Let

$$x = \text{the length of the longer rafter in feet.}$$

Then

$$x - 10 = \text{the length of the rafter on the side of the collector.}$$

We use the Pythagorean theorem:

$$x^2 + (x - 10)^2 = 50^2 \quad \text{or} \quad x^2 + x^2 - 20x + 100 = 2{,}500$$

or

$$2x^2 - 20x - 2{,}400 = 0$$

$$x^2 - 10x - 1{,}200 = 0 \quad \text{(We divided both sides by 2.)}$$

$$(x + 30)(x - 40) = 0$$

$$\begin{array}{c|c} x + 30 = 0 & x - 40 = 0 \\ x = -30 & x = 40. \end{array}$$

Because the length is always positive, the lengths of the rafters are 40 feet and $40 - 10 = 30$ feet.

Check $30^2 + 40^2 = 900 + 1600 = 2500 = 50^2.$

Example 3 A rectangle has a length twice its width. If the width is increased by 5 feet and the length is decreased by 5 feet, the new rectangle has an area of 225 square feet. Find the dimensions of the original rectangle.

Solution Let

$$x = \text{the width of the original rectangle.}$$

Then we can summarize the information in the problem as follows:

	Original Rectangle	New Rectangle
Width	x	$x + 5$
Length	$2x$	$2x - 5$

Using the formula $A = lw$, we see that

$$(2x - 5)(x + 5) = 225$$
$$2x^2 + 5x - 25 = 225$$
$$2x^2 + 5x - 250 = 0$$
$$(2x + 25)(x - 10) = 0$$

$2x + 25 = 0$	$x - 10 = 0$
$2x = -25$	$x = 10.$
$x = -\dfrac{25}{2}$	

Since the width is positive. $x = 10$. Thus, the width of the original rectangle is 10 feet and the length is $2 \cdot 10 = 20$ feet.

Check $(20 - 5)(10 + 5) = (15)(1) = 225.$

Example 4 A car traveled 100 miles at a uniform speed. If the speed had been 10 miles per hour faster, the trip would have been completed one-half hour sooner. Find the speed of the car.

Solution Let

$$x = \text{the speed of the car in miles per hour.}$$

Then

$$x + 10 = \text{the car's speed at 10 miles per hour faster.}$$

Using the formula $t = d/r$ (time equals distance divided by rate of speed), we have the following information:

	d	r	$t = \dfrac{d}{r}$
Actual trip	100	x	$\dfrac{100}{x}$
Trip at 10 miles per hour faster	100	$x + 10$	$\dfrac{100}{x + 10}$

The difference of the two times is $\frac{1}{2}$ hour. Therefore, we can write the equation as follows:

$$\frac{1}{2} = \frac{100}{x} - \frac{100}{x + 10}.$$

To solve this equation, we first multiply each side by the LCD, $2x(x + 10)$,

$$2x(x + 10)\left(\frac{1}{2}\right) = 2x(x + 10)\left(\frac{100}{x} - \frac{100}{x + 10}\right)$$

$$= 2x(x + 10)\left(\frac{100}{x}\right) - 2x(x + 10)\left(\frac{100}{x + 10}\right)$$

or

$$x(x + 10) = 200(x + 10) - 200x$$

$$x^2 + 10x = 2{,}000$$

$$x^2 + 10x - 2{,}000 = 0$$

$$(x + 50)(x - 40) = 0$$

$$x + 50 = 0 \quad | \quad x - 40 = 0$$

$$x = -50 \quad | \quad x = 40.$$

Since the car's speed must be positive, $x = 40$. Therefore, the car traveled at 40 miles per hour.

PROBLEM SET 4.3

In problems 1–23. use a quadratic equation to solve each word problem.

1. Find two numbers whose sum is 11 and whose product is 30.
2. Find two numbers whose sum is 18 and whose product is 72.
3. The sum of the reciprocals of two consecutive positive odd integers is $\frac{8}{15}$. What are the integers?
4. The difference of the reciprocals of two consecutive positive even integers is $\frac{1}{12}$. What are the integers?
5. If twice the reciprocal of a number is subtracted from the number, the difference is 1. Find the number.
6. If the reciprocal of a number is added to the number, the sum is $\frac{26}{5}$. Find the number.
7. A hiker walks 5 miles along straight railroad tracks that cut diagonally through a rectangular plot. The length of the plot is 1 mile greater than its width. Find the dimensions of the plot.
8. The longer leg of a right triangle is 2 feet more than twice the shorter leg. If the hypotenuse is 13 feet, what is the length of the shorter leg?
9. A surveyor marking off a field in the shape of a right triangle finds that the hypotenuse is 30 feet more than twice one leg. If the other leg is 120 feet, how long is the hypotenuse?

C 10. The horizon is 1,600 miles away from a space capsule orbiting the earth. Assuming that the distance from the surface to the center of the earth is 4.000 miles. find how far the space capsule is above the earth. Approximate this distance to two decimal places.

11. A rectangular garden is fenced in with 160 feet of fence. If the area of the garden is 1,500 square feet. find the dimensions of the garden.
12. A swimming pool 30 feet long and 20 feet wide has a cement walk of uniform width around it. If the area of the walk is the same as the area of the surface of the pool, find the width of the walk.

C 13. A picture measuring 7 inches by 9 inches has a frame surrounding it of uniform width. If the area of the picture is the same as the area of the frame, how wide is the frame? Express your answer in radical form. and then approximate it by rounding off to two decimal places.

14. What is the length of the side of a square if the numerical value of its perimeter is the same as the numerical value of its area?
15. Two joggers leave at the same time from the intersection of two straight roads that form a right angle. Each jogger runs on a different road. After 1 hour they are 10 miles apart. If one jogger runs 2 miles per hour faster than the other. how fast is each running?

16. A rectangle has a length 2 yards greater than its width. If the width is doubled and the length is decreased by 2 yards, the area is increased by 63 square yards. Find the dimensions of the rectangle.

17. A rectangular backyard is 20 feet longer than it is wide. A flower bed 10 feet wide is dug along the width of the yard. If the area of the remaining part of the yard is 3,000 square feet, how wide is the yard?

18. Mary is 2 years older than Jane, and in 3 years the sum of the squares of their ages will be 130. How old is each now?

19. A carpenter working alone can build a cottage in 5 days less time than his apprentice can do the job working alone. If they can build the cottage together in 6 days, how long does it take the carpenter to do the job alone?

20. A plumber can do a certain job in 9 days less time than his helper. If they can do the job together in 20 days, how long would it take each to do the job alone?

21. A train travels 300 miles at a uniform rate. If the train had traveled 5 miles per hour faster, the trip would have taken 2 hours less. How fast did the train travel?

22. A boat takes 1 hour longer to travel 12 miles up a stream than it takes to make the return trip downstream. If the rate of the current is 1 mile per hour, how fast can the boat go in still water?

23. An object is dropped from the top of a building that is 400 feet tall. As it is falling, its distance d above the ground at time t (in seconds) is given by

$$d = 400 - 16t^2.$$

How long does the object take to hit the ground?

4.4 Complex Solutions to Quadratic Equations (Optional)

So far we have solved quadratic equations that give rise to real solutions. In order to solve quadratic equations such as

$$x^2 = -9$$

we introduce a new system of numbers called the *complex numbers*. This system is based on the following definition: The number i is a number such that

$$i = \sqrt{-1} \quad \text{and} \quad -i = -\sqrt{-1}.$$

If we square both sides of the equation $i = \sqrt{-1}$, we have:

$$i^2 = -1.$$

If we extend the product property for radicals so that it also applies when one of the radicands is negative, we can write square roots of negative numbers in terms of i. For example,

$$\sqrt{-9} = \sqrt{9 \cdot (-1)} = \sqrt{9} \cdot \sqrt{-1} = 3i$$

$$\sqrt{-25} = \sqrt{25 \cdot (-1)} = \sqrt{25} \cdot \sqrt{-1} = 5i$$

$$\sqrt{-7} = \sqrt{7 \cdot (-1)} = \sqrt{7} \cdot \sqrt{-1} = \sqrt{7}i$$

$$\sqrt{-12} = \sqrt{12 \cdot (-1)} = \sqrt{12} \cdot \sqrt{-1} = 2\sqrt{3}i.$$

We shall use i to define a new type of number, the *complex number*.

4.4 COMPLEX SOLUTIONS TO QUADRATIC EQUATIONS (OPTIONAL)

DEFINITION

A number of the form

$$a + bi,$$

where a and b are real numbers and $i = \sqrt{-1}$, is called a **complex number**.

For example,

$$5 + 3i, \quad 2 - 7i, \quad 3 + \sqrt{5}i, \quad \text{and} \quad \frac{2}{3} + 2\sqrt{11}i$$

are complex numbers.

The complex numbers

$$a + bi \quad \text{and} \quad a - bi$$

are called **conjugates** of each other. If $a + bi$ is a complex number, then its conjugate is denoted by $\overline{a + bi}$ and

$$\overline{a + bi} = a - bi.$$

For example,

$$\overline{4 + 3i} = 4 - 3i$$

$$\overline{2 - 7i} = 2 + 7i$$

and

$$\overline{-3 - 5i} = -3 + 5i.$$

Complex numbers are used extensively in electrical and aeronautical engineering and the space sciences, as well as in chemistry and in physics. Our use of complex numbers will be limited to the solutions of quadratic equations such as

$$x^2 + 9 = 0 \quad \text{and} \quad x^2 - 6x + 13 = 0.$$

For example, to solve the equation

$$x^2 + 9 = 0.$$

we have

$$x^2 = -9$$

$$x = \pm\sqrt{-9} = \pm 3i.$$

The two solutions $3i$ and $-3i$ satisfy the original equation:

$$(3i)^2 + 9 \stackrel{?}{=} 0 \quad \text{or} \quad (-3i)^2 + 9 \stackrel{?}{=} 0$$

$$9i^2 + 9 \stackrel{?}{=} 0 \quad\quad\quad 9i^2 + 9 \stackrel{?}{=} 0$$

$$9(-1) + 9 \stackrel{?}{=} 0 \quad\quad\quad 9(-1) + 9 \stackrel{?}{=} 0$$

$$-9 + 9 = 0 \quad\quad\quad -9 + 9 = 0.$$

In Examples 1–3, solve each quadratic equation.

Example 1 $x^2 - 6x + 13 = 0$

Solution Substituting $a = 1$, $b = -6$, and $c = 13$ in the quadratic formula

$$x = \frac{-b \pm \sqrt{b^2 - 4ac}}{2a},$$

we have:

$$x = \frac{-(-6) \pm \sqrt{(-6)^2 - 4(1)(13)}}{2(1)}$$

$$= \frac{6 \pm \sqrt{36 - 52}}{2}$$

$$= \frac{6 \pm \sqrt{-16}}{2} = \frac{6 \pm 4i}{2}$$

$$= \frac{2(3 \pm 2i)}{2} = 3 \pm 2i.$$

Therefore, the solutions are $3 + \sqrt{2}i$ and $3 - \sqrt{2}i$.

Example 2 $x^2 - 6x + 11 = 0$

Solution Substituting $a = 1$, $b = -6$, and $c = 11$ in the quadratic formula

$$x = \frac{-b \pm \sqrt{b^2 - 4ac}}{2a},$$

we have:

$$x = \frac{-(-6) \pm \sqrt{(-6)2 - 4(1)(11)}}{2(1)}$$

$$= \frac{6 \pm \sqrt{36 - 44}}{2}$$

$$= \frac{6 \pm \sqrt{-8}}{2}$$

$$= \frac{6 \pm 2\sqrt{2}i}{2} = \frac{2(3 \pm \sqrt{2}i)}{2}$$

$$= 3 \pm 2i.$$

Therefore, the solutions are $3 + 2i$ and $3 - 2i$.

Example 3 $2x^2 - 4x + 3 = 0$

Solution Substituting $a = 2$, $b = -4$, and $c = 3$ in the quadratic formula

$$x = \frac{-b \pm \sqrt{b^2 - 4ac}}{2a},$$

we have

$$x = \frac{-(-4) \pm \sqrt{(-4)^2 - 4(2)(3)}}{2(2)}$$

$$= \frac{4 \pm \sqrt{16 - 24}}{4} = \frac{4 \pm \sqrt{-8}}{4}$$

$$= \frac{4 \pm 2\sqrt{2}i}{4} = \frac{2(2 \pm \sqrt{2}i)}{4}$$

$$= \frac{2 \pm \sqrt{2}i}{2}$$

$$= 1 \pm \frac{\sqrt{2}}{2}i.$$

Therefore, the solutions are $1 + \dfrac{\sqrt{2}}{2}i$ and $1 - \dfrac{\sqrt{2}}{2}i$.

Complex numbers can be manipulated as if they were binomial forms in real-number algebra, with the exception that i^2 is to be replaced with -1. The following example illustrates the mechanics of carrying out addition, subtraction, multiplication, and division.

Example 4 Perform the indicated operations and write each answer in the form $a + bi$.

(a) $(2 + 7i) + (3 - 4i)$

(b) $(6 + 5i) - (3 - 7i)$

(c) $(3 + 4i)(2 - i)$

(d) $\dfrac{4 + 5i}{3 - 2}i$

Solution (a) We remove parentheses and combine like terms, so that

$$(2 + 7i) + (3 - 4i) = 2 + 7i + 3 - 4i$$
$$= 2 + 3 + 7i - 4i$$
$$= 5 + 3i.$$

(b) We remove parentheses and combine like terms, so that

$$(6 + 5i) - (3 - 7i) = 6 + 5i - 3 + 7i$$
$$= 6 - 3 + 5i + 7i$$
$$= 3 + 12i.$$

(c) We multiply and replace i^2 by -1:

$$(3 + 4i)(2 - i) = 6 - 3i + 8i - 4i^2$$
$$= 6 - 3i + 8i + 4$$
$$= 10 + 5i.$$

(d) In order to eliminate i from the denominator, we multiply the numerator and denominator by the conjugate of $3 - 2i$; that is, $3 + 2i$, so that

$$\dfrac{4 + 5i}{3 - 2i} = \dfrac{4 + 5i}{3 - 2i} \cdot \dfrac{3 + 2i}{3 + 2i} = \dfrac{12 + 8i + 15i + 10i^2}{3^2 - (2i)^2}$$
$$= \dfrac{12 + 23i + 10(-1)}{9 - 4i^2}$$
$$= \dfrac{12 + 23i - 10}{9 + 4}$$
$$= \dfrac{2 + 23i}{13} = \dfrac{2}{13} + \dfrac{23}{13}i.$$

PROBLEM SET 4.4

In problems 1–12, convert the given square roots of negative numbers to the complex form $a + bi$.

1. $\sqrt{-36}$
2. $\sqrt{-49}$
3. $-\sqrt{-121}$
4. $-\sqrt{-144}$
5. $5 - \sqrt{-9}$
6. $3 + \sqrt{-4}$
7. $12 + \sqrt{-25}$
8. $-2 + \sqrt{-49}$
9. $3 - \sqrt{-32}$
10. $5 - \sqrt{-48}$
11. $5 + \sqrt{-72}$
12. $3 - \sqrt{-11}$

In problems 13–30, solve each equation.

13. $x^2 - 4x + 8 = 0$
14. $x^2 - 2x + 3 = 0$
15. $x^2 - 4x + 13 = 0$
16. $x^2 - \sqrt{7}x + 2 = 0$
17. $2x^2 - 3x + 4 = 0$
18. $2x^2 - 5x + 5 = 0$
19. $x^2 - 2x - 5 = 0$
20. $2x^2 - 6x + 7 = 0$
21. $5t^2 - 2t + 7 = 0$
22. $7y^2 - 8y + 3 = 0$
23. $3u^2 - u + 7 = 0$
24. $5m^2 - 8m + 17 = 0$
25. $4t^2 + 7t + 4 = 0$
26. $2u^2 - u + 5 = 0$
27. $4y^2 - 6y + 9 = 0$
28. $3t^2 - 6t + 5 = 0$
29. $2x^2 - 4x + 11 = 0$
30. $2y^2 + 6y + 5 = 0$

31. Verify that $2 + i$ is a solution of the equation $x^2 - 4x + 5 = 0$, and find another solution.

32. Verify that $\frac{2}{3} - i$ is a solution of the equation $9y^2 - 12y + 13 = 0$, and find another solution.

In problems 33–48, perform the indicated operation and write the answer in the form $a + bi$.

33. $(3 + 6i) + (2 - 3i)$
34. $(5 - 2i) + (7 + 3i)$
35. $(2 + 3i) + (2 - 3i)$
36. $(25 - 7i) + (25 + 7i)$
37. $(-2 - 3i) - (-3 + 4i)$
38. $(7 - 5i) - (3 + 2i)$
39. $(3 + 5i) - (7 - 2i)$
40. $(7 - 3i) - (7 + 3i)$
41. $(1 + 3i)(2 + 5i)$
42. $(-2 + 3i)(4 + 5i)$
43. $(5 - 12i)(5 + 12i)$
44. $(7 - 11i)(7 + 11i)$
45. $\dfrac{1}{3 + 2i}$
46. $\dfrac{3 + 4i}{-2 + 5i}$
47. $\dfrac{2 - 6i}{1 - 2i}$
48. $\dfrac{5 - 2i}{11 + 2i}$

Chapter 5

GRAPHING AND SYSTEMS OF LINEAR EQUATIONS

5.1 The Cartesian Coordinate System
5.2 Graphs of Linear Equations
5.3 Intercepts and Slope of a Line
5.4 Graphing Linear Inequalities

5.1 The Cartesian Coordinate System

In Chapter 2, we saw that a point on a number line, or a coordinate axis, can be specified by a number x called the *coordinate* of that point (Figure 1).

Figure 1

This number line was then used to *graph* a solution of an equation (or inequality) in one variable. For instance, to graph the solution $x = 3$ of the equation $2x - 1 = 5$, we place a *dot* at the point whose coordinate is 3 (Figure 2). In this section, we see how to graph the solutions of equations containing two variables.

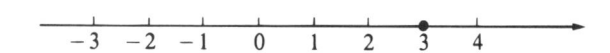

Figure 2

From *Beginning Algebra, 4th edition* by M.A. Munem and W. Tschirhart. Copyright © 2000 by Kendall/Hunt Publishing Company. Reprinted by permission.

Like an equation with one variable, an equation with two variables is either true or false, depending upon which particular pair of numbers is substituted for the variables. If a true statement results from such a substitution, we say that the substitution **satisfies** the equation. For instance, the substitution $x = 1$ and $y = 4$ satisfies the equation $3x + y = 7$, since $3(1) + 4 = 7$ is true. If a substitution $x = a$ and $y = b$ satisfies an equation with two unknowns x and y, we say that the **ordered pair** (a, b) is a **solution** of the equation. Thus, the ordered pair $(1, 4)$ is a solution of the equation $3x + y = 7$. The notation $(1, 4)$ is called an ordered pair because it is a pair of numbers written in a specific order. The first number in the ordered pair is to be substituted for x and the second number in the ordered pair is to be substituted for y. Thus,

$$(1, 4) \neq (4, 1)$$

Notice that the ordered pair $(1, 4)$ is not the only solution of the equation $3x + y = 7$. Another solution is $(2, 1)$, because when we replace x by 2 and y by 1, we obtain

$$3(2) + 1 = 6 + 1 + 7.$$

We have seen that both $(1, 4)$ and $(2, 1)$ are solutions of the equation $3x + y = 7$. Now we ask: How many solutions does this equation have? The answer is that we can find as many solutions as we want by selecting different values of x and calculating the corresponding values of y. It is impossible to actually list all the solutions to most equations in two variables. However, we can use a *graph* to represent these solutions.

To graph a solution of an equation with two variables, we use two perpendicular number lines L_1 and L_2, which meet at a common point called the **origin** (Figure 3). L_1 and L_2 are scaled as real number lines by using the origin as the 0 point for each of the two lines. The two number lines are called the **coordinate axes**. The coordinate axes L_1 and L_2 are often referred to as the **horizontal axis** or ***x* axis** and the **vertical axis** or ***y* axis**. Together, the x and y axes from the **Cartesian coordinate plane** or the ***xy* plane**. The numbers on the x axis are *positive to the right of the vertical axis* and *negative to the left of the vertical axis*. The numbers on the y axis are *positive above the horizontal axis* and **negative** *below the horizontal axis* (Figure 4).

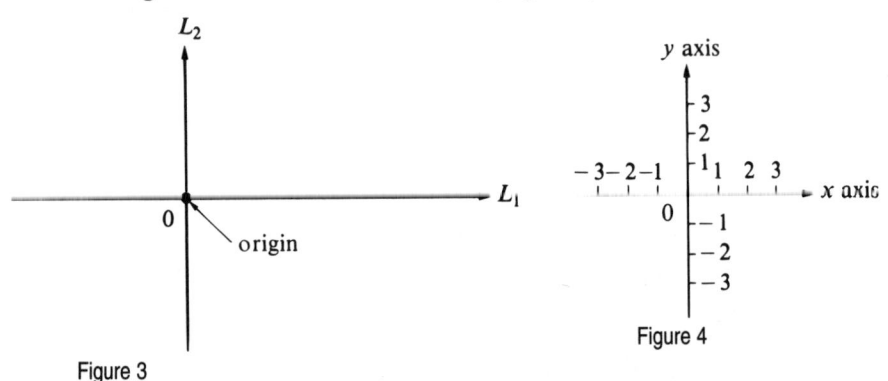

Figure 3

Figure 4

Each point P in the plane can be labeled according to its position. For instance, to label the point P in Figure 5, we drop a perpendicular line from P to the x axis. Note that this line intersects the x axis at a (the x coordinate of P). We also drop a perpendicular line from P to the y axis and see that it intersects the y axis at b (the y coordinate of P).

The x coordinate is called the **abscissa** of P and the y coordinate is called the **ordinate** of P. The abscissa and the ordinate are usually called the **coordinates** of the point P. These coordinates are written as an ordered pair (a, b),

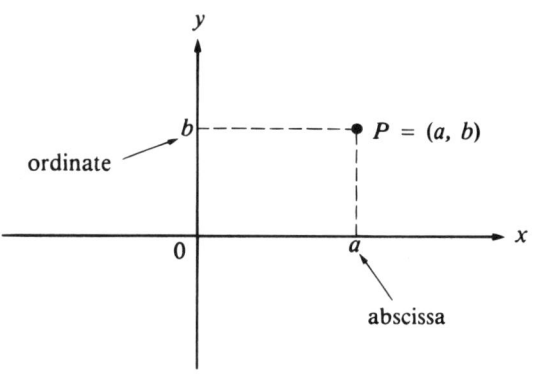

Figure 5

enclosed in parentheses, with the abscissa first and the ordinate second. To **plot** or **graph** the point P with coordinates (a, b) means to draw Cartesian coordinate axes and to place a dot representing P at the point with abscissa a and b. We can think of the ordered pair (a, b) as the numerical "address" of P. The correspondence between P and (a, b) helps us to identify the point P with its "address" (a, b) by writing $P = (a, b)$. With this identification in mind, we can call an ordered pair of real numbers (a, b) *a* **point.**

To graph the point $(1, 1)$, we start at the origin and move one unit to the right along the x axis, then one unit up from the x axis. Similarly, $(7, -5)$ is graphed by moving seven units to the right of the y axis and five units down. We locate $(-\frac{1}{2}, 0)$ by moving one-half unit to the left of the y axis, and no units up or down from the x axis (Figure 6).

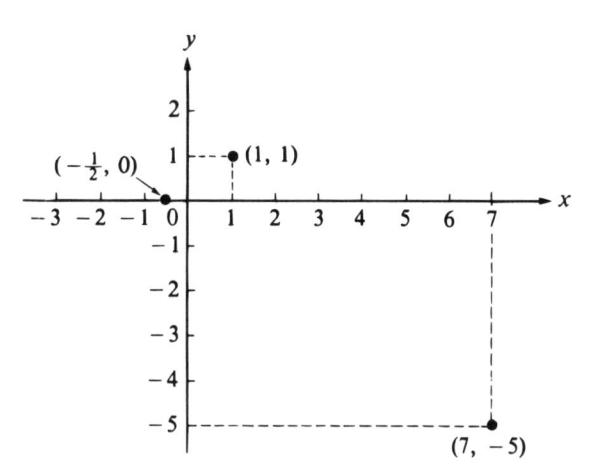

Figure 6

The x and y axis divide the plane into four regions, called **quadrants I, II, III,** and **IV** (Figure 7). Note that the quadrants do not include the coordinate axes.

Figure 7

Example 1 Plot each point and indicate which quadrant or coordinate axis contains the point.
(a) (2, 2) (b) (−2, 3) (c) (−3, −4) (d) (−4, 1)
(e) (1, 0) (f) (3, −4) (g) $(0, -\frac{3}{2})$ (h) (0, 0)

Solution The points are plotted in Figure 8:
(a) (2,2) lies in quadrant I.
(b) (−2, 3) lies in quadrant II.
(c) (−3, −4) lies in quadrant III.
(d) (−4, 1) lies in quadrant II.
(e) (1, 0) lies on the x axis.
(f) (3, −4) lies in quadrant IV.
(g) $(0, -\frac{3}{2})$ lies on the y axis.
(h) (0, 0), the origin, lies on both axes.

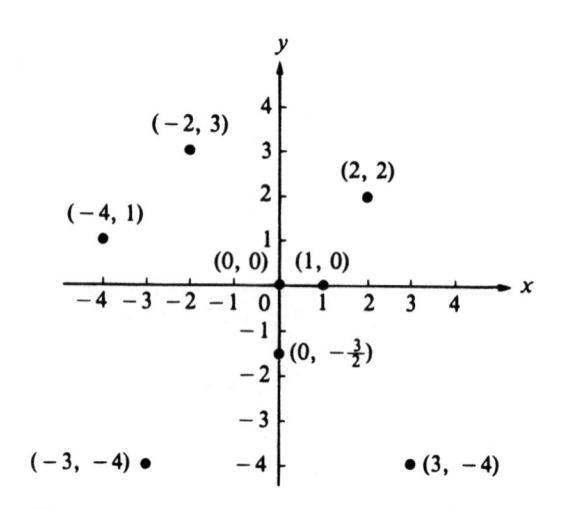

Figure 8

Example 2 Find the coordinates of each of the points described and graph the points.
(a) A: four units to the right of the y axis, and two units up from the x axis.
(b) B: three units to the left of the y axis, and one unit up from the x axis.
(c) C: two units to the left of the y axis, and four units down from the x axis.
(d) D: five units to the right of the y axis, and two units down from the x axis.

Solution
(a) $A = (4, 2)$ (b) $B = (-3, 1)$
(c) $C = (-2, -4)$ (d) $D = (5, -2)$

The graphs of these points are shown in Figure 9.

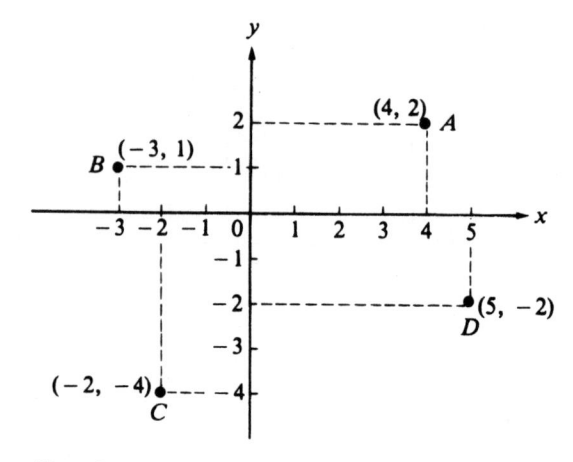

Figure 9

Example 3   Find the coordinates of each of the points shown in Figure 10.

Solution   From Figure 10, we see that

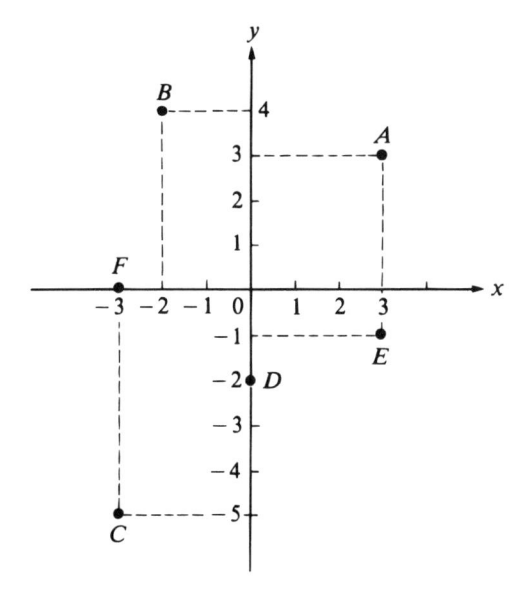

$$A = (3, 3), \quad B = (-2, 4),$$
$$C = (-3, -5), \quad D = (0, -2),$$
$$E = (3, -1), \quad F = (-3, 0).$$

Figure 10

Example 4   Which of the following ordered pairs are solutions of $3x + 2y = 13$?

$$(1, 2), \quad \left(0, \frac{15}{2}\right), \quad (3, 2), \quad \left(\frac{13}{3}, 0\right), \quad \text{and} \quad (-1, 8).$$

Plot the points that are solutions of the equation.

Solution   We test $(1, 2)$ in the equation $\quad 3x + 2y = 13$:
$$3(1) + 2(2) \stackrel{?}{=} 13$$
$$7 \neq 13.$$

Therefore, $(1, 2)$ is not a solution.
  Testing $(0, \frac{15}{2})$, we have:
$$3(0) + 2\left(\frac{15}{2}\right) \stackrel{?}{=} 13$$
$$15 \neq 13.$$

Therefore, $(0, \frac{15}{2})$ is not a solution.
  For $(3, 2)$ we have:
$$3(3) + 2(2) \stackrel{?}{=} 13$$
$$13 = 13.$$

Therefore, $(3, 2)$ is a solution.
  For $(\frac{13}{3}, 0)$ we have:
$$3\left(\frac{13}{3}\right) + 2(0) \stackrel{?}{=} 1$$
$$13 = 13.$$

Therefore, $(\frac{13}{3}, 0)$ is a solution.

For $(-1, 8)$ we have:
$$3(-1) + 2(8) \stackrel{?}{=} 13$$
$$13 \neq 13.$$

Therefore, $(-1, 8)$ is a solution.

The points $(3, 2)$, $(\frac{13}{3}, 0)$, and $(-1, 8)$ are plotted in Figure 11.

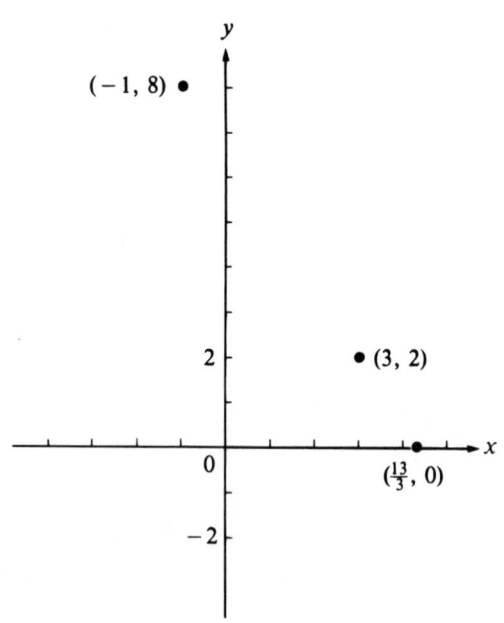

Figure 11

Example 5 Complete the following table for the equation $y = 2x - 4$:

x	-4	2		
y			4	-8

Plot the resulting points.

Solution We substitute the given values in the equation $y = 2x - 4$:

If $x = -4$, then $y = 2(-4) - 4 = -12$.
If $x = 2$, then $y = 2(2) - 4 = 0$.
If $y = 4$, then $4 = 2x - 4$, or $x = 4$.
If $y = -8$, then $-8 = 2x - 4$, or $x = -2$.

The completed table is:

x	-4	2	4	-2
y	-12	0	4	-8

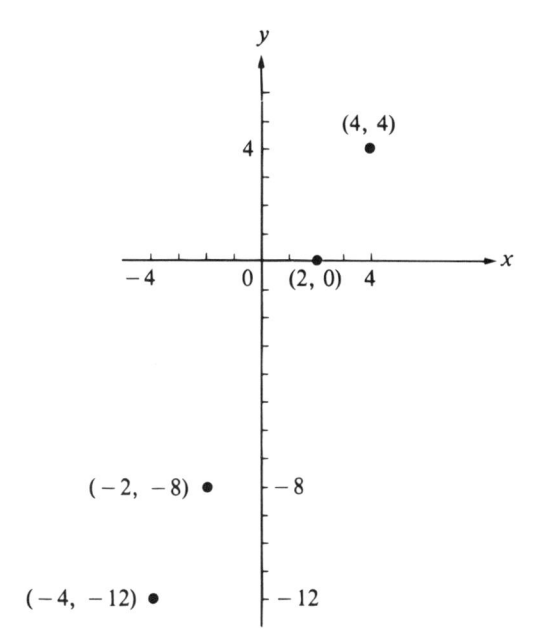

Now we plot the points $(-4, -12)$, $(2, 0)$, $(4, 4)$ and $(-2, -8)$ in Figure 12.

Figure 12

PROBLEM SET 5.1

In problems 1–20, plot each point and indicate which quadrant or coordinate axis contains the point.

1. $(2, -1)$
2. $(-3, -3)$
3. $(3, 2)$
4. $(2, \frac{1}{2})$
5. $(5, 0)$
6. $(-5, 4)$
7. $(-4, 3)$
8. $(\frac{7}{2}, \frac{5}{2})$
9. $(-\frac{1}{2}, -4)$
10. $(4, -\frac{3}{2})$
11. $(0, 2)$
12. $(0, 0)$
13. $(6, 7)$
14. $(3, 3)$
15. $(-3, 0)$
16. $(-7, 6)$
17. $(-2, -2)$
18. $(0, -5)$
19. $(2, -\frac{4}{3})$
20. $(-1, -1)$

In problems 21–30, find the coordinates of the points described and graph the points.

21. A point four units to the right of the origin and three units down from the x axis.
22. A point five units to the left of the y axis and two units up from the x axis.
23. A point seven units to the left of the y axis and on the x axis.
24. A point $3\frac{1}{2}$ units to the right of the y axis and six units down from the x axis.
25. A point eight units to the right of the y axis and ten units up from the x axis.
26. A point one unit to the left of the y axis and three units down from the x axis.
27. A point two units to the left of the y axis and four units up from the x axis.
28. A point eleven units to the right of the y axis and two units up from the x axis.
29. A point three units below the x axis and five units to the left of the y axis.
30. A point three units below the x axis and on the y axis.
31. Find the coordination of each of the points shown in Figure 13.

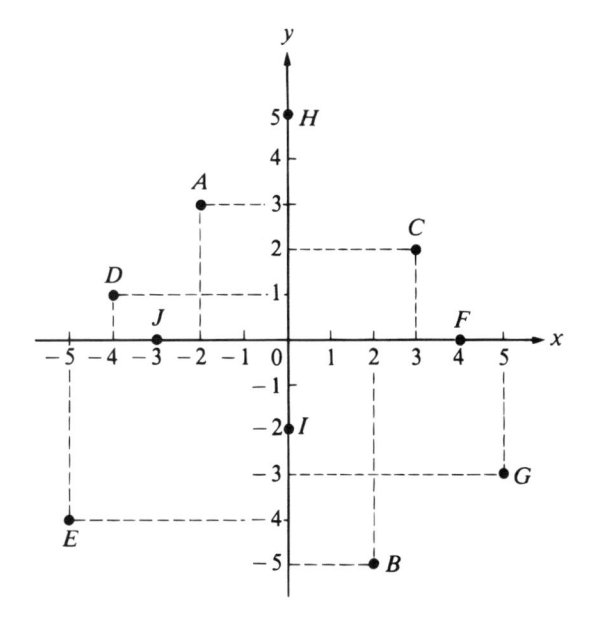

Figure 13

32. Which of the following pairs are solution of the equation $4x - 3y = 8$?
   $(1, -1)$, $(2, 0)$, $(-1, -4)$,
   $(1, \frac{4}{3})$,   and $(5, 4)$.

   Plot the points that are solutions of the equation.

33. Which of the following ordered pairs are solution of the equation $2x + y = 5$?
   $(2, 1)$ $(1, 2)$ $(4, -3)$,
   $(0, 5)$,   and $(2, \frac{3}{2})$.

   Plot the points that are solutions of the equation.

34. Complete the following table for the equation $y = -4x + 7$, and plot the resulting points:

x	1		0	$\frac{3}{4}$
y		-5	0	

35. Complete the following table for the equation $y = 3x - 2$, and plot the resulting points:

x	2	-1	0
y		7	-11

5.2 Graphs of Linear Equations

As we mentioned in Section 5.1, we can use a graph to represent the solutions of an equation containing two variables. The **graph of an equation** in two variables is the graph of all the solutions of the given equation. If an equation has a graph that is a straight line, we can graph the line by plotting a few points whose coordinates are solutions of the equation, and then drawing the line that contains these points.

For example, some solutions of the equation $3x + y = 7$ are $(1, 4)$, $(2, 1)$, $(3, -2)$, and $(4, -5)$. Graphing these points, we obtain a partial graph of the equation $3x + y = 7$ (Figure 1). Note that the points in the partial graph appear to be on a straight line. In fact, all solutions of $3x + y = 7$ do lie on a single straight line.

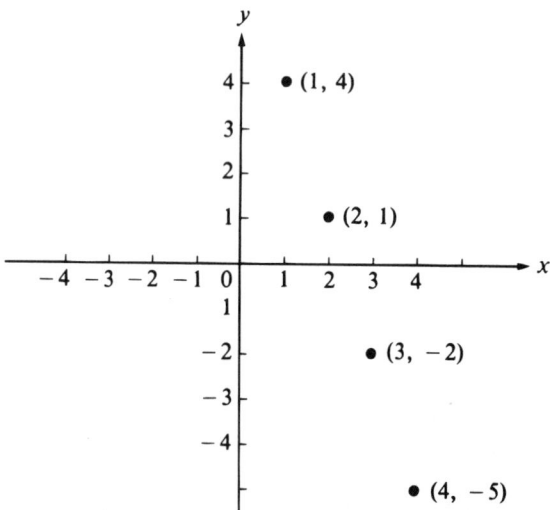

Figure 1

In general, any equation of the form
$$Ax + By = C,$$
where A, B, and C are constants (A and B cannot both equal 0) and x and y are variables has a graph that is a straight line.

For this reason, such equations are called **linear equations.** A straight line is completely determined by any two distinct (different) points of the line. We can graph a linear equation by finding any two different points whose coordinates are solutions of the equation, and then drawing a straight line through these points. However, to avoid mistakes, we suggest the following steps:

> Step 1. Find three points on the graph of the given equation. (Solve the equation for one of the variables. Choose three different values of one variable, and calculate the corresponding values of the other variable).
> Step 2. Plot the three points on a Cartesian coordinate system.
> Step 3. Draw a straight line through the three points.

Note: If you are unable to draw a straight line through the three points, then you probably made a mistake in step 1.

In examples 1–8, graph each equation.

Example 1 $y = x - 2$

Solution We choose three values for x and determine the corresponding values for y. This gives us three points (x, y). If we choose the values -1, 0, and 2 for x, we obtain the points in the following table:

x	$y = x - 2$	Point (x, y)
-1	$y = -1 - 2 = -3$	$(-1, -3)$
0	$y = 0 - 2 = -2$	$(0, -2)$
2	$y = 2 - 2 = 0$	$(2, 0)$

We plot these points and draw a straight line containing them (Figure 2).

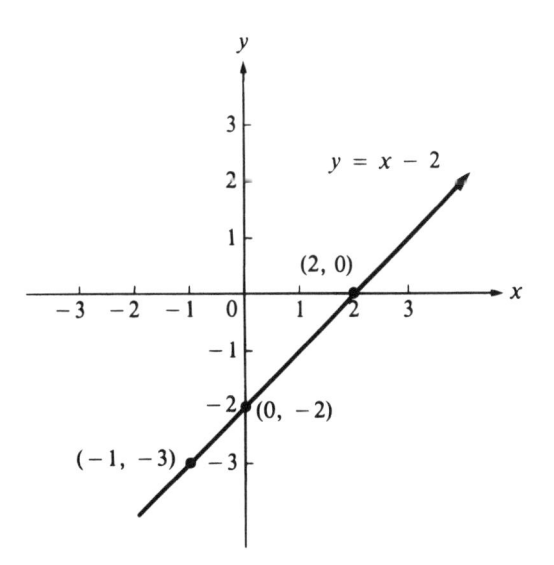

Figure 2

Example 2 $y = -2x + 4$

Solution We choose three values for x and determine the corresponding values for y to obtain three points of the form (x, y). If we choose the values $-2, 0,$ and 3 for x, we obtain the points shown in the following table:

x	$y = -2x + 4$	Point (x, y)
-2	$y = -2(-2) + 4 = 8$	$(-2, 8)$
0	$y = -2(0) + 4 = 4$	$(0, 4)$
3	$y = -2(3) + 4 = -2$	$(3, -2)$

We plot these points and draw a straight line containing them (Figure 3).

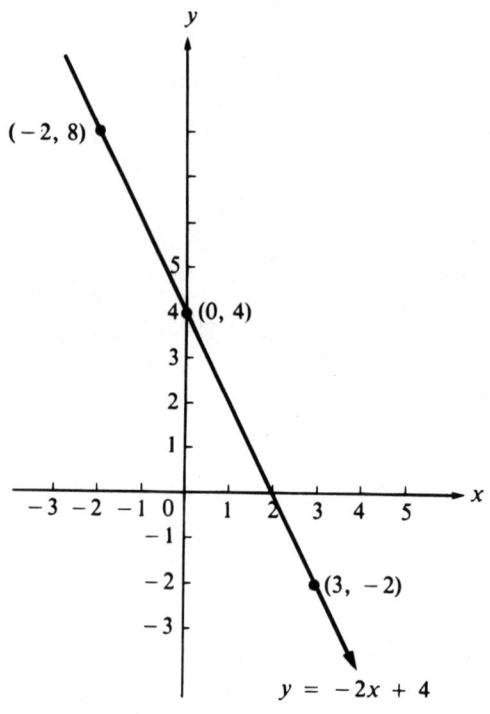

Figure 3

Example 3 $y = -\frac{3}{5}x + 2$

Solution We choose the values $-5, 0,$ and 5 for x and find the corresponding values for y:

x	$y = -\frac{3}{5}x + 2$	Point (x, y)
-5	$y = -\frac{3}{5}(-5) + 2 = 5$	$(-5, 5)$
0	$y = -\frac{3}{5}(0) + 2 = 2$	$(0, 2)$
5	$y = -\frac{3}{5}(5) + 2 = -1$	$(5, -1)$

We plot these points and draw a straight line containing them (Figure 4).

5.2 GRAPHS OF LINEAR EQUATIONS 227

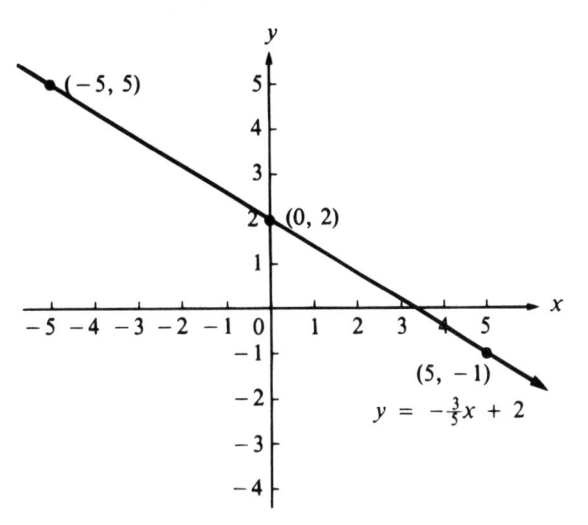

Figure 4

Example 4 $y = 3x$

Solution We choose the values -2, 0, and 1 for x and find the corresponding values for y:

x	$y = 3x$	Point (x, y)
-2	$y = 3(-2) = -6$	$(-2, -6)$
0	$y = 3(0) = 0$	$(0, 0)$
1	$y = 3(1) = 3$	$(1, 3)$

We plot these points and draw a straight line containing them (Figure 5).

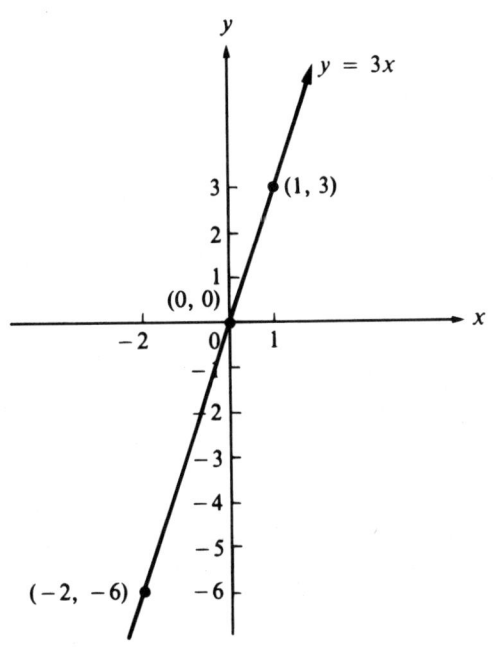

Figure 5

Example 5 $x - 2y = 6$

Solution   We choose three values for x and calculate the corresponding values for y. The calculations will be easier if first we solve the equation for y in terms of x:

$$-2y = -x + 6, \qquad y = \tfrac{1}{2}x - 3.$$

We choose the values of 6, 0, and 2 for x and calculate the values for y:

x	$y = \tfrac{1}{2}x - 3$	Point (x, y)
6	0	$(6, 0)$
0	-3	$(0, -3)$
2	-2	$(2, -2)$

We plot these points and draw a straight line containing them (Figure 6).

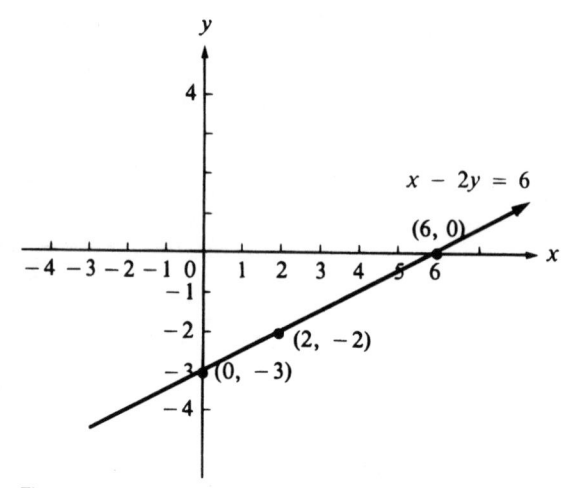

Figure 6

Example 5 $3x + 2y - 5 = 0$

Solution   We solve the equation for y in terms of x:

$$2y = -3x + 5, \qquad y = -\frac{3}{2}x + \frac{5}{2}.$$

We choose the values -1, 1, and 3 for x and calculate the y values:

x	$y = -\tfrac{3}{2}x + \tfrac{5}{2}$	Point (x, y)
-1	4	$(-1, 4)$
1	1	$(1, 1)$
3	-2	$(3, -2)$

We plot these points and draw a straight line containing them (Figure 7).

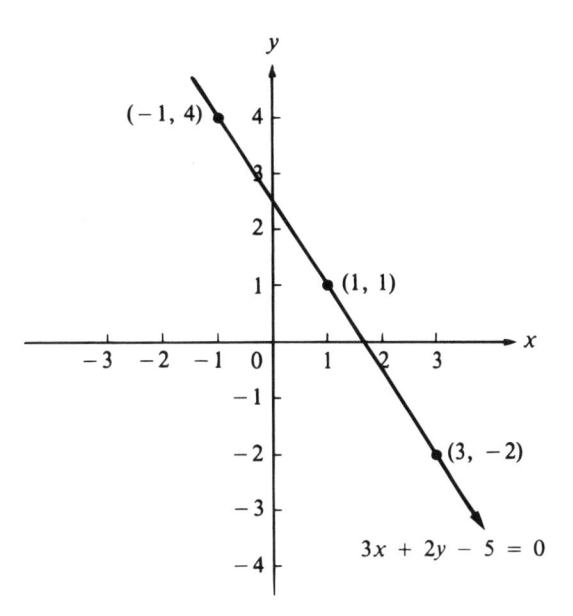

Figure 7

$3x + 2y - 5 = 0$

Example 7 $3y = 6$

Solution The equation $3y = 6$ is equivalent to $3y = 0x + 6$ or $y = 0x + 2$. We choose the values -4, 0, and 3 for x, and we obtain 2 in each case as a value for y. In fact, no matter what value is used for x, we obtain $y = 2$, since $0x = 0$. The graph is a *horizontal line* crossing the y axis at 2 (Figure 8).

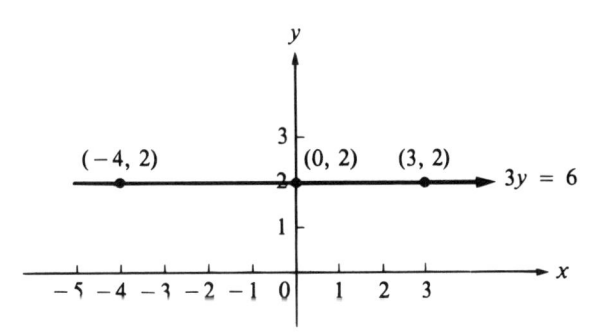

Figure 8

Example 8 $x + 3 = 0$

Solution The equation $x + 3 + 0$ is equivalent to $x = 0y - 3$ or $x = 0y - 3$. If we choose the values $-2, 0,$ and 3 for y, x will be -3. In fact, for any value of y we find that $x = -3$. The graph of $x + 3 = 0$ is a *vertical line* crossing the x axis at -3 (Figure 9).

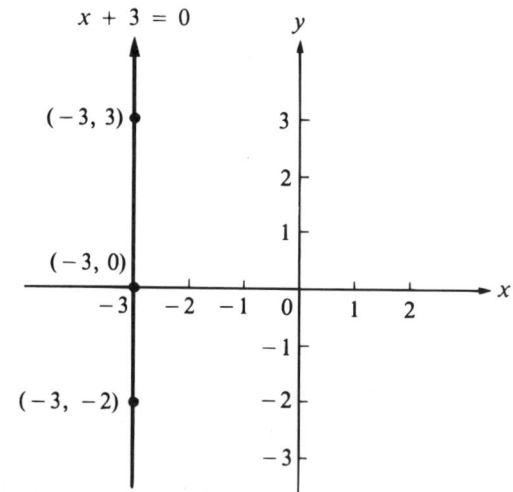

Figure 9

PROBLEM SET 5.2

In problems 1–32, graph each linear equation.

1. $y = x + 1$
2. $y = x - 3$
3. $y = -x + 4$
4. $y = -x + 5$
5. $y = 2x$
6. $y = \frac{2}{3}x$
7. $y = -\frac{3}{4}x$
8. $y = -2x$
9. $y = 3x + 2$
10. $y = 4x - 3$
11. $y = 3x - 1$
12. $y = -5x + 7$
13. $y - 5 - \frac{2}{5}x$
14. $y = 7 + \frac{3}{4}x$
15. $2x + 3y = 6$
16. $4x - 5y = 20$
17. $x - 4y = 5$
18. $-3x + 4y = 7$
19. $-6x + 5y - 15 = 0$
20. $7x - 4y + 8 = 0$
21. $4x - 16 = 0$
22. $-2x + 5 = 0$
23. $2y - 3 = 0$
24. $3y + 10 = 0$
25. $y = -\frac{2}{3}$
26. $y = \frac{3}{5}$
27. $x = 2$
28. $x = -\frac{1}{2}$
29. $x = -\frac{3}{4}$
30. $x = \frac{3}{7}$
31. $x = 0$
32. $y = 0$

33. A jogger's heartbeat N (in beats per minute) is related to her speed V (in feet per second) by the linear equation

$$N = \frac{5}{2}V + 50.$$

Sketch the graph of the equation, then determine the jogger's heartbeat if her speed is 16 feet per second.

34. A manufacturer has determined that the relationship between the profit P earned and the number x of items produced is given by the equation

$$P = 30x - 340.$$

Sketch the graph of this equation, and determine the profit P if the manufacturer produced 200 items.

35. In electronics, it was found that if the resistance of a simple electric circuit is 28 ohms, then the relationship between the electromotive force E (in volts) and the current I (in amperes) is given by the equation

$$E = 28I.$$

Sketch the graph of this equation for $0 \leq I \leq 1$, and determine E when $I = 0.5$ ampere.

36. In a certain income tax bracket, the amount y (in dollars) of income tax that an individual must pay the federal government is related to the amount x (in dollars) of income after deductions by the equation

$$y = 0.28x - 2{,}116.$$

Sketch the graph of the equation for $14{,}000 \leq x \leq 25{,}000$ and find y when $x = \$21{,}000$.

5.3 Intercepts and Slope of a Line

In Section 5.2, we considered the problem of graphing linear equations of the form

$$Ax + By = C.$$

Now we examine three special quantities associated with straight lines and their graphs. One is the slope of the line. The other two are the x and y intercepts of the graph. Consider the equation

$$3x - 2y = 12.$$

In order to graph this equation using the method of Section 5.2, we would plot three points and connect them by a straight line. However, we can graph a linear equation by finding any two points, and the easiest points to find are those where the graph crosses the x axis and the y axis. To find where the line crosses the x axis, let $y = 0$ and solve the equation for x. We have:

$$3x = 12 \quad \text{or} \quad x = 4.$$

To find where the line crosses the y axis, we let $x = 0$ and solve the equation for y. We have:

$$-2y = 12 \quad \text{or} \quad y = -6.$$

The numbers 4 and -6 are called the x intercept and the y intercept, respectively, so the graph of the equation $3x - 2y = 12$ is found by drawing the straight line that contains these two points (Figure 1).

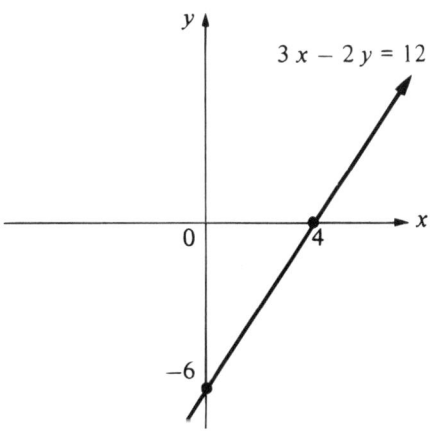

Figure 1

In general:

DEFINITION 1

The **x intercept** of a line is the x coordinate of the point where the graph crosses the x axis.
The **y intercept** is the y coordinate of the point where the graph crosses the y axis.

To determine the x intercept of line L, set $y = 0$ in any equation of L and solve for x. To determine the y intercept, set $x = 0$ and solve for y.

In Examples 1 and 2, find the x and y intercepts of the line with the given equation, and sketch the graph.

Example 1 $3x + 2y = 6$

Solution   To determine the x intercept, we set $y = 0$, so that
$$3x + 2(0) = 6$$
$$3x = 6$$
$$x = 2.$$

To determine the y intercept, we set $x = 0$, so that
$$3(0) + 2y = 6$$
$$2y = 6$$
$$y = 3.$$

We plot the x intercept and the y intercept and draw the straight line determined by these two points (Figure 2).

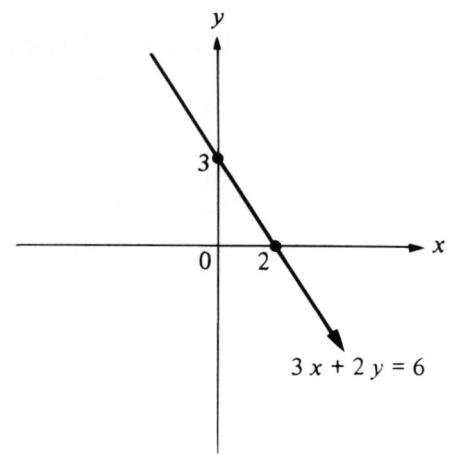

Figure 2

Example 2 $2x - 3y = 10$

Solution   We determine the x and y intercepts. Let $y = 0$,
$$2x - 3(0) = 10$$
$$2x = 10$$
$$x = 5.$$

Let $x = 0$,
$$2(0) - 3y = 10$$
$$-3y = 10$$
$$y = -\tfrac{10}{3}.$$

We plot these intercepts and draw the straight line through them (Figure 3).

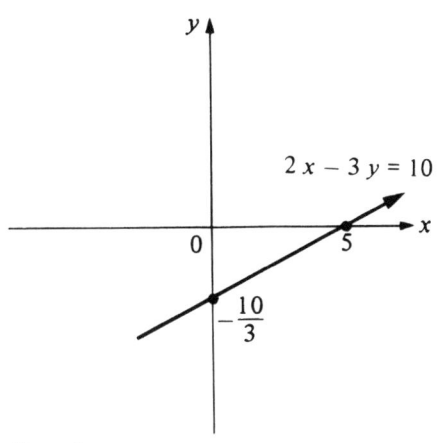

Figure 3

The Slope of a Line

We commonly use the word "slope" to refer to steepness of an incline, or some deviation from the horizontal. For instance, we speak of a ski slope, or the slope of a hill or the slope of a roof. In mathematics, we express the slope (or steepness) of a line between two points as a ratio of the change in vertical distance, called the rise (Figure 4a), to the change in horizontal distance, called the run (Figure 4b).

In Figure 5, the vertical distance $|\overline{BC}|$ is the rise, and the horizontal distance $|\overline{AC}|$ is the run. The ratio of the *rise* to the *run* is called the **slope** of the line segment $\overline{AB}$, and is represented by the letter m; that is,

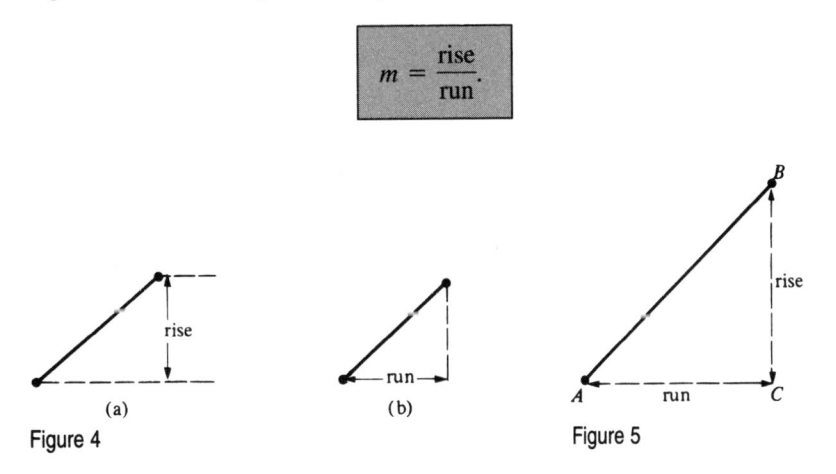

$$m = \frac{\text{rise}}{\text{run}}.$$

Figure 4

Figure 5

If the line segment $\overline{AB}$ is horizontal, its rise is zero, and its slope $m =$ rise/run $= 0$ (Figure 6a). If $\overline{AB}$ slants upward to the right, its rise is considered to be positive, and the slope $m =$ rise/run is positive (Figure 6b). If $\overline{AB}$ slants downward to the right, its rise is considered to be negative; hence, its slope $m =$ rise/run is negative (Figure 6c). The run is always considered to be nonnegative.

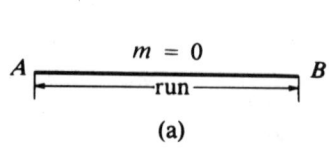

234 CHAPTER 5 GRAPHING AND SYSTEMS OF LINEAR EQUATIONS

Figure 6

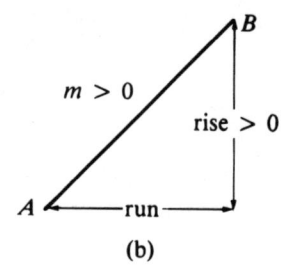

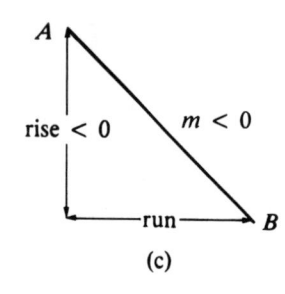

Now let's consider the line segment $\overline{AB}$ in which $A = (x_1, y_1)$ and $B = (x_2, y_2)$ (Figure 7). Note in Figure 7 that B is above and to the right of A. The line segment $\overline{AB}$ has rise $= y_2 - y_1$ and run $= x_2 - x_1$. Thus the slope is given by the formula

$$m = \frac{y_2 - y_1}{x_2 - x_1}$$

provided that $x_1 \neq x_2$.

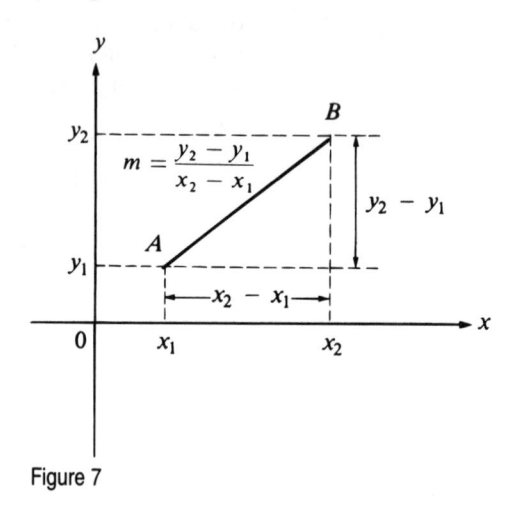

Figure 7

Example 3 Sketch each line segment $\overline{AB}$ and find its slope m.
(a) $A = (4, 1)$ and $B = (7, 6)$
(b) $A = (3, 9)$ and $B = (7, 4)$
(c) $A = (-1, 4)$ and $B = (3, 4)$
(d) $A = (2, -3)$ and $B = (2, 2)$

Solution The line segments are sketched in Figure 8.

(a) $m = \dfrac{y_2 - y_1}{x_2 - x_1} = \dfrac{6 - 1}{7 - 4} = \dfrac{5}{3}$

(b) $m = \dfrac{y_2 - y_1}{x_2 - x_1} = \dfrac{4 - 9}{7 - 3} = \dfrac{-5}{4}$

(c) $m = \dfrac{y_2 - y_1}{x_2 - x_1} = \dfrac{4 - 4}{3 - (-1)} = \dfrac{0}{4} = 0$

(d) m is undefined because $x_2 - x_1 = 2 - 2 = 0$.

5.3 INTERCEPTS AND SLOPE OF A LINE

$B = (7, 6)$, $A = (4, 1)$, $m = \frac{5}{3}$

(a)

$A = (3, 9)$, $B = (7, 4)$, $m = -\frac{5}{4}$

(b)

$A = (-1, 4)$, $B = (3, 4)$, $m = 0$

(c)

$B = (2, 2)$, $A = (2, -3)$, m is undefined

(d)

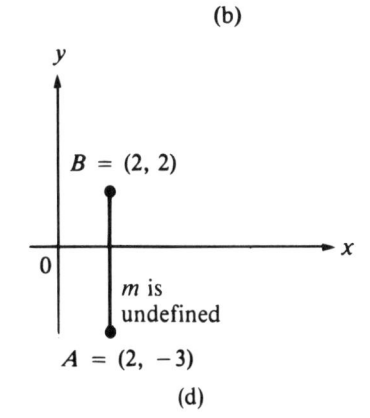

Figure 8

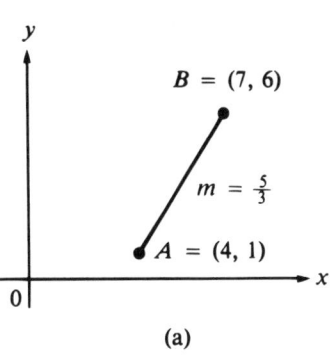

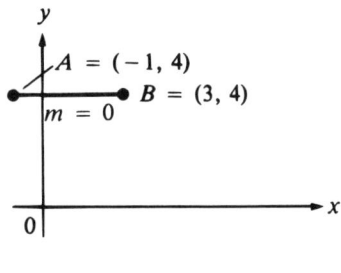

If two line segments $\overline{AB}$ and $\overline{CD}$ lie on the same line L, then they have the same slope (Figure 9). In fact, the slope of a line L is the same no matter which two points of the line are considered. The common slope of all line segments lying on a line L is called the slope of L.

Figure 9

Equation of a Line

Suppose that we are given a point on a line and the slope of the line. We ask: Can we find an equation to describe all the points on the line? To answer this question, consider a nonvertical line L having slope m and containing the point $P_1 = (x_1, y_1)$ (Figure 10.) If $P = (x, y)$ is any other point on L, then the slope m is given by

$$m = \frac{y - y_1}{x - x_1}.$$

We multiply both sides of this equation by $x - x_1$:

$$y - y_1 = m(x - x_1).$$

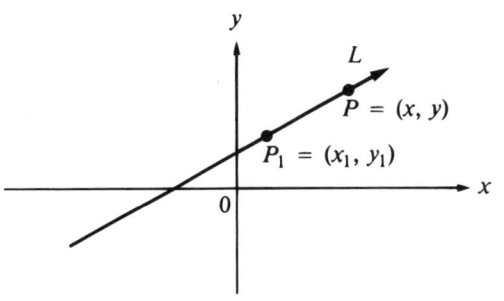

Figure 10

The latter equation is known as the **point-slope form** of an equation for the line L, which contains the point $P_1 = (x_1, y_1)$ and has the slope m.

For example, the equation of the line that contains the point $P_1 = (x_1, y_1) = (2, 3)$ and whose slope is $m = 5$ is given by

$$y - 3 = 5(x - 2).$$

If the point $P_1 = (x_1, y_1) = (0, b)$, then an equation of the line L is given by

$$y - b = m(x - 0),$$

which simplifies to

$$y - b = mx$$

or

$$y = mx + b.$$

This is called the **slope-intercept form** of an equation for L. *In the equation $y = mx + b$ the coefficient of x is the slope and the constant term b is the y intercept.* In other words, when a linear equation in two variables is solved for y, the coefficient of x is the slope, and the constant term is the y intercept.

Example 4 Find the slope and y intercept of the line $3x + 2y - 12 = 0$. Sketch the graph.

Solution First, we solve the equation for y in terms of x to put it in slope-intercept form:

$$3x + 2y - 12 = 0$$
$$2y = -3x + 12$$
$$y = -\frac{3}{2}x + 6$$

Therefore the slope is $-\frac{3}{2}$ and the y intercept is 6 (Figure 11).

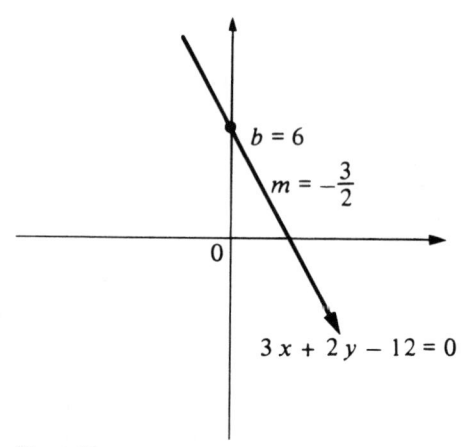

Figure 11

Example 5 Find the slope-intercept form of an equation of a line L having slope $-\frac{2}{3}$ and y intercept $b = 5$.

Solution We substitute $m = -\frac{2}{3}$ and $b = 5$ in the equation $y = mx + b$ to obtain

$$y = -\frac{2}{3}x + 5.$$

Example 6 Find an equation for the line that contains the point $(-2, 3)$ and has slope $m = 4$. Express the equation in slope-intercept form.

Solution We substitute $x_1 = -2$, $y_1 = 3$, and $m = 4$ in $y - y_1 = m(x - x_1)$:

$$y - 3 = 4[x - (-2)]$$

or

$$y - 3 = 4(x + 2)$$
$$y = 4x + 8 + 3$$
$$y = 4x + 11.$$

PROBLEM SET 5.3

In problems 1–20, find the x intercept and the y intercept of each line, and sketch the graph.

1. $2x - y = 8$
2. $3x - y = 3$
3. $2x - 3y + 6 = 0$
4. $3x - 5y - 15 = 0$
5. $2x + y - 4 = 0$
6. $2x + 4y = -9$
7. $-2x + 3y - 12 = 0$
8. $5x + y = 20$
9. $-3x + y = 0$
10. $5x + 3y = -10$
11. $7x + 2y - 14 = 0$
12. $-3x - y = 7$
13. $\frac{1}{2}x - y = -3$
14. $x - \frac{5}{3}y = 5$
15. $y = \frac{2}{3}x + 2$
16. $y = -\frac{1}{3}x + 7$
17. $y + 1 = -\frac{3}{4}(x - 2)$
18. $y + 2 = -\frac{2}{3}(x - 1)$
19. $0.3x - 0.6y = 0.18$
20. $0.5x + 0.2y = 0.54$

In problems 21–30, find the slope of the line segment joining the given points.

21. $(3, 2)$ and $(5, 6)$
22. $(1, 3)$ and $(-7, 5)$
23. $(-12, 1)$ and $(5, -10)$
24. $(10, 5)$ and $(-1, 2)$
25. $(-2, 1)$ and $(-3, 4)$
26. $(4, -1)$ and $(4, 3)$
27. $(1, 3)$ and $(-2, 3)$
28. $(-1, 5)$ and $(2, 5)$
29. $(3, 2)$ and $(3, -5)$
30. $(4, 3)$ and $(-7, 3)$

In problems 31–40, write each equation in the slope-intercept form, find the slope and the y intercept, then sketch the graph.

31. $5x + 2y + 10 = 0$
32. $x - 5y - 9 = 0$
33. $5x - 3y - 15 = 0$
34. $x + 2y = 2$
35. $2y + 3x = 0$
36. $4y - 7x = 0$
37. $-5x - 2y = -10$
38. $-5x + 4y = -20$
39. $y + 5 = 0$
40. $y - 3 = 0$

In problems 41–56, find an equation of the line L.

41. L contains $P = (-1, 2)$ and has slope $m = 5$.
42. L contains $P = (0, 3)$ and has slope $m = -7$.
43. L contains $P = (7, 3)$ and has slope $m = -3$.
44. L contains $P = (-1, 4)$ and has slope $m = \frac{2}{5}$.
45. L contains $P = (5, -1)$ and has slope $m = \frac{3}{7}$.
46. L contains $P = (-4, 6)$ and has slope $m = -\frac{3}{4}$.
47. L contains $P = (0, 0)$ and has slope $m = -\frac{3}{8}$.
48. L contains $P = (0, 0)$ and has slope $m = \frac{4}{9}$.
49. L contains $P = (-1, -5)$ and has slope $m = 0$.
50. L contains $P = (-\frac{1}{2}, 4)$ and has slope $m = 0$.
51. L contains $P_1 = (-3, 2)$ and $P_2 = (3, 5)$.
52. L contains $P_1 = P_1 = (-2, 4)$ and $P_2 = (0, 1)$.
53. L has slope $m = -3$ and y intercept $b = 5$.
54. L has slope $m = -7$ and y intercept $b = 2$.
55. L has slope $m = -\frac{3}{7}$ and y intercept $b = 0$.
56. L has slope $m = -\frac{5}{11}$ and y intercept $b = 0$.

In problems 57 and 58, use the fact that two different nonvertical lines in the plane with slopes m_1 and m_2 are **parallel** if and only if $m_1 = m_2$ (Figure 12) to show that the line containing A and B is parallel to the line containing C and D.

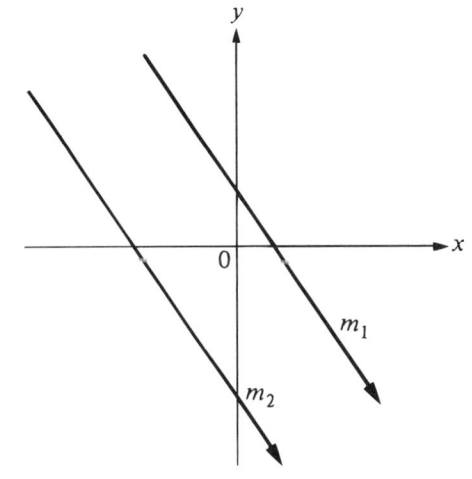

Figure 12

57. $A = (-2, 5)$, $B = (2, -1)$, $C = (-4, 1)$, and $D = (0, -5)$.
58. $A = (2, 4)$, $B = (3, 8)$, $C = (5, 1)$, and $D = D = (4, -3)$.

In problems 59 and 60, use the fact that two different nonvertical lines in the plane with slopes m_1, and m_2 are

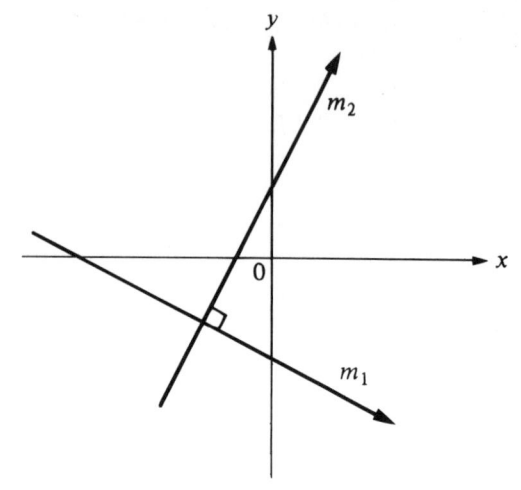

Figure 13

perpendicular if and only if $m_1 m_2 = -1$ (Figure 13) to show that the line containing A and B is perpendicular to the line containing C and D.

59. $A = (-1, 1)$, $B = (1, 5)$, $C = (-2, -3)$, and $D = (2, -5)$.
60. $A = (2, 4)$, $B = (3, 8)$, $C = (8, -2)$, and $D = (-4, 1)$.

In problems 61–63, first find the slope of the line containing the two points, then use the equation $y = mx + b$ to find b.

61. Water freezes at 32° Fahrenheit and 0° Celsius, and boils at 212° Fahrenheit and 100° Celsius. Find a linear relationship between the Fahrenheit and Celsius scales.

62. In a psychological experiment, it was found that the relationship between the scores on two types of personality tests—test A and test B—was linear. A student who scored a 55 on test A would score a 95 on test B, and a student who scored 35 on test A would score 55 on test B. What score on test B would a student get if she scored a 60 on test A?

63. The annual simple interest earned I is related to the amount P invested in a bank by a linear equation. If you invest $650, you earn $45.50 in 1 year, and if you invest $1,375, you earn $96.25 in a year.
    (a) Write an equation that expresses the relationship between the annual interest earned I and the amount P invested.
    (b) How much money should you invest to earn $515.55 in one year?

OBJECTIVES

1. Graph Linear Inequalities in Two Variables
2. Graph Compound Inequalities
3. Solve Applied Problems
4. Use Technology Exploration

5.4 Graphing Linear Inequalities

Graphing Linear Inequalities in Two Variables

Recall that the graph of a linear equation in two variables is a straight line. Here we graph linear inequalities in two variables that can be written in the standard form $Ax + By < C$ where A, B and C are real numbers and A and B are not both zero. The symbol $<$ may be replaced by $\geq$, $\leq$, or $>$. Examples of linear inequalities are:

$$3x - y < 6, \quad 5x - y > 7, \quad 0x + y \leq 1, \quad \text{and} \quad x + 0y \leq 2$$

A **solution** of a linear inequality in two variables is an ordered pair of numbers (x, y) that makes the inequality true. All the ordered pairs that make the inequality true is called the **solution set**. The **graph** of an inequality in two variables represents its solution set. To graph a linear inequality in x and y, we first graph the **boundary line** of the related equation. This line divides the xy plane into two regions called **half planes,** one above the line and one below it (assuming this line is not vertical). To determine which half-plane satisfies the inequality, we test a point in either region.

For example, to solve the inequality $y \leq 3x + 6$, first we graph the boundary line $y = 3x + 6$ (Figure 1). This line divides the xy plane into an upper half-plane A and a lower half-plane C. The set of all points on the line is represented by B. The point $(0, 0)$ is in the half-plane C. It makes the inequality true:

$$0 \leq 3(0) + 6.$$

The solution of the inequality $y \leq 3x + 6$ consists of all points in the half-plane C and all points on the boundary line B.

The following step procedure may be used to graph linear inequalities in two variables:

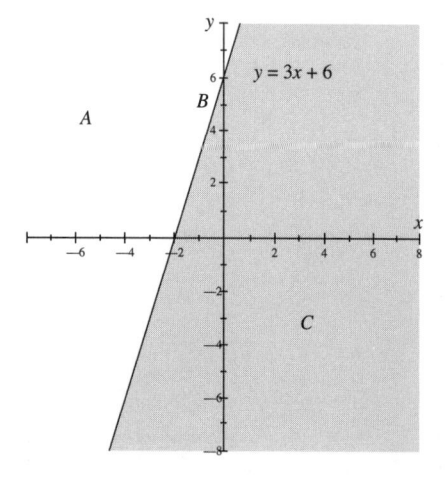

Figure 1

Graphing Linear Inequalities in Two Variables

Step 1. Sketch the graph of the boundary line obtained by replacing the inequality sign with an equal sign and graph this related equation. If the inequality has the symbol < or >, draw a dashed line. A dashed line indicates that the boundary line is not part of the solution set. If the inequality has the symbols ≤ or ≥, draw a solid line to indicate that points on the line are solutions of the inequality.

Step 2. Determine which half-plane corresponds to the inequality. To do this, select any convenient **test point** (x, y) not on the boundary line. Substitute the coordinates of this test point into the original inequality.

   (i) If the point makes the original inequality true, shade the half-plane containing the test point.

   (ii) If the point makes the original inequality false, shade the other half-plane.

Example 1 Graphing a Linear Inequality

Sketch the graph of the inequality $y < x + 2$

Solution **Step 1.** The boundary line for this inequality is the graph of the related equation $y = x + 2$. We draw the graph of a dashed boundary line because the inequality contains the symbol < (which is the strict inequality, less than but not equal to).

Step 2. To determine which half-plane is the solution, we select the test point $(1, 2)$ not on the boundary line. Replacing x by 1 and y by 2 in the original inequality leads to the following statement

$$y < x + 2$$
$$2 < 1 + 2$$
$$2 < 3 \text{ which is true.}$$

Because $(1, 2)$ satisfies the inequality, so does every point on the same side of the boundary line as $(1, 2)$. Therefore, we shade the lower half-plane that contains the point $(1, 2)$ (Figure 2).

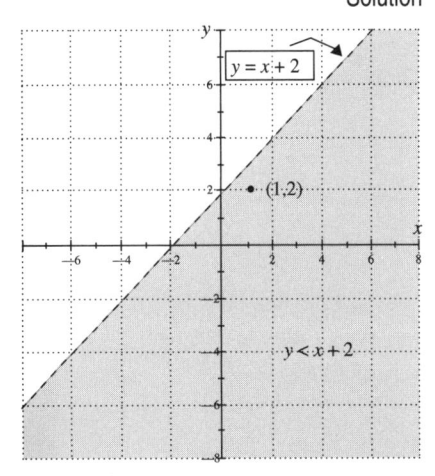

Figure 2

Example 2 Graphing a Linear Inequality

Sketch the graph of the inequality $2x - y \le 1$

Solution **Step 1.** Because the inequality contains the symbol ≤ (meaning less than or equal to), we graph the related equation as a solid boundary line

$$2x - y = 1$$

Step 2. We test the inequality at the point $(1, -3)$ not on the boundary to determine which half-plane satisfies the inequality

$$2x - y \le 1$$

When $x = 1$ and $y = -3$, we have

$$2(1) - (-3) \le 1$$
$$5 \le 1 \text{ False}$$

This point does not satisfy the inequality, so the correct half-plane to shade is the upper half-plane that does not contain the point $(1, -3)$ (Figure 3).

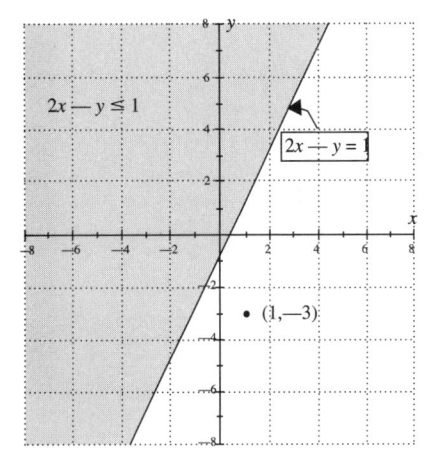

Figure 3

Example 3 Graphing Special Linear Inequalities

Sketch the graph of each inequality

(a) $y \geq 3$ \hspace{2cm} (b) $x < -1$

Solution (a) First we draw the boundary line as a solid horizontal line $y = 3$ since the symbol $\geq$ is used. Next we choose $(-2, 1)$ as a test point and find that $1 \geq 3$ is false. Therefore, we shade the upper half-plane that does not contain the point $(-2, 1)$ (Figure 4a).

(b) We begin by drawing the boundary line as the dashed line $x = -1$, since equality is not included. Then we choose $(0, 3)$ as a test point and find that $0 < -1$ is false. Therefore, we shade the half-plane (to the left of the vertical boundary line) which does not contain the point $(0, 3)$ (Figure 4b).

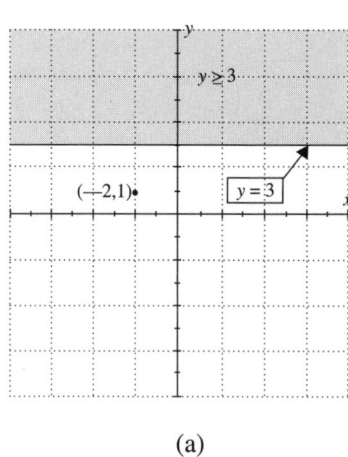

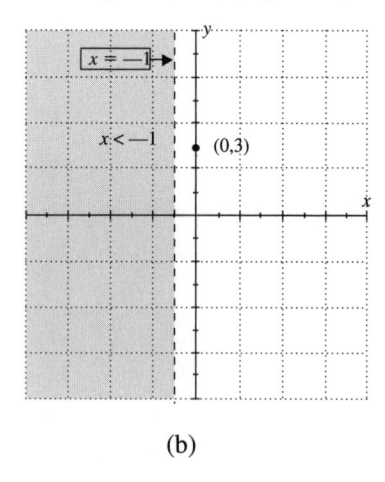

Figure 4

Graphing Compound Inequalities

In section 2.4 we solved compound inequalities containing one variable. Linear inequalities containing two variables can also be connected with the words "and" or "or" to form compound inequalities.

Graphing a Compound Inequality

Example 4 Sketch the graph of the compound inequality

$$x \geq 2 \text{ and } y \geq 4x + 3$$

Solution Figure 5a shows the graph of $x \geq 2$ to be the region to the right of and including the solid line $x = 2$. Figure 5b shows the region above and including the solid boundary line $y \geq 4x + 3$. Because the inequalities are connected with the word "and", the graph of the compound inequality is the area common to both or the intersection of the two graphs (Figure 5c).

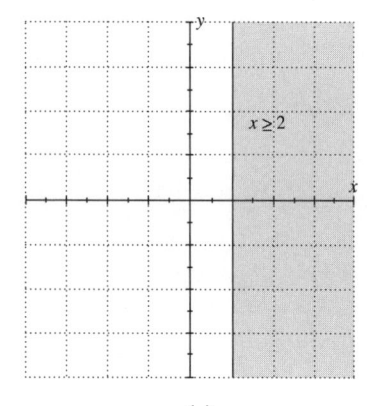

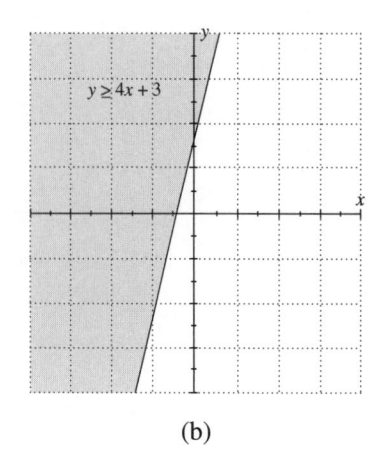

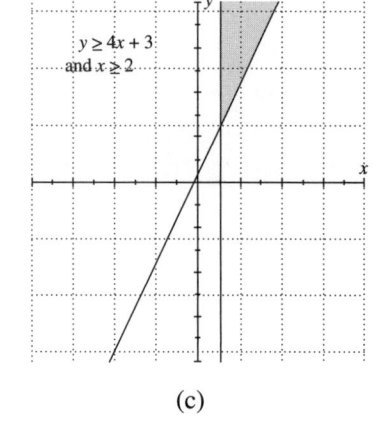

Figure 5

5.4 GRAPHING LINEAR INEQUALITIES 241

Example 5 Graphing a Compound Inequality

Sketch the graph of the compound inequality
$$y \geq x - 2 \text{ or } y \leq 1$$

Solution Figure 6a shows the graph of $y \geq x - 2$ as the region above and including the solid boundary line $y = x - 2$. Figure 6b is the graph of $y \leq 1$ as the region below and including the boundary line $y = 1$. This compound inequality uses the connector "or" so that the graph is all the points in one graph or the other as shown in Figure 6c.

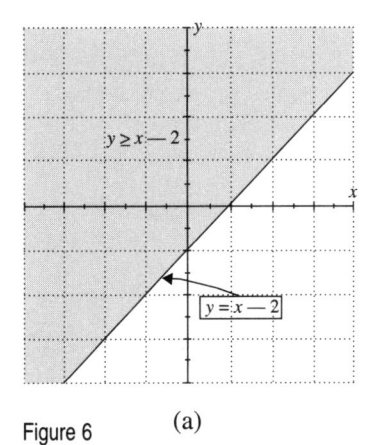

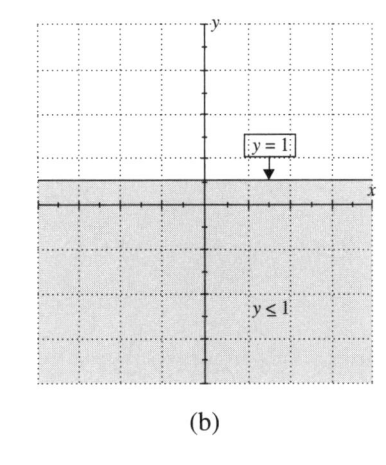

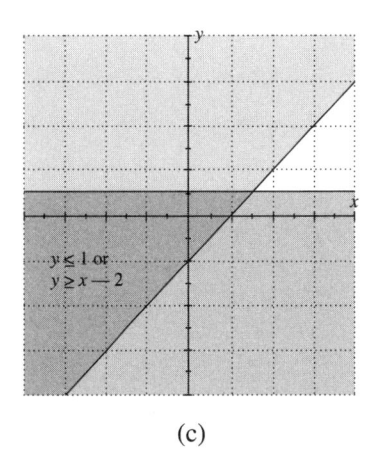

Figure 6 (a) (b) (c)

Example 6 Identifying a Region by Using Linear Inequalities

Identify the region that satisfies all the conditions.
$$\begin{cases} x + 2y \leq 8 \\ x \geq 0 \\ y \geq 0 \end{cases}$$

Solution The graph of the inequality $x + 2y \leq 8$ consists of all points in the plane on or below the solid line $x + 2y = 8$. The graph of $x \geq 0$ consists of all points on or to the right of the y axis, while the graph of $y \geq 0$ includes all points in the plane on or above the x axis. Therefore, the graph of these inequalities is the shaded region in the first quadrant including that part of the coordinate axes (Figure 7).

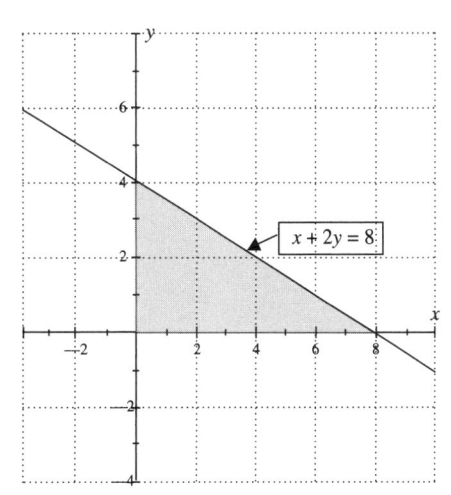

Figure 7

Solving Applied Problems

Linear inequalities in two variables are used in a variety of applications and models as the next example shows.

Example 7 **Solving a Nutrition Problem**

A nutritionist determines that an apple contains 100 calories and a pear contains 120 calories. Suppose that a person eats an apple and a pear for a mid-morning snack several times a week. The person wishes to keep the total number of calories per week for the snack to be at most 1,180. If x and y represent the number of apples and pears consumed per week respectively, write a linear inequality to model this situation and graph it.

Solution Since x and y represent the number of apples and pears consumed per week, the total number of calories consumed per week for the snack is modeled by the following linear inequality:

$$100x + 120y \leq 1180$$

with the restriction that $x \geq 0$ and $y \geq 0$. Figure 8 shows the shaded region which describes the solution set of this inequality.

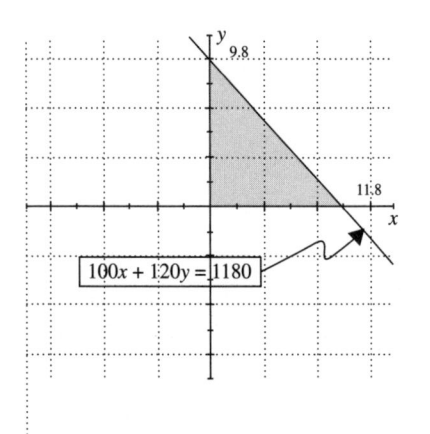

Figure 8

Using Technology Exploration

Linear inequalities can be graphed using a grapher with a SHADE feature.

Example 8 **Using a Grapher to Graph a Linear Inequality**

Use a grapher with a SHADE feature to graph the inequality $3x + 4y \leq 12$.

Solution To use a grapher the inequality must first be solved for y in terms of x. That is,

$$3x + 4y \leq 12$$
$$4y \leq -3x + 12$$
$$y \leq -\frac{3}{4}x + 3$$

and graph the above inequality (Figure 9).

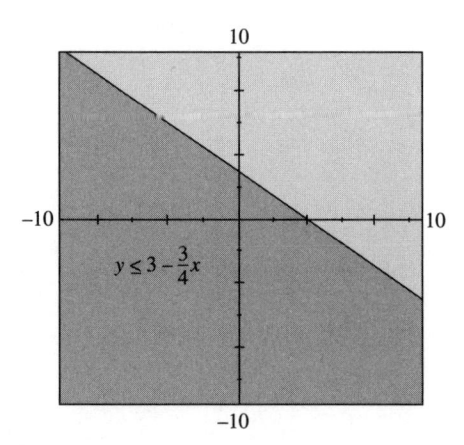

Figure 9

PROBLEM SET 5.4

Mastering the Concepts

In Problems 1 and 2, match the shaded region with the appropriate inequality.

(i) $y > 2x - 3$ (ii) $x + y > 4$ (iii) $10y > -x + 5$
(iv) $y > x$ (v) $y < 3$ (vi) $x \geq 2$

1. (a)

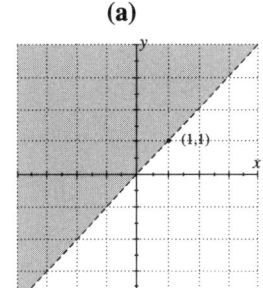

2. (a)

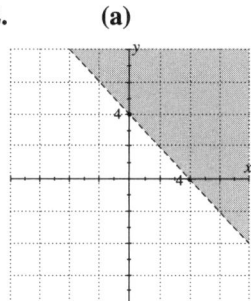

1. (b)

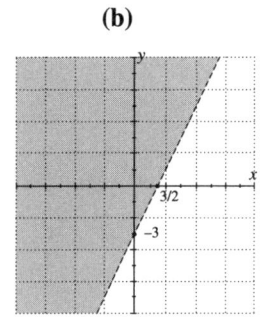

2. (b)

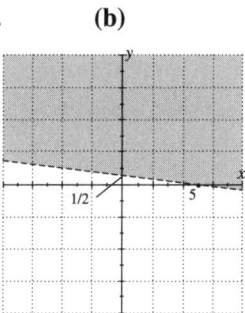

1. (c)

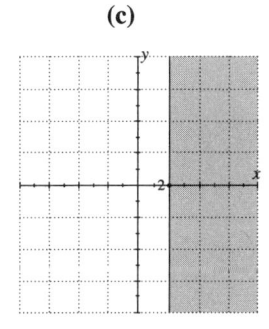

2. (c)

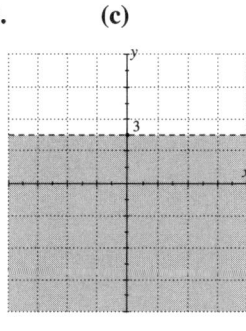

In problems 3 and 4 write an inequality whose solution set is the given graph.

3. (a)

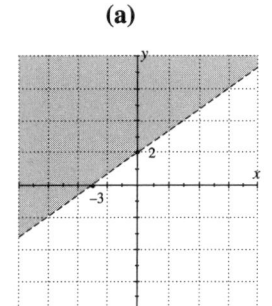

4. (a)

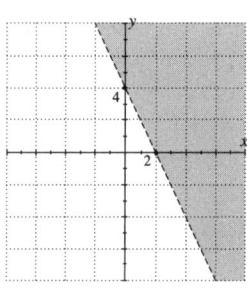

3. (b)

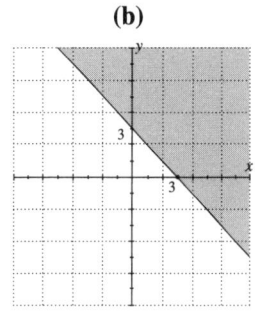

4. (b)

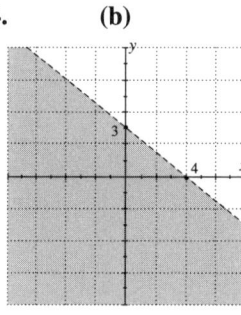

3. (c)

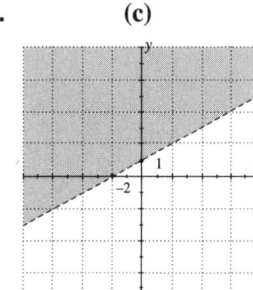

4. (c)

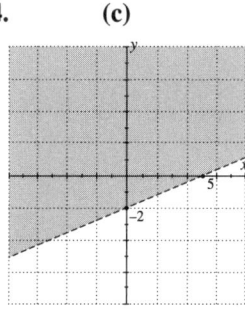

In problems 5 and 6, the boundary line for the graph of the given inequality is shown. Shade the correct half-plane that indicates the solution of the inequality.

5. (a) $y \leq 2x - 3$ (b) $y \geq -3x$

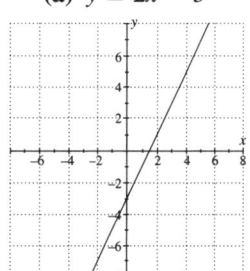

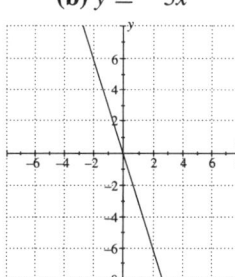

6. (a) $-2x + y < 6$ (b) $y \geq 2$

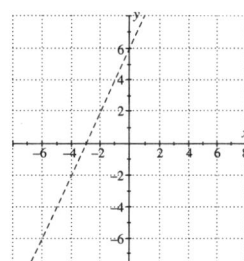

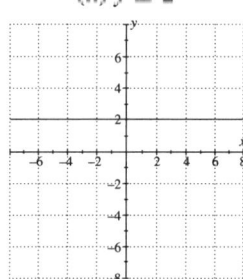

In Problems 7–20, sketch the graph of the given inequality.

7. $y \leq 2x + 5$
8. $y \geq 3x + 4$
9. $y > 3x$
10. $y \leq -4x$
11. $y > -2x + 3$
12. $2y > 3x + 7$
13. $2x + y \leq 3$
14. $5y \geq 7$

15. $3y \leq -4$
16. $y \leq -1$
17. $x \geq 2$
18. $2x - 6 < 0$
19. $x \leq 2$
20. $x < -2$

In Problems 21 and 22. write a compound inequality whose solution is described in the given graph.

21. (a) 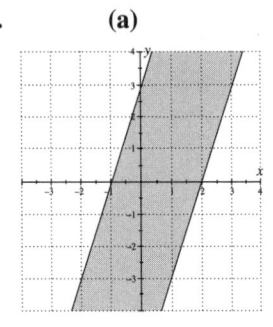
22. (a)

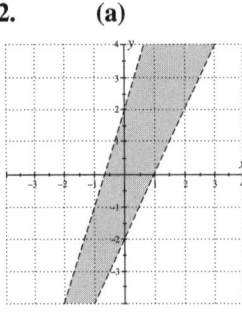

21. (b) 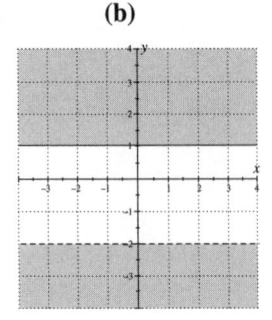
22. (b)

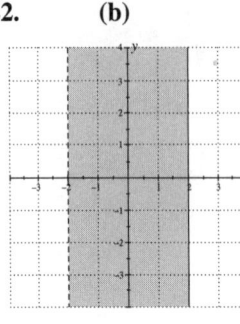

21. (c) 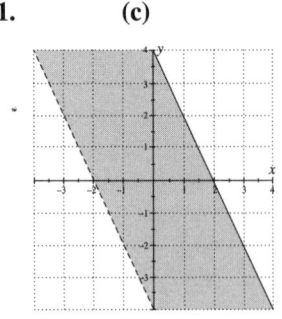
22. (c)

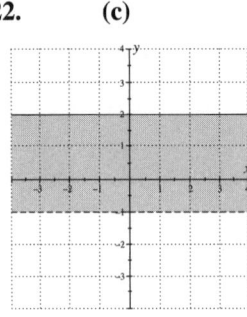

In Problems 23–32, sketch the graph of each compound inequality.

23. $y \geq -2$ and $x \leq 1$
24. $y \leq 2$ or $x \geq -1$
25. $y \leq -3$ or $x \geq 2$
26. $y \geq -2$ and $x \leq 1$
27. $2x + y < 3$ and $x + y < 1$
28. $-3x + y > 2$ and $x - 2y \leq 4$
29. $2x + y < 4$ or $x - y < 2$
30. $y \geq 3$ or $y < -2x + 1$

31. $\begin{cases} 2x + y \leq 10 \\ x \geq 0 \\ y \geq 0 \end{cases}$

32. $\begin{cases} 2x + 3y \leq 18 \\ x \geq 0 \\ y \geq 0 \end{cases}$

Applying the Concepts

33. **Inventory Levels:** An electronic store sells two brands of VCRs, A and B. Customer demand indicates that it is necessary to stock at least twice as many units of brand A as of brand B. Due to limitation of space, there is room for no more than 100 units in the store. Write a linear inequality that describes all possibilities for stocking the two brands and produce its graph.

34. **Investments:** A retirement fund invests in two mutual funds, X and Y. The X fund pays 5% annual simple interest and the Y fund pays 6% annual simple interest. The total interest from both investments for one year must be at least $10,600. Write a linear inequality that describes this situation and produce its graph.

35. **Ticket Prices:** For an upcoming event in an auditorium, tickets are priced at $8 and $5. The total sales for that event must be at least $3000. Write a linear inequality that describes this situation and produce its graph.

36. **Fuel Economy:** A certain make of car gets 16 miles per gallon in city driving and 24 miles per gallon in highway driving. The car is driven at most 380 miles on a full tank of gasoline. Write a linear inequality that describes this situation and produce its graph.

Developing and Extending the Concepts

37. Figure 10 shows the graph of the equation $|x + y| = 2$. Use the graph to describe the region which indicates the solution of each inequality.

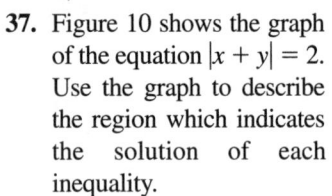

    (a) $|x + y| \leq 2$
    (b) $|x + y| \geq 2$

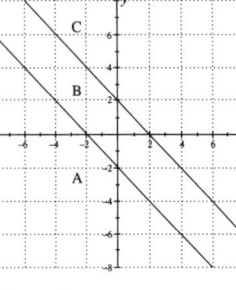

Figure 10

38. Figure 11 shows the graph of the equation $|y| = 1$. Use the graph to describe the region which indicates the solution of each inequality.

    (a) $|y| \geq 1$
    (b) $|y| \leq 1$

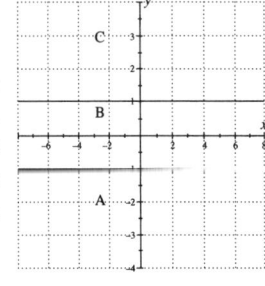

Figure 11

In Problems 39 and 40, sketch the graph of each absolute value inequality.

39. $|x - y| < 0$
40. $|y - 3x| \geq 0$

G In Problems 41–44, use a grapher with a SHADE feature to graph each inequality.

41. $y \leq 6 - 3x$
42. $y \geq \frac{2}{3}x - 3$
43. $2x + y > 3$
44. $x - 2 < -y$

In Problems 45 and 46, write inequalities that describe the indicated quadrant.

45. (a) Quadrant II (b) Quadrants I and IV
46. (a) Quadrant IV (b) Quadrants III and IV

In Problems 47 and 48, sketch the graph of each inequality that fits the description.

47. A fast food restaurant has 20 Level I and 15 Level II employees. For a 7 hour day, each Level I employee is paid $50 and each Level II employee is paid $40. The daily payroll of the restaurant must never exceed $1,400.

48. A pen company manufactures two types of pens, A and B, which it sells for $2.75 and $1.50 respectively. Their total production is at most 15,000 pens per week. In order to remain profitable, their gross receipts must be at least $20,000.

Solutions to Selected Problems

Chapter 0

PROBLEM SET 0.1 PAGE 5–6

1. $10 = 3 + 7$ **3.** $12 \neq 16$ **5.** $11 - 2 = 9$ **7.** $16 \div 2 = 8$ **9.** $3 < 10$ **11.** $27 > 6 + 9$
13. $1 + 2 < 7 - 2$ **15.** True **17.** True **19.** False **21.** True **23.** True **25.** 36 **27.** 35 **29.** 34
31. 19 **33.** 4 **35.** 34 **37.** 62 **39.** 12 **41.** 6 **43.** 2 **45.** $\frac{26}{25}$ **47.** 1 **49.** 75 **51.** 20

PROBLEM SET 0.2 PAGE 11–12

1. $\{-2, -1, 0, 1, 2, 3, 4\}$ **3.** $\{6, 8, 10, 12\}$ **5.** $\{-49, -48, -47, \ldots, -3, -2, -1\}$ **7.** $\{6, 9, 12, 15, 18\}$
9. $\{4, 8, 12, 16, \ldots\}$ **11.** ⟵•—•—•—•—•—•—•—•—•—•—•—•⟶ **13.** ⟵•—•—•—•—•—•—•—•—•—•—•—•⟶
-6 -5 -4 -3 -2 -1 0 1 2 3 4 5 -6 -5 -4 -3 -2 -1 0 1 2 3 4 5
15. -3.5 **17.** -0.8 **19.** 1.2 **21.** 0.875 **23.** 0.35 **25.** -0.68 **27.** $0.533\ldots$ **29.** $\frac{43}{100}$ **31.** $\frac{66}{25}$
33. $-\frac{1}{125}$ **35.** $-\frac{3,663}{100}$ **37.** $\frac{347}{500}$ **39.** 0.16 **41.** 0.147 **43.** 0.065 **45.** 0.29 **47.** 0.303 **49.** 1.8 **51.** $\frac{19}{100}$
53. $\frac{19}{80}$ **55.** $\frac{37}{500}$ **57.** $\frac{193}{1,000}$ **59.** $\frac{249}{1,000}$ **61.** $\frac{5}{4}$ **63.** 24 **65.** 180 **67.** 72.5 **69.** 35.7 **71.** 92.125
73. \$343,000 **75.** 7,200 **77.** 7.03 grams

PROBLEM SET 0.3 PAGE 17

1. 15 **3.** 45 **5.** 30 **7.** 14 **9.** 224 **11.** Commutative property for addition **13.** Additive inverse
15. Distributive property **17.** Associative property for multiplication **19.** Multiplicative inverse
21. Commutative property for multiplication **23.** Distributive property **25.** Identity element for multiplication
27. Associative property for addition **29.** Identity element for addition **31.** $-3, \frac{1}{3}$ **33.** $-6, \frac{1}{6}$ **35.** $-\frac{1}{5}, 5$
37. 0, none **39.** $-\frac{3}{4}, \frac{4}{3}$ **41.** 60 **43.** 24 **45.** 65 **47.** 23 **49.** 161 **51.** 260 **53.** 20 **55.** 24
57. $\frac{14}{17}$

PROBLEM SET 0.4 PAGE 21–22

1. 9 **3.** 11 **5.** -5 **7.** -8 **9.** 8 **11.** 15 **13.** 38 **15.** -11 **17.** -35 **19.** 11 **21.** -8 **23.** -20
25. 12 **27.** -15 **29.** -43 **31.** 5 **33.** 7 **35.** -8 **37.** -16 **39.** 27 **41.** 36 **43.** -29 **45.** -80
47. 26 **49.** 20 **51.** -47 **53.** 0 **55.** 33 **57.** -60 **59.** 15 **61.** $20°$ **63.** \$265 **65.** (a) positive;
(b) negative

PROBLEM SET 0.5 PAGE 26

1. 18 **3.** 20 **5.** −16 **7.** −33 **9.** 75 **11.** −63 **13.** 100 **15.** −1 **17.** 15 **19.** 30 **21.** 5 **23.** −2
25. −7 **27.** −3 **29.** −3 **31.** 4 **33.** 3 **35.** −7 **37.** −36 **39.** 3 **41.** 0 **43.** undefined **45.** 20
47. 27 **49.** −14 **51.** 18 **53.** 15 **55.** −27 **57.** 15 **59.** −74

PROBLEM SET 0.6 PAGE 34–35

1. $\frac{3}{4}$ **3.** $-\frac{3}{4}$ **5.** $\frac{2}{3}$ **7.** $\frac{1}{3}$ **9.** $-\frac{1}{4}$ **11.** $\frac{8}{3}$ **13.** $\frac{5}{7}$ **15.** $\frac{6}{11}$ **17.** $-\frac{7}{5}$ **19.** $\frac{18}{13}$ **21.** $-\frac{3}{4}$ **23.** $\frac{2}{5}$ **25.** $\frac{3}{10}$
27. $\frac{23}{20}$ **29.** $\frac{4}{35}$ **31.** $\frac{37}{24}$ **33.** $\frac{23}{80}$ **35.** $\frac{199}{210}$ **37.** $\frac{149}{1080}$ **39.** $\frac{19}{78}$ **41.** $-\frac{11}{210}$ **43.** $\frac{17}{24}$ **45.** $\frac{7}{12}$ **47.** $\frac{7}{30}$ **49.** $-\frac{13}{16}$
51. $\frac{2}{5}$ **53.** $-\frac{3}{5}$ **55.** $\frac{4}{3}$ **57.** $-\frac{2}{3}$ **59.** $\frac{40}{99}$ **61.** $\frac{10}{27}$ **63.** $-\frac{16}{135}$ **65.** $\frac{4}{9}$ **67.** $-\frac{5}{12}$ **69.** $\frac{16}{11}$ **71.** $\frac{3}{2}$ **73.** $\frac{5}{4}$ **75.** $\frac{13}{24}$
77. $\frac{23}{48}$ **79.** $\frac{35}{6}$ **81.** $\frac{188}{315}$ **83.** $\frac{11}{16}$ **85.** $\frac{21}{10}$ feet

PROBLEM SET 0.7 PAGE 40–41

1. $\frac{2}{100}$ **3.** $\frac{1}{10}$ **5.** 4.9 **7.** 1.386 **9.** 5.029 **11.** 4700 **13.** 0.67 **15.** 0.2 **17.** $12,700,000 **19.** 0.56

PROBLEM SET 0.8 PAGE 46–47

1. 6.99 **3.** 160.21 **5.** 627.139 **7.** 14.972 **9.** 205.77 **11.** 44.419 **13.** 94.4 **15.** 19.7
17. (a) Kotowski: $1,398.27; Dohner: $1,715.69; Humbert: $1,619.43 **(b)** Mon.: $824.18; Tue.: $713.83; Wed.: $953.35; Thur.: $1,090.05; Fri.: $1,151.98 **19.** 61.499 **21.** 63.7 **23. (a)** $x = 2.3; y = 6.4$ **(b)** 45

PROBLEM SET 0.9 PAGE 52–53

1. 0.045789 **3.** 59.438 **5.** 13.34 **7.** 26.718 **9.** 0.01196 **11.** 5.4 **13.** 25.8 **15.** 10.08 **17.** $1610 **19.** 0.0625 **21.** $X = 0.255; Y = 1.11$ **23.** 1.05 **25. (a)** 46.556 **(b)** 22.644 **27.** 494.91 **29.** $89.31
31. 3487.26 **33.** 85.428

PROBLEM SET 0.10 PAGE 58–60

1. 54.8 **3.** 4.98 **5.** 3.9 **7.** 4.58 **9.** 0.035 **11.** 5.3 **13.** 0.64 **15.** 2673.913 **17.** 1.4 **19.** 300
21. $0.018 **23. (a)** 0.9947 **(b)** 0.0053 **25. (a)** 0.13 **(b)** 0.13 **27.** 1.43 **29.** 67.4
31. (a) $1092.48 **(b)** $5644.48 **(c)** $156.79

PROBLEM SET 0.11 PAGE 64–65

1. 0.857 **3.** 0.846 **5.** 3.25 **7.** $\frac{37}{50}$ **9.** $\frac{7}{20}$ **11.** $1\frac{1}{5}$ **13.** 0.327 **15.** $\frac{3}{8}$ **17.** 0.375; 0.625; 0.5625
19. (a) 0.25 **(b)** $5.94 **21. (a)** $14\frac{3}{4}; 9\frac{2}{3}$ **(b)** 14.75; 9.67

PROBLEM SET 0.12 PAGE 68

1. 15 square inches **3.** 25.42 square centimeters **5.** 40.88 square feet **7.** 64 square centimeters
9. 39.69 square feet **11.** 94.09 square yards **13.** 17 square feet **15.** 62.01 square meters
17. 26 square inches **19.** 9 square feet **21.** 18.615 square meters **23.** 18.69 square inches
25. 40 square feet **27.** 24.51 square inches **29.** $40\frac{3}{8}$ square feet **31.** 78.54 square feet
33. 30.19 square meters **35.** 60.82 square inches

PROBLEM SET 0.13 PAGE 71–72

1. 16 inches **3.** 20.6 centimeters **5.** 25.8 feet **7.** 32 centimeters **9.** 25.2 feet **11.** 38.8 yards
13. 16.8 feet **15.** 34 meters **17.** $20\frac{2}{3}$ inches **19.** 12 inches **21.** 17.7 meters **23.** $10\frac{3}{4}$ inches
25. 24 inches **27.** 50.4 meters **29.** $1\frac{1}{3}$ feet **31.** 43.98 feet **33.** 32.67 inches **35.** 103.04 centimeters

PROBLEM SET 0.14 PAGE 76–78

1. $V = 60$ cubic inches; $S = 94$ square inches   **3.** $V = 80\frac{8}{9}$ cubic yards; $S = 114\frac{7}{9}$ square yards
5. $V = 919.275$ cubic feet; $S = 578.14$ square feet
7. $V = 250\pi$ cubic inches; $LS = 100\pi$ square inches; $S = 150\pi$ square inches
9. $V = 189.953\pi$ cubic meters; $LS = 92.66\pi$ square meters; $S = 126.28\pi$ square meters
11. $V = 87.856\pi$ cubic feet; $LS = 51.687\pi$ square feet; $S = 74.8\pi$ square feet
13. $V = 400$ cubic inches; $LS = 260$ square inches
15. $V = 92.38$ cubic feet; $LS = 123.6$ square feet
17. $V = 743.4$ cubic inches; $LS = 366$ square inches
19. $V = 1671$ cubic inches; $LS = 20\pi$ square inches
21. $V = 194.672\pi$ cubic feet; $LS = 105.8\pi$ square feet
23. $V = 8.192\pi$ cubic centimeters; $LS = 12.8\pi$ square centimeters
25. $V = \dfrac{256\pi}{3}$ cubic feet; $S = 64\pi$ square feet
27. $V = 234.155\pi$ cubic meters; $S = 125.44\pi$ square meters
29. $V = \dfrac{4{,}000\pi}{3}$ cubic inches; $S = 400\pi$ square inches

REVIEW PROBLEM SET PAGE 78–79

1. $18 = 6 \times 3$ **3.** $7 \neq 2 + 4$ **5.** 19 **7.** 43 **9.** 23 **11.** 15 **13.** 3 **15.** (number line from −6 to 5 with points at −4, −2, 0, 2, 4)
17. (number line from 0 to 9 with points at 2, 4, 6, 8) **19.** 0.182 **21.** 0.238 **23.** $\frac{11}{100}$ **25.** $\frac{29}{100}$ **27.** 45 **29.** 255.5
31. Commutative property for addition   **33.** Commutative property for multiplication
35. Associative property for addition   **37.** Distributive property   **39.** Identity element for addition
41. Associative property for multiplication   **43.** Identity element for multiplication   **45.** Multiplicative inverse
47. $-12, \frac{1}{12}$ **49.** $17, -\frac{1}{17}$ **51.** $-\frac{5}{3}, \frac{3}{5}$ **53.** 105 **55.** 155 **57.** 22 **59.** 18 **61.** -14 **63.** 25 **65.** -32
67. -4 **69.** 13 **71.** 40 **73.** -55 **75.** 1 **77.** -5 **79.** -30 **81.** 0 **83.** -24 **85.** 42 **87.** -16
89. 54 **91.** -2 **93.** 5 **95.** -3 **97.** -3 **99.** -5 **101.** 6 **103.** 0 **105.** 2 **107.** $\frac{1}{2}$ **109.** $\frac{9}{16}$
111. $\frac{21}{160}$ **113.** $\frac{4}{15}$ **115.** $\frac{27}{10}$ **117.** -31 **119.** 9 **121.** 0 **123.** $\frac{3}{2}$ **125.** $-\frac{28}{45}$ **127.** 17th floor **129.** 14°C

CHAPTER 0 TEST PAGE 80–81

1. (a) $35 = 5 \times 6 + 5$; **(b)** $7 \times 4 > 9$ **2. (a)** 48; **(b)** 3 **3.** (number line from −5 to 4 with points at −3, 0, 2, 4) **4.** 0.173 **5.** $\frac{3}{25}$ **6.** 6.6
7. (a) Additive inverse $= -4$; multiplicative inverse $= \frac{1}{4}$; **(b)** Additive inverse $= -\frac{2}{3}$; multiplicative inverse $= \frac{3}{2}$
8. 45 **9. (a)** 13; **(b)** -5 **10. (a)** -15; **(b)** -5; **(c)** 4; **(d)** 20; **(e)** -3; **(f)** 19
11. (a) 21; **(b)** -18; **(c)** -3; **(d)** 7; **(e)** -60 **12. (a)** $\frac{5}{7}$; **(b)** $-\frac{3}{8}$ **13. (a)** $\frac{8}{17}$; **(b)** $\frac{1}{3}$; **(c)** $\frac{23}{30}$; **(d)** $\frac{46}{231}$
14. (a) $\frac{3}{25}$; **(b)** $-\frac{1}{4}$; **(c)** $\frac{2}{3}$; **(d)** $-\frac{1}{5}$; **(e)** $-\frac{3}{2}$ **15.** b **16.** a **17.** c **18.** d **19.** a **20.** e **21.** c **22.** e **23.** a
24. a **25.** c **26.** d **27.** e **28.** a **29.** d **30.** d **31.** c **32.** c **33.** d **34** b

Chapter 1

PROBLEM SET 1.1 PAGE 87–88

1. Base 2; exponent 4 **3.** Base -3; exponent 5 **5.** Base 7; exponent 2 **7.** Base x; exponent 10
9. Base y; exponent 8 **11.** 3^4 **13.** $(-2)^4$ **15.** u^7 **17.** x^3y^2 **19.** $-8r^2s^3 + t^4$ **21.** 8 **23.** -81 **25.** -64
27. 1 **29.** $\frac{1}{64}$ **31.** 2^5 **33.** 3^6 **35.** -5^{11} **37.** m^8 **39.** $-x^9$ **41.** 3^6 **43.** 2^8 **45.** x^{12} **47.** m^{10} **49.** z^{30}
51. 3^2x^2 **53.** u^6v^6 **55.** -4^3t^3 **57.** 2^4x^4 **59.** $3^2x^2y^2$ **61.** 3.782×10^3 **63.** 3.84×10^5 **65.** 7.8×10^8
67. 210 **69.** 750,000 **71.** 31,200,000 **73.** 8.7×10^5 miles **75.** 5.98×10^{24} tons

PROBLEM SET 1.2 PAGE 94

1. $\frac{1}{16}$ **3.** $\frac{1}{1000}$ **5.** $-\frac{1}{8}$ **7.** $-\frac{1}{81}$ **9.** $\frac{7}{x^3}$ **11.** 1 **13.** 1 **15.** -1 **17.** 1, for $x \neq -3$ **19.** 4 **21.** -27
23. z^3 **25.** $\frac{1}{x^3}$ **27.** 1 **29.** $-u^3$ **31.** 1 **33.** $\frac{1}{r^4}$ **35.** 1,000 **37.** $\frac{1}{y^6}$ **39.** $\frac{1}{x}$ **41.** $\frac{8}{27}$ **43.** $\frac{x^4}{y^4}$
45. $-\frac{u^5}{v^5}$ **47.** $-\frac{125}{t^3}$ **49.** $\frac{z^4}{w^4}$ **51.** $\frac{125}{64}$ **53.** $\frac{z^2}{9}$ **55.** $\frac{1}{3,125}$ **57.** $\frac{1}{y^9}$ **59.** p^{12} **61.** $\frac{625}{z^8}$ **63.** p^2y^6 **65.** $\frac{x^{12}}{y^{28}}$
67. $170.56 **69.** $438.49 **71.** $152.88 **73.** $649.64 **75.** 3.8×10^{-2} **77.** 6.9×10^{-5} **79.** 1.9401×10^{-2}
81. 1.7900×10^{-9}

PROBLEM SET 1.3 PAGE 99–100

1. -3 **3.** 13 **5.** -9 **7.** -2 **9.** 4 **11.** -2 **13.** -2 **15.** Not a real number **17.** Not a real number
19. 1 **21.** -15 **23.** 17 **25.** 0.2 **27.** -0.5 **29.** 0.2 **31.** x^4 **33.** z^{25} **35.** $-u$ **37.** $2x$ **39.** $3w$
41. $-3wt^2$ **43.** 1.4142 **45.** 2.2361 **47.** 3.3166 **49.** 10.6301 **51.** 400 tons **53.** 1.68 seconds
55. $P = 1,062$

PROBLEM SET 1.4 PAGE 104–105

1. $4\sqrt{2}$ **3.** $3\sqrt{5}$ **5.** $4\sqrt{10}$ **7.** $5\sqrt{6}$ **9.** $2\sqrt[3]{9}$ **11.** $4\sqrt[3]{5}$ **13.** $-5\sqrt[3]{2}$ **15.** $2\sqrt[4]{5}$ **17.** $3\sqrt[4]{7}$
19. $2\sqrt[5]{5}$ **21.** $-3\sqrt[5]{2}$ **23.** $x^2\sqrt{x}$ **25.** $2u\sqrt{2u}$ **27.** $5x\sqrt{2x}$ **29.** $xy^2\sqrt{xy}$ **31.** $2u^2v^4\sqrt{5u}$
33. $2z^3\sqrt[3]{2z}$ **35.** $-5xy\sqrt[3]{2xy^2}$ **37.** $2uv\sqrt[4]{3uv^3}$ **39.** $2wz\sqrt[5]{7wz^2}$ **41.** $\frac{\sqrt{3}}{5}$ **43.** $\frac{\sqrt{5}}{4}$ **45.** $\frac{\sqrt{7}}{11}$
47. $\frac{7}{8}$ **49.** $\frac{2}{5}$ **51.** $-\frac{\sqrt[3]{11}}{2}$ **53.** $\frac{\sqrt[4]{5}}{2}$ **55.** $\frac{\sqrt[4]{23}}{3}$ **57.** $\frac{\sqrt[5]{19}}{2}$ **59.** $-\frac{\sqrt[5]{171}}{3}$ **61.** $\frac{\sqrt{11}}{t}$ **63.** $\frac{\sqrt{3}}{2x}$
65. $\frac{a}{9}$ **67.** $\frac{\sqrt[3]{13}}{3a}$ **69.** $-\frac{\sqrt[3]{17z}}{4x}$ **71.** $\frac{x^2\sqrt{x}}{2z^4}$ **73.** $\frac{\sqrt[4]{37}}{3u^2v^3}$ **75.** $-\frac{\sqrt[5]{19}}{2x^2}$ **77.** 8 volts **79.** 20%

PROBLEM SET 1.5 PAGE 107

1. $5\sqrt{2}$ **3.** $8\sqrt{5}$ **5.** $10\sqrt[3]{6}$ **7.** $10\sqrt{x}$ **9.** $-9\sqrt[3]{y}$ **11.** $10\sqrt{z} - 3\sqrt{w}$ **13.** $20\sqrt[3]{x^2} - \sqrt[3]{x}$ **15.** $5\sqrt{2}$
17. $5\sqrt{2}$ **19.** $43\sqrt{5}$ **21.** $2\sqrt{2}$ **23.** $3\sqrt{3}$ **25.** $5\sqrt{2}$ **27.** $11\sqrt{2} - 4\sqrt{3}$ **29.** $5\sqrt[3]{2}$ **31.** $24\sqrt[3]{3}$
33. $6\sqrt{2x}$ **35.** $\sqrt{5y}$ **37.** $\sqrt[3]{2x^2}$ **39.** $(6y + 3)\sqrt{y}$ **41.** $71t\sqrt{3t}$ **43.** $13x\sqrt{2x}$ **45.** $11m\sqrt[3]{4m}$

PROBLEM SET 1.6 PAGE 109–110

1. 8 **3.** $5\sqrt{2}$ **5.** 5 **7.** $4x$ **9.** $4mn$ **11.** 18 **13.** $49y$ **15.** $\sqrt{6}+3$ **17.** 2 **19.** $3x+5\sqrt{x}$
21. 4 **23.** 8 **25.** 29 **27.** $u-v$ **29.** $28+10\sqrt{3}$ **31.** $21-4\sqrt{5}$ **33.** $m+2\sqrt{mn}+n$
35. $x-6\sqrt{xy}+9y$ **37.** $9+5\sqrt{15}$ **39.** $8x-10\sqrt{xy}-3y$

PROBLEM SET 1.7 PAGE 112–113

1. 3 **3.** 2 **5.** 2 **7.** $4x^2$ **9.** $5u^3v^2$ **11.** $10x$ **13.** $3t^3$ **15.** $\dfrac{5\sqrt{3}}{3}$ **17.** $\dfrac{\sqrt{14}}{7}$ **19.** $\dfrac{\sqrt{15}}{3}$ **21.** $\dfrac{10\sqrt{t}}{t}$
23. $\dfrac{9\sqrt{2x}}{2x}$ **25.** $\dfrac{5\sqrt{5}+5}{2}$ **27.** $\dfrac{13\sqrt{3}-13}{4}$ **29.** $\dfrac{4\sqrt{t}-4}{t-1}$ **31.** $\dfrac{x+2\sqrt{x}}{x-4}$ **33.** $3\sqrt{7}+3\sqrt{5}$

PROBLEM SET 1.8 PAGE 115–116

1. 5 **3.** 12 **5.** 15 **7.** 16 **9.** 26 **11.** 11.40 **13.** 17.53 **15.** 18.96 **17.** 30.03 **19.** 217.19
21. 5 **23.** 13 **25.** $3\sqrt{29}$ **27.** 10 **29.** $3\sqrt{5}$ **31.** $2\sqrt{5}$ **33.** $\sqrt{41}$ **35.** $6\sqrt{5}$ **37.** 60 feet
39. 125 feet **41.** $2\sqrt{209}$ feet

REVIEW PROBLEM SET PAGE 116–118

1. 10 **3.** -8 **5.** 5 **7.** 3 **9.** -3 **11.** z^3 **13.** $7w^4$ **15.** $-3m^3n^{10}$ **17.** $2x^4y^6$ **19.** $2t^7s^5$
21. 2.8284 **23.** 3.8730 **25.** 13.0767 **27.** $3\sqrt{7}$ **29.** $6\sqrt{3}$ **31.** $3\sqrt[3]{5}$ **33.** $-4\sqrt[3]{3}$ **35.** $2\sqrt[4]{2}$
37. $2\sqrt[5]{7}$ **39.** $z^4\sqrt{z}$ **41.** $4x^2\sqrt{2x}$ **43.** $5uv^5\sqrt{5uv}$ **45.** $-3m^3n^3\sqrt[3]{4m}$ **47.** $2t^3s^8\sqrt[4]{3t^3s^2}$
49. $\dfrac{\sqrt{11}}{3}$ **51.** $\dfrac{4}{5}$ **53.** $\dfrac{\sqrt[3]{10}}{3}$ **55.** $-\dfrac{\sqrt[3]{13}}{5}$ **57.** $\dfrac{\sqrt{15}}{2u^2}$ **59.** $\dfrac{2x\sqrt{2}}{5y^2}$ **61.** $\dfrac{\sqrt[3]{10}}{3u^2v^3}$ **63.** $\dfrac{\sqrt[4]{7}}{2x^2y}$ **65.** $8\sqrt{7}$
67. $10\sqrt{3}$ **69.** $10\sqrt[3]{2t}$ **71.** $60\sqrt{2}$ **73.** $12\sqrt{2}$ **75.** $\sqrt[3]{2}$ **77.** $3\sqrt{5z}$ **79.** $16x\sqrt[3]{4x}$ **81.** $6\sqrt{5}$
83. $3x^2\sqrt{5}$ **85.** $5\sqrt{3}+5$ **87.** $z-9$ **89.** $9+2\sqrt{14}$ **91.** $2x+7\sqrt{xy}+3y$ **93.** $3\sqrt{2}$ **95.** 3
97. $4x^2$ **99.** $-3uv^2$ **101.** $\dfrac{13\sqrt{7}}{7}$ **103.** $\dfrac{8\sqrt{x}}{x}$ **105.** $\dfrac{4\sqrt{7}-8}{3}$ **107.** $\dfrac{3z\sqrt{z}+3z}{z-1}$ **109.** $\dfrac{19+7\sqrt{10}}{43}$
111. 50 **113.** 2 **115.** $\dfrac{36}{5}$ **117.** $\dfrac{11}{2}$ **119.** 7 **121.** 9 **123.** 2 **125.** 25 **127.** 20 **129.** $\sqrt{130}$
131. 5 **133.** 17 **135.** $\sqrt{137}$ **137.** $5\sqrt{5}$ feet **139.** $5-\sqrt{13}$ miles

Chapter 2

PROBLEM SET 2.1 PAGE 122

1. Polynomial; monomial **3.** Polynomial; monomial **5.** Polynomial; binomial **7.** Not a polynomial
9. Polynomial; trinomial **11.** Polynomial; none of these **13.** Polynomial; binomial **15.** Not a polynomial
17. Degree = 1; 5, 7 **19.** Degree = 2; 1, 7, -3 **21.** Degree = 2; 1, -7, 3 **23.** Degree = 0; 5
25. Degree = 3; $\frac{1}{2}, \frac{-1}{3}, \frac{2}{5}, \frac{-3}{2}$ **27.** Degree = 7; $-7, 3, 4$ **29.** Degree = 5; 3, -2, 11 **31.** 17 **33.** -29
35. 14 **37.** 10 **39.** -6 **41.** 12 **43.** 43 **45.** 5 **47.** -9 **49.** -23
51. (a) 13 hours; **(b)** 11 hours; **(c)** 9 hours **53.** $9.50 **55. (a)** $287.50; **(b)** $587.50

PROBLEM SET 2.2 PAGE 124

1. $P(0) = 7, P(1) = 8, P(2) = 9, P(3) = 10$
3. $S(-1) = -14, S(0) = -9, S(1) = -6, S(2) = -5$
5. $R(2) = 7, R(0) = 1, R(-3) = -8$
7. $Q(-2.3) = 6.675, Q(-1.3) = 6.425, Q(0) = 6.1, Q(1.3) = 5.775$
9. $P(3) + Q(-2.3) - S(0.15) = 13.66925$
11. $P(0) = 3/4, P(1) = 5/4, P(2) = 7/4, P(3) = 9/4$
13. $S(-1) = -4.875, S(0) = -19/2, S(1) = -13.375, S(2) = -16.5$
15. $R(2) = -25/12, R(0) = 43/12, R(3) = -59/12$
17. They have a loss of $20.00. $P(40) = R(40) - C(40)$
$P(40) = 600 - 620$ dollars
$P(40) = \$ -20.00$

PROBLEM SET 2.3 PAGE 127–128

1. $21x^2$ 3. $15s^4$ 5. $5y^3$ 7. $-2z^5$ 9. $8x^3y$ 11. $11u^2v^3 - 7u^3v^2$ 13. $8t^2 - 4$ 15. $5z^2 + z - 6$
17. $2s^2 + 10s + 7$ 19. $2z^3 - 8z^2 + 5z - 10$ 21. $13x^3y - 7x^2y^2 + 7xy^3$ 23. $7x^2 + 7x + 5$
25. $8s^3 - 6s^2 + 5s - 6$ 27. $3z^2 - 5z - 4$ 29. $10u^3 + u^2 + 12$ 31. $-7y^3 + 4y^2 + 9y + 5$
33. $15m^3n - 13 mn^3$ 35. $4u^4v^2 + 13u^2v - 16uv^5 - u^4v^4$ 37. $3x^3y^2 - 15x^2y + 5xy^2 + 2$ 39. $6x^2 - 5x + 8$
41. $4t^2 - 6t - 3$ 43. $6s^3 + 3s^2 - 6s + 10$ 45. $z^4 - 5z^2 + 3$ 47. $10n^2 - 7n - 9$ 49. $2x^2y^3 - 13x^3y + 4xy^2$
51. $2x^2 - 18x + 5$ 53. $-14y^2 - 14y + 3$ 55. $-5x^3 + 3x^2 - 4x + 2$ 57. $15x - 3$

PROBLEM SET 2.4 PAGE 131–134

1. $8x^7$ 3. $-6t^4$ 5. $15u^3v^5$ 7. $-40x^8$ 9. $2m^4n^5$ 11. $6y^2 + 3y$ 13. $15s^2 - 10s$ 15. $-12z^2 + 18z$
17. $2x^3 - 2x^2 - 2x$ 19. $4a^3b^3 - 8a^2b^3 + 12ab^4$ 21. $15m^2n + 10mn^2$ 23. $-x^5y^3 + 2x^4y^3 - 4x^4y^4$
25. $2x^2 + 7x + 3$ 27. $3u^2 + u - 10$ 29. $-2t^2 + 17t - 21$ 31. $4z^4 + 3z^2 - 1$ 33. $15x^2 - 14xy - 8y^2$
35. $6r^2 - 19rs + 3s^2$ 37. $2x^3 - 7x^2 + 11x - 4$ 39. $4z^4 + 10z^3 - 5z^2 + 2z - 3$ 41. $3x^3 + 5x^2y + 2xy^2 + 8y^3$
43. $4m^3 - 19m^2n + 23mn^2 - 6n^3$ 45. $2x^4 - 3x^2 + 2x + 3$ 47. $6x^4 + 7x^3y - 2x^2y^2 + 7xy^3 - 2y^4$
49. $15y^5 + 11y^4 - 8y^3 + 9y^2 - 11y + 2$ 51. $12x^2 - 7x - 12$ 53. $\dfrac{6x^2 + 19x + 15}{2}$

PROBLEM SET 2.5 PAGE 135–136

1. $x^2 + 3x + 2$ 3. $z^2 + 2z - 15$ 5. $y^2 - 21y + 110$ 7. $v^2 - v - 56$ 9. $z^2 - 121$ 11. $2x^2 + 5x + 3$
13. $8y^2 - 6y - 5$ 15. $7s^2 - 17s + 6$ 17. $6u^2 - 19u - 20$ 19. $4x^2 + 4x + 1$ 21. $3y^2 + 5yz - 2z^2$
23. $4m^2 - 20mn + 25n^2$ 25. $6x^2 - 5xy - 6y^2$ 27. $54t^2 - 15t - 56$ 29. $50y^2 - 35yz + 6z^2$
31. $2x^2 + 15x + 7$ 33. $2z^2 + 11z + 12$ 35. $20y^2 - 47y + 24$ 37. $35m^2 + 24mn - 35n^2$
39. $18x^2 - 3xy - 10y^2$ 41. $x^2 + 10x + 25$ 43. $y^2 - 4y + 4$ 45. $m^2 + 12m + 36$ 47. $4x^2 + 28x + 49$
49. $9z^2 - 24z + 16$ 51. $121 + 44t + 4t^2$ 53. $25x^2 - 10xy + y^2$ 55. $16t^2 + 24ts + 9s^2$
57. $49u^2 - 42uv + 9v^2$ 59. $36w^2 + 120wz + 100z^2$ 61. $4x^2 + 2x + \frac{1}{4}$ 63. $9m^2n^2 - 6mnp + p^2$
65. $25x^4y^2 + 20x^2yz + 4z^2$ 67. $x^2 - 16$ 69. $u^2 - 100$ 71. $z^2 - 64$ 73. $9x^2 - y^2$ 75. $4t^2 - 49$
77. $36x^2 - 25y^2$ 79. $9m^2 - 121n^2$ 81. $9x^2 - \frac{1}{16}$ 83. $25t^4 - 9s^4$ 85. $49x^4y^2 - 16z^6$
87. $49x^2 + 154x + 121$

PROBLEM SET 2.6 PAGE 141

1. $4x^3$ **3.** $-3m^2$ **5.** $-5\dfrac{y^2}{z^3}$ **7.** $2x^2 + 5x$ **9.** $5t - 7t^3 + 3t^5$ **11.** $2m^2n^2 - \dfrac{3m}{n} + \dfrac{5n}{m}$ **13.** $-\dfrac{7z}{w^2} + \dfrac{2}{w} - \dfrac{3}{z}$
15. $x + 3$ **17.** $m - 6$ **19.** $1 + y$ **21.** $x^2 + 2x + 1$ **23.** $x^2 + 2x + 3$; R = 2 **25.** $m - 3$; R = -5
27. $3x^2 + 6x - 6$; $R = -17x + 8$ **29.** $2x^2 - 3x + 2$ **31.** $x^2 - 4xy + y^2$ **33.** $m^3 - 8n^3$ **35.** $y^2 + 2$; R = 10
37. $4u^2 + 6uv + 9v^2$ **39.** $p^4 - p^3q + p^2q^2 - pq^3 + q^4$

PROBLEM SET 2.7 PAGE 144

1. $2 \cdot 3^2$ **3.** $3 \cdot 17$ **5.** $3^2 \cdot 7$ **7.** $2^4 \cdot 5$ **9.** $2^2 \cdot 3 \cdot 7$ **11.** $2^3 \cdot 5^2$ **13.** $13 \cdot 17$ **15.** $2 \cdot 29$
17. $2^2 \cdot 3 \cdot 11$ **19.** $3 \cdot 11 \cdot 31$ **21.** $2^2 \cdot 11$ **23.** $2 \cdot 3^3$ **25.** 5^3 **27.** $2^3 \cdot 3 \cdot 7$ **29.** $3^3 \cdot 7$ **31.** prime
33. $2 \cdot 59$ **35.** $2^2 \cdot 3^2 \cdot 5$ **37.** $2 \cdot 3 \cdot 5^2$ **39.** $7 \cdot 5^2$ **41.** 7 **43.** 6 **45.** 4 **47.** 5 **49.** 6

PROBLEM SET 2.8 PAGE 149–150

1. $2x^2$ **3.** $4x^2y^2$ **5.** xy **7.** $5(x + 2)$ **9.** $3(r - 2)$ **11.** $5(f - 1)$ **13.** $12(u^2 + 2)$ **15.** $x(3x - 1)$
17. $7z(3z - 1)$ **19.** $32(2a^2 + 3)$ **21.** $3u^2(u + 2)$ **23.** $5m^2(2 - m)$ **25.** $x^3(4x^5 + 9)$ **27.** $a(x^2 + y^2)$
29. $\pi rh(r + 2)$ **31.** $6rs(s + 5r)$ **33.** $ab^2(5a + 7b)$ **35.** $6pq(p + 4q)$ **37.** $2(t^2 + 2t + 3)$
39. $w(w^2 + 3w + 5)$ **41.** $r^3(r^2 + 2r + 3)$ **43.** $3x^2(x - 2x^3 + 5)$ **45.** $4A^2(4A + 3 - 2A^2)$
47. $xy(4x + 12 - 7y)$ **49.** $-5rt(7r^2 + 5rt + 3t^2)$ **51.** $5xy^2(x^2y^4 - 2x - 5)$ **53.** $-2c^2d^3(2d + 4c - cd)$
55. $ab^2(8ac^2 - 4 - 3b)$ **57.** $(m + 4)(a + b)$ **59.** $(a + 1)(t - b)$ **61.** $(u + 2)(v + 8)$ **63.** $(2x + y)(3 - z)$
65. $(c + 1)(x + y)$ **67.** $(a + 3)(b - 5)$ **69.** $(s - 2)(t + 7)$ **71.** $(a + 2b)(3c - d)$ **73.** $(x^2 + 4)(a^2 - 5)$
75. $(b + 1)(c - 1)$ **77.** $(x + 2w)(5y - z)$ **79.** $-3, 0$ **81.** $0, 7$ **83.** $0, 4$ **85.** $0, 3$ **87.** $0, 3$ **89.** $0, 16$

PROBLEM SET 2.9 PAGE 153

1. $(x - 3)(x + 3)$ **3.** $(2 - t)(2 + t)$ **5.** $(2c - 1)(2c + 1)$ **7.** $(1 - 10r)(1 + 10r)$ **9.** $(5 - 9c)(5 + 9c)$
11. $(3f - 5)(3f + 5)$ **13.** $(2k - 3)(2k + 3)$ **15.** $(10 - 11w)(10 + 11w)$ **17.** $(ab - 1)(ab + 1)$
19. $(3xy - 4)(3xy + 4)$ **21.** $(xy - 25)(xy + 25)$ **23.** $(3m - 7n)(3m + 7n)$ **25.** $(3xyz - 4w)(3xyz + 4w)$
27. $(9abc - 5uv)(9abc + 5uv)$ **29.** $(b - 10cdf)(b + 10cdf)$ **31.** $(c - d)(c + d)(c^2 + d^2)$
33. $(1 - 3t)(1 + 3t)(1 + 9t^2)$ **35.** $(2a - 3b)(2a + 3b)(4a^2 + 9b^2)$ **37.** $5x(x - y)(x + y)$
39. $27ab(b - 2a)(b + 2a)$ **41.** $3(a - 5)(a + 5)$ **43.** $2v(5u - 6)(5u + 6)$ **45.** $2d(2c - 3)(2c + 3)$
47. $3r(r - 4s)(r + 4s)$ **49.** $(x + 1)^2$ **51.** $(y - 5)^2$ **53.** $(2u - 3v)^2$ **55.** $(8t + 3)^2$ **57.** $(9i + 7)^2$
59. $-5, 5$ **61.** $-\dfrac{1}{3}, \dfrac{1}{3}$ **63.** $-11, 11$ **65.** $-\dfrac{5}{7}, \dfrac{5}{7}$ **67.** $-\dfrac{11}{4}, \dfrac{11}{4}$ **69.** 9 **71.** 7, 8

PROBLEM SET 2.10 PAGE 157

1. $(x + 1)(x + 2)$ **3.** $(r + 1)(r + 5)$ **5.** $(t - 3)(t - 7)$ **7.** $(m - 2)(m - 5)$ **9.** $(x - 1)(x + 3)$
11. $(u + 2)(u - 6)$ **13.** $(a + 5)(a - 6)$ **15.** $(c - 3)(c + 15)$ **17.** $(w - 6)(w + 8)$ **19.** $(a - 3)(a - 16)$
21. $(m - 3)(m - 4)$ **23.** $(x - 10)(x + 11)$ **25.** $(w - 2)(w - 2)$ **27.** $(x + 4)(x + 4)$ **29.** $(y + 5)(y + 5)$
31. $(y - 7z)(y + 9z)$ **33.** $(a - 3b)(a + 10b)$ **35.** $(x + 2y)(x - 22y)$ **37.** $(w - 3u)(w - 12u)$
39. $(x + 9y)(x + 9y)$ **41.** $(a + 5b)(a + 8b)$ **43.** $(y - 3)(y - 4)$ **45.** Not factorable **47.** $(m - 1)(m - 6)$
49. Not factorable **51.** $-3, -4$ **53.** $-2, 6$ **55.** $-11, 2$ **57.** $-11, 6$ **59.** $-7, 5$ **61.** 8

PROBLEM SET 2.11 PAGE 162–163

1. $(2x + 3)(x + 1)$ **3.** $(3r + 1)(r + 2)$ **5.** $(2t - 3)(t - 2)$ **7.** $(5z - 2)(z + 1)$ **9.** $(3c + 2)(2c + 1)$
11. $(4z - 1)(z + 2)$ **13.** $(2y - 5)(y - 3)$ **15.** $(3r + 4)(2r - 3)$ **17.** $(7c - 2)(6c + 1)$ **19.** $(4f - 5)(6f - 1)$
21. $(3y - 5)(3y - 4)$ **23.** Not factorable **25.** Not factorable **27.** $(2x + 5)(7x - 3)$ **29.** $(5h - 4)(4h + 5)$
31. $(5x - 7y)(3x + 2y)$ **33.** $(4a - 5b)(5a + 4b)$ **35.** $(2y - 9z)(3y - 8z)$ **37.** $(7x + 3y)(x + 4y)$
39. Not factorable **41.** $(8a - 7b)(4a + 3b)$ **43.** Not factorable **45.** $a(x - 3)(x - 5)$ **47.** $c(d + 4)(d - 2)$
49. $2y(3y - 1)(y + 5)$ **51.** $5x^2(2z + 1)(z - 1)$ **53.** $4y(5w + 3z)(5w + 3z)$ **55.** $5pq^4(4p - 7q)(p + 2q)$
57. $-\frac{5}{2}, \frac{1}{5}$ **59.** $\frac{4}{3}, \frac{1}{2}$ **61.** $\frac{5}{6}, \frac{1}{2}$ **63.** $-3, \frac{1}{2}$ **65.** $-\frac{1}{2}, \frac{9}{2}$

REVIEW PROBLEM SET PAGE 163–164

1. 64 **3.** $-\frac{1}{27}$ **5.** -1 **7.** $5u^3v^2$ **9.** $-2t^4s^3$ **11.** $4t^4y - 8xy^3$ **13.** Polynomial **15.** Not a polynomial
17. Polynomial **19.** Binomial; degree = 1; 3, −7 **21.** Trinomial; degree = 2; 4, −5, 13
23. Monomial; degree = 5; −8 **25.** Degree = 3; 8, 9, −1, 7 **27.** 21 **29.** 19 **31.** 12 **33.** $960 **35.** 20°
37. $37,500 **39.** $19w^2$ **41.** $4x^3$ **43.** $-22t$ **45.** $4y^3 - 5y^2 + 3y + 4$ **47.** $p^3 + 3p^2 + 7p + 3$
49. $5x^2y + 4xy + 7xy^3$ **51.** $5y^2 + 4y - 6$ **53.** $5z^2 - 8z - 9$ **55.** $-2u^2v + 4uv + 4uv^2$ **57.** 5.892×10^3
59. 7.92×10^{-4} **61.** 37,200 **63.** 0.0123 **65.** 4.3281×10^{-1} **67.** 8^{15} **69.** 9^7 **71.** x^5y^5 **73.** $-u^7$
75. $\frac{a^6}{b^6}$ **77.** z^3 **79.** x^{12} **81.** t^6 **83.** $8a^{12}$ **85.** $-\frac{1}{y^3}$ **87.** $-9x^{12}y^{14}$ **89.** $\frac{2p^{12}}{q^2}$ **91.** x^6 **93.** $\frac{x^4}{y^6}$
95. $-21x^4y^2z^3$ **97.** $6t^5 + 15t^3 - 12t^2$ **99.** $v^3 + v^2 + v + 6$ **101.** $6x^3 + x^2y + xy^2 + y^3$
103. $m^4 - m^3 + 2m + 8$ **105.** $x^2 + 13x + 42$ **107.** $t^2 - 10t + 9$ **109.** $2y^2 - 9y - 56$
111. $3x^2 + 7xy - 20y^2$ **113.** $30u^2 - 43uv + 15v^2$ **115.** $z^2 + 8z + 16$ **117.** $4x^2 - 49$ **119.** $25t^2 - 20t + 4$
121. $16x^2 - 8xy + y^2$ **123.** $100 - 9v^2$ **125.** $5xy^2$ **127.** $4uv^4 - 5u^3v$ **129.** $4t^4 - 3 + \frac{1}{t^2}$
131. $x + 9$ **133.** $m^2 - 2m + 3$ **135.** $x^2 + 7x + 3$ **137.** $x^3 + 2x^2y - xy^2 + y^3$

CHAPTER 2 TEST PAGE 165

1. (a) -81; (b) 1; (c) $\frac{1}{16}$ (d) 4 **2.** (a) $3x^3y^4$; (b) $2xy^2 + 3x^3y$
3. (a) Binomial; degree = 2; 3, 5; (b) Monomial; degree = 5; −7; (c) Trinomial; degree = 3; 2, −4, 7
4. -12 **5.** (a) 15°; (b) $28.80 **6.** (a) $4x^2 + 3x - 4$; (b) $6y^2 + 2y - 3$; (c) $2m^2n + 4mn + 3mn^2$
7. (a) 4.385×10^2; (b) 7.29×10^{-4} **8.** (a) 27,800; (b) 0.00198 **9.** 1.2613×10^{-8}
10. (a) $-12x^3y^5$; (b) $-6t^6 + 8t^5 - 4t^3$; (c) $2x^3 - 7x^2y + 5xy^2 - y^3$
11. (a) $x^2 + 11x + 24$; (b) $x^2 - 3x - 10$; (c) $6x^2 - 2xy - 20y^2$
12. (a) $x^2 - 49$; (b) $9z^2 + 12z + 4$; (c) $u^2 - 8uv + 16v^2$ **13.** (a) $-4u^2vw^2$; (b) $4x^2 - 2xy$; (c) $x^2 + 3x - 4$
14. (a) x^{12}; (b) y^8; (c) $\frac{2}{x^3}$; (d) $-\frac{8x^3}{y^3}$; (e) $\frac{1}{4x^6y^4}$; (f) $\frac{y^{14}}{x^8}$ **15.** $9x^2 + 24x + 16$

Chapter 3

PROBLEM SET 3.1 PAGE 172–174

1a. 8 **3a.** 10 **5a.** 13/4 **7a.** $-17/3$ **9.** 1 **13.** $-13/43$ **17.** 3
1b. 0 **3b.** 22/3 **5b.** -2 **7b.** 1/2 **11.** 5 **15.** 8/3 **19.** $-11/2$
21. no solution **23.** 210 **27.** 59/8 **31.** 1 **35.** $-63/10$ **39.** 5 **43.** -5
 25. 2 **29.** 1 **33.** $-10/19$ **37.** $-4/29$ **41.** 2 **45.** 35/6

47. 800   **51a.** identity   **53a.** inconsistent   **55.** 86 nickels   **57a.** $135   **59a.** $130,200
49. 0.25   **51b.** identity   **53b.** identity   91 dimes   **57b.** 600 mi   **59b.** $160,000
                                                                                         43 quarters

61. −1.5   **63b.**   **65b.**   **67.** If $a = b$,
                                                                                                                                        then $a + c = b + c$
63a. 2   **69.** 9°
65a. 2

PROBLEM SET 3.2 PAGE 178–180

1. 5c   **3.** $\dfrac{c - b}{a}$ **5.** $-\dfrac{b}{2}$ **7.** $-\dfrac{5a}{3}$ **9.** $-c/2$
                                                                                                                       11. $7d/2m$
                                                                                                                       13. $-7b$

15. $(12 - 4x)/3$ **17.** $(99 - 11x)/9$ **19.** $(63 - 27x)/14$ **21.** $100x, + 3$ **23.** $(4x - 90)/23$
25. $\dfrac{A}{rt + 1}$ **27.** $\dfrac{Ty}{x}$ **29.** $\dfrac{10P}{7}$ **31.** $\dfrac{y - b}{m}$ **33.** $\dfrac{C}{2\pi}$

35. $\dfrac{P - 2w}{2}$ **37a.** $3V/h$ **41a.** $bx/(b - y)$ **43.** 240 ft
                              37b. $3V/B$ **41b.** $ay/(a - x)$ **45.** $22.52
                              39a. $V + 32t$ **39b.** $\dfrac{V_0 - V}{32}$

47a. 113.04 ft³   **49c.**   **55a.**   **55b.**
47b. $V/\pi r^2$
47c. 6 ft
49a. $T/0.15 + 1000$
49b. $2680
49d. $114

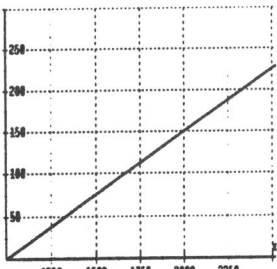

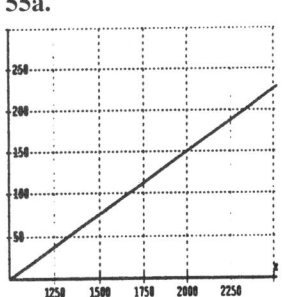

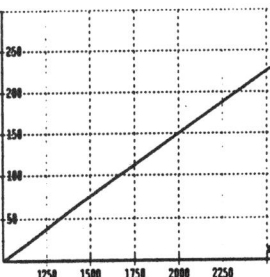

                       53a. $C = 0.33 + 0.22(w - 1)$ **55a.** $y = 3 - 2x, 3/2$ **61a.** $0.5N(N - 3)$
51a. $\dfrac{V}{\pi r^2} - \dfrac{4}{3}r$ **53b.** $w = \dfrac{C - 0.33}{0.22} + 1$ **55b.** $y = 4x, 0$ **61b.** 9 diagonals
                                                                                               57. $7/a$ **63a.** 1.60 in
51b. 15.61m   **59.** $-17/a^2$ **63b.** 3.60 in
                       53c. 11 oz

PROBLEM SET 3.3 PAGE 187–190

1. 18, 20, 22   **3.** 4, 6, 8   **5.** 2, 4, 6   **7.** 32 ft by 15 ft
9. 199m by 399m   **11.** 8 in by 10 in   **13.** 11 ft from one end   **15.** 27 ft, 29 ft, 31 ft
17. 32 cm, 96 cm, 52 cm   **21.** $45,000
19. $20,000 at 9%, $12,000 at 8.5%   **23.** $1067.86 @ 4.25%; $2382.14 @ 7.75%
25. 25 gal   **27.** 300 lbs @ 30%, 100 lbs @ 10%   **29.** 15 tons sand; 30 tons gravel
31. 350 mi @ 35 mph; 850 mi @ 85 mph   **33.** 20 mph
35. bus: 52.5 mph; train: 83.5 mph   **37.** 30 tens, 90 fives   **39.** son 11, father 36
41. $1029   **43.** $50   **45.** $23.54   **47.** 135

49. 9% **51.** 22 weeks **53.** 2500 m **55.** 2.7 minutes
57. $9.00 **59a.** 59, 60, 61 **61.** 14 m by 26 m **63.** 128 in³
59b. 58, 60, 62
59c. not possible

PROBLEM SET 3.4 PAGE 197–198

1a. $(-1, 1]$ **1b.** $[-1, 1)$ **3a.** $[-1, \infty)$ **3b.** $(-\infty, -1)$

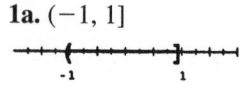

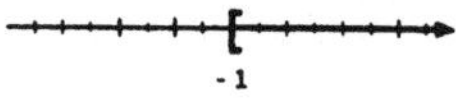

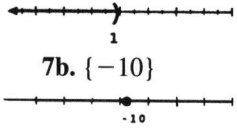

5a. $(-\infty, 2]$ or $[4, \infty)$ **5b.** $(-\infty, \infty)$ **7a.** $(-\infty, \infty)$ **7b.** $\{-10\}$

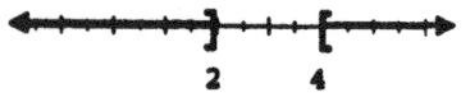

9a. addition **11.** $x \geq 1, [1, \infty)$ **13.** $x > -2, (-2, \infty)$ **15.** $x \geq 1, [1, \infty)$
9b. multiplication
9c. addition
9d. multiplication

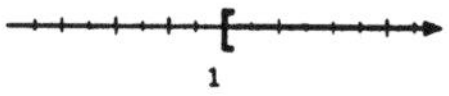

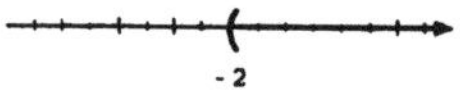

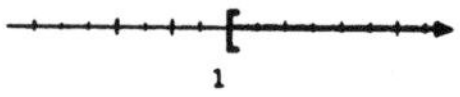

17. $x < 5, (-\infty, 5)$ **19.** $x < 0\ (-\infty, 0)$ **21.** $x \geq 3/4, [3/4, \infty)$ **23.** $x \geq 5, [5, \infty)$

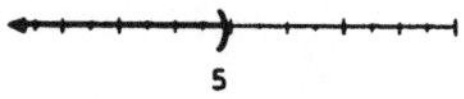

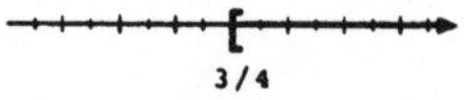

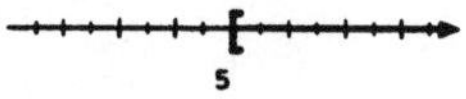

25. $x < 13, (-\infty, 13)$ **27.** $x \geq -3, [-3, \infty)$ **29.** $x < 1/3, (-\infty, 1/3)$ **31.** $x < -24/25, (-\infty, -24/25)$

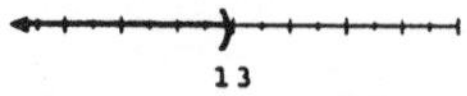

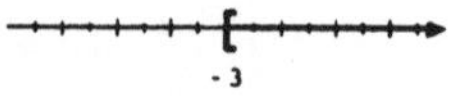

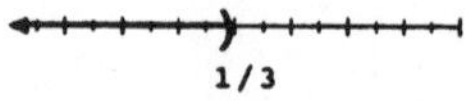

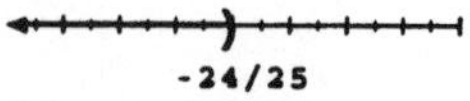

33. $x \leq 24, (-\infty, 24]$  **35.** $x > 3, (3, \infty)$ 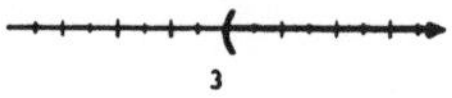 **37.** $x \geq -2, [-2, \infty)$ **39.** $x > 1, (1, \infty)$

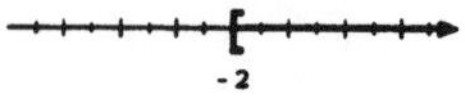

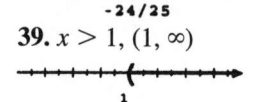

41. $2 \leq x \leq 4, [2, 4]$  **43.** $-3 < x < 2, (-3, 2)$ **45.** $-1/2 \leq x \leq 3/2, [-1/2, 3/2]$ **47.** $x < 1/2$ or $x > 2$, $(-\infty, 1/2)$ or $(2, \infty)$

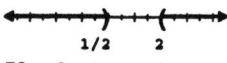

49. $x < 3/4$ or $x > 7/4$ **51.** $x < -2$ or $x > 8/3$ **53.** 15 or more **59a.** $3 \leq x \leq 4$
$(-\infty, 3/4)$ or $(7/4, \infty)$ $(-\infty, -2)$ or $(8/3, \infty)$ **55.** 15 mins or less **59b.** $21 \leq x \leq 28$ per week
  59c. $1095 \leq x \leq 1460$ per yr
57. 82 or more **61.** 33 cm
63. $x \geq 6, (6, \infty)$ **65a.** $(1, \infty)$ **67a.** $(5, \infty)$ **67b.** $[-\infty, 19/4]$
65b. $(-\infty, 1)$

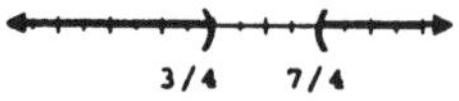

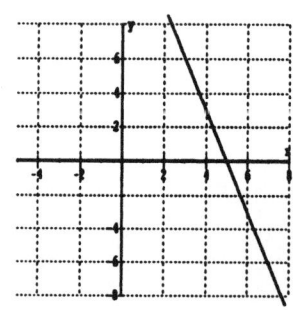

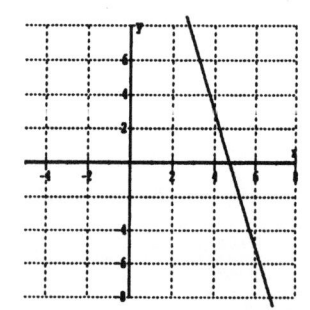

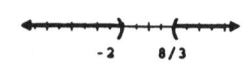

Chapter 4

PROBLEM SET 4.1 PAGE 204

1. $-3, 7$ **3.** $-6, -5$ **5.** $-4, 11$ **7.** $-2, \frac{7}{3}$ **9.** $-\frac{2}{3}, \frac{1}{7}$ **11.** $\frac{3}{5}, \frac{2}{3}$ **13.** $-\frac{7}{2}, \frac{1}{9}$ **15.** $-4, 4$

17. $-9, 9$ **19.** $-\frac{5}{3}, \frac{5}{3}$ **21.** $-\frac{3}{4}, \frac{3}{4}$ **23.** $-\frac{\sqrt{7}}{2}, \frac{\sqrt{7}}{2}$ **25.** $-\frac{5\sqrt{2}}{3}, \frac{5\sqrt{2}}{3}$ **27.** $-1, 1$ **29.** $-\frac{11}{2}, \frac{11}{2}$

31. $-\dfrac{3\sqrt{2}}{5}, \dfrac{3\sqrt{2}}{5}$ **33.** 1, 5   **35.** −9, −1   **37.** $4 - \sqrt{7}, 4 + \sqrt{7}$ **39.** $-6 - \sqrt{3}, -6 + \sqrt{3}$ **41.** 0, 3
43. $\dfrac{-1 - 2\sqrt{2}}{3}, \dfrac{-1 + 2\sqrt{2}}{3}$ **45.** $-\dfrac{5}{2}, -\dfrac{1}{2}$ **47.** 1, 4   **49.** $\dfrac{-5 - \sqrt{37}}{6}, \dfrac{-5 + \sqrt{37}}{6}$
51. $\dfrac{1 - \sqrt{6}}{5}, \dfrac{1 + \sqrt{6}}{5}$ **53.** $\dfrac{-5 - 2\sqrt{2}}{3}, \dfrac{-5 + 2\sqrt{2}}{3}$ **55.** 40 centimeters   **57.** 2   **59.** 1, 3   **61.** 2, 3

◆ PROBLEM SET 4.2 PAGE 207–208

1. $\dfrac{-5 - \sqrt{17}}{4}, \dfrac{5 + \sqrt{17}}{4}$ **3.** $3 - \sqrt{7}, 3 + \sqrt{7}$ **5.** $\dfrac{-3 - \sqrt{21}}{2}, \dfrac{-3 + \sqrt{21}}{2}$ **7.** $-\dfrac{1}{2}, \dfrac{5}{3}$ **9.** $-\dfrac{2}{3}, \dfrac{3}{2}$
11. $-1, -\dfrac{1}{3}$ **13.** $-5, \dfrac{3}{2}$ **15.** $-\dfrac{4}{5}, \dfrac{2}{3}$ **17.** $-2, -\dfrac{2}{3}$ **19.** $-\dfrac{4}{5}, \dfrac{3}{2}$ **21.** $-\dfrac{7}{2}, \dfrac{2}{3}$ **23.** −7, 9   **25.** $-\dfrac{5}{6}, 1$
27. $\dfrac{1}{2}, 1$ **29.** $-\dfrac{3}{4}, \dfrac{7}{8}$ **31.** $\dfrac{6 - \sqrt{57}}{7}, \dfrac{6 + \sqrt{57}}{7}$ **33.** $\dfrac{-5 - \sqrt{229}}{6}, \dfrac{-5 + \sqrt{229}}{6}$
35. $\dfrac{-1 - \sqrt{33}}{4}, \dfrac{-1 + \sqrt{33}}{4}$ **37.** $\dfrac{-7 - \sqrt{29}}{2}, \dfrac{-7 + \sqrt{29}}{2}$ **39.** $\dfrac{6 - \sqrt{26}}{5}, \dfrac{6 + \sqrt{26}}{5}$
41. −0.135, 2.468   **43.** −2.581, 0.581   **45.** $0, -\dfrac{1}{2}, 4$ **47.** 6.93%

◆ PROBLEM SET 4.3 PAGE 211–212

1. 5, 6   **3.** 3, 5   **5.** −1 or 2   **7.** 3 miles by 4 miles   **9.** 130 feet   **11.** 30 feet by 50 feet
13. $\dfrac{-8 + \sqrt{127}}{2} \approx 1.63$ inches   **15.** 6 mph, 8 mph   **17.** 50 feet   **19.** 10 days   **21.** 25 mph   **23.** 5 seconds

Chapter 5

◆ PROBLEM SET 5.1 PAGE 223–224

1. IV   **3.** I   **5.** x axis   **7.** II   **9.** III   **11.** y axis   **13.** I   **15.** x axis   **17.** III   **19.** IV   **21.** $(4, -3)$
23. $(-7, 0)$ **25.** $(8, 10)$ **27.** $(-2, 4)$ **29.** $(-5, -3)$ **31.** $A = (-2, 3); B = (2, -5); C = (3, 2); D = (-4, 1);$
$E = (-5, -4); F = (4, 0); G = (5, -3); H = (0, 5); I = (0, -2); J = (-3, 0)$ **33.** $(2, 1); (4, -3); (0, 5)$

35.

x	2	−1	0	3	−3
y	4	−5	−2	7	−11

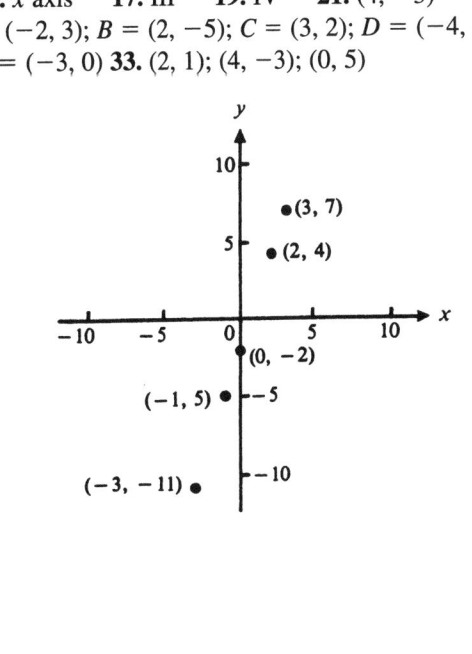

258 SOLUTIONS TO SELECTED PROBLEMS

PROBLEM SET 5.2 PAGE 230

1. $y = x + 1$; $(-1, 0)$, $(0, 1)$

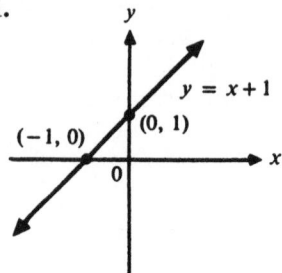

3. $y = -x + 4$; $(0, 4)$, $(4, 0)$

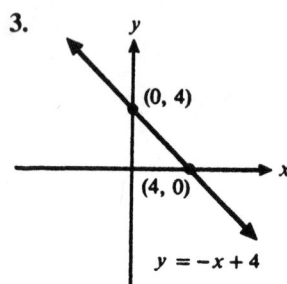

5. $y = 2x$; $(0, 0)$, $(1, 2)$

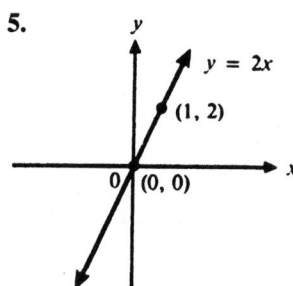

7. $y = -\frac{3}{4}x$; $(0, 0)$, $(1, -\frac{3}{4})$

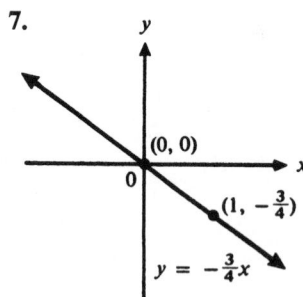

9. $y = 3x + 2$; $(0, 2)$, $(-\frac{2}{3}, 0)$

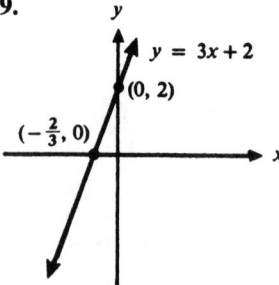

11. $y = 3x - 1$; $(0, -1)$, $(1, 2)$

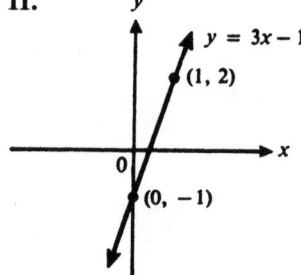

13. $y = 5 - \frac{2}{5}x$; $(0, 5)$, $(\frac{25}{2}, 0)$

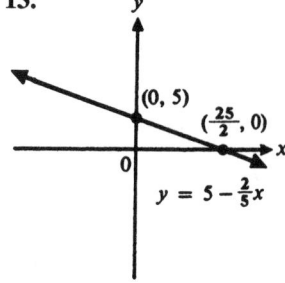

15. $2x + 3y = 6$; $(0, 2)$, $(3, 0)$

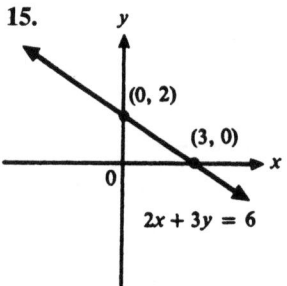

17. $x - 4y = 5$; $(5, 0)$, $(0, -\frac{5}{4})$

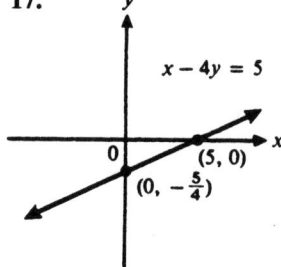

19. Graph of $-6x + 5y - 15 = 0$ with intercepts $(-\frac{5}{2}, 0)$ and $(0, 3)$.

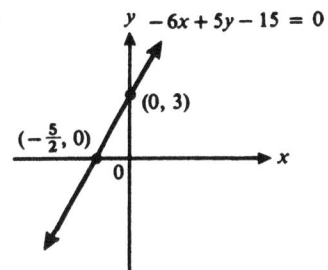

21. Graph of $4x - 16 = 0$ (vertical line) through $(4, 0)$.

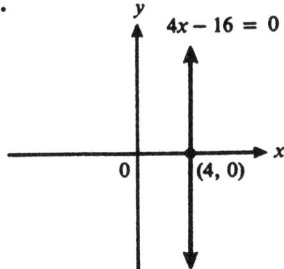

23. Graph of $2y - 3 = 0$ (horizontal line) through $(0, \frac{3}{2})$.

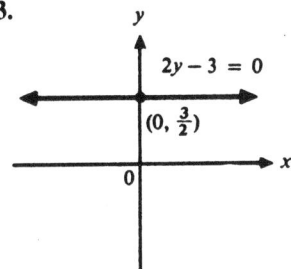

25. Graph of $y = -\frac{2}{3}$ (horizontal line) through $(0, -\frac{2}{3})$.

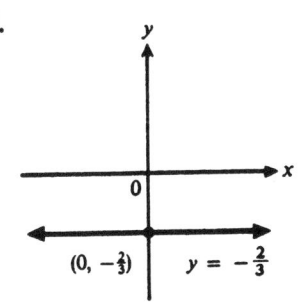

27. Graph of $x = 2$ (vertical line) through $(2, 0)$.

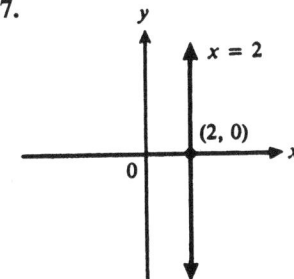

29. Graph of $x = -\frac{3}{4}$ (vertical line) through $(-\frac{3}{4}, 0)$.

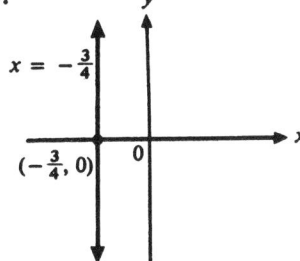

31. Graph of $x = 0$ (the y-axis).

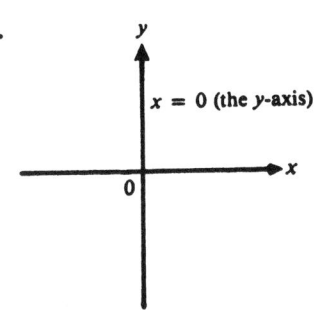

33. 90 beats per minute

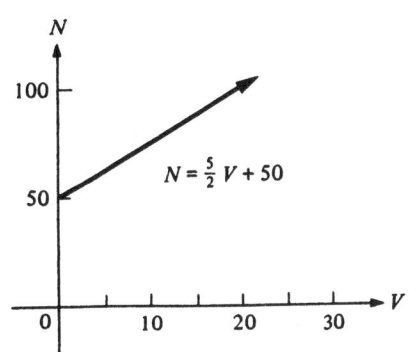

Graph of $N = \frac{5}{2}V + 50$, with N-intercept at 50.

35. 14 volts

Graph of $E = 28I$, passing through origin, reaching about 30 near $I = 1$.

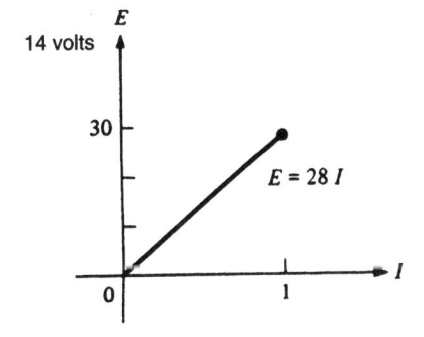

PROBLEM SET 5.3 PAGE 237–238

1. Graph of $2x - y = 8$ with intercepts $(4, 0)$ and $(0, -8)$.

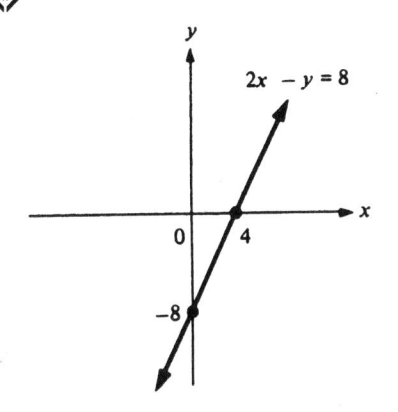

3. Graph of $2x - 3y + 6 = 0$ with intercepts $(-3, 0)$ and $(0, 2)$.

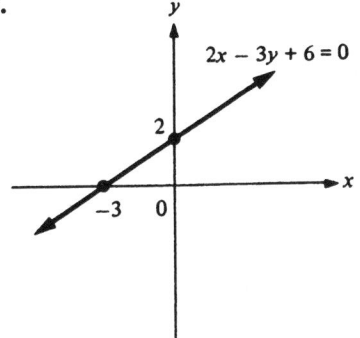

5. Graph of $2x + y - 4 = 0$ with intercepts $(2, 0)$ and $(0, 4)$.

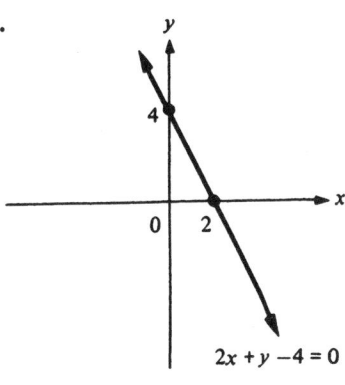

260 SOLUTIONS TO SELECTED PROBLEMS

7.

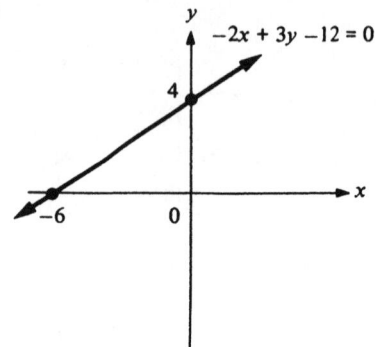

9.

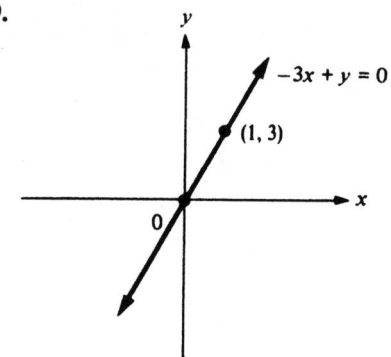

11.

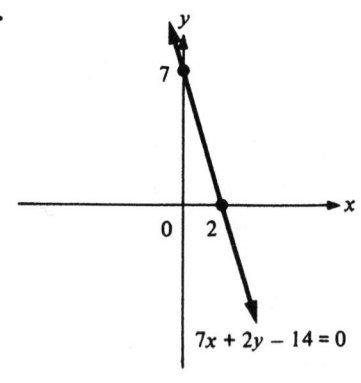

13.

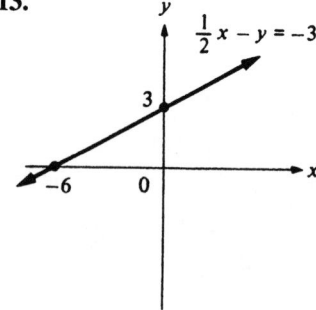

15.

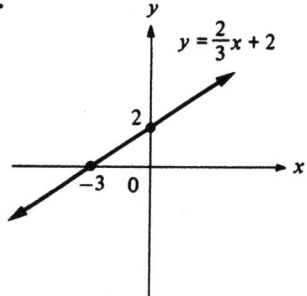

17.

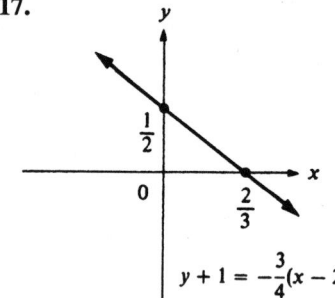

19.

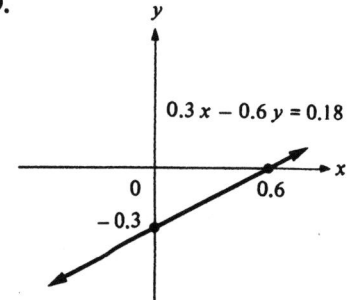

21. 2   **23.** $\frac{-11}{17}$ **25.** -3
27. 0   **29.** Undefined

31. $y = \frac{-5}{2}x - 5$

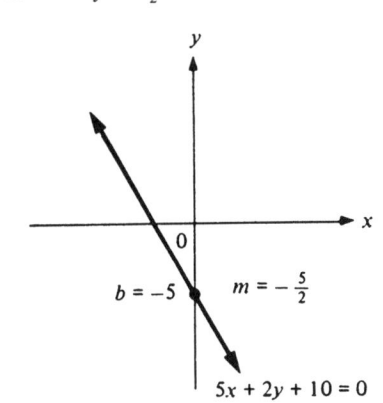

[Graph: line with $b = -5$, $m = -\frac{5}{2}$, equation $5x + 2y + 10 = 0$]

33. $y = \frac{5}{3}x - 5$

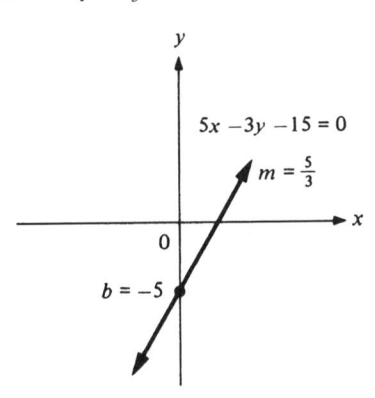

[Graph: line with $5x - 3y - 15 = 0$, $m = \frac{5}{3}$, $b = -5$]

35. $y = \frac{-3}{2}x$

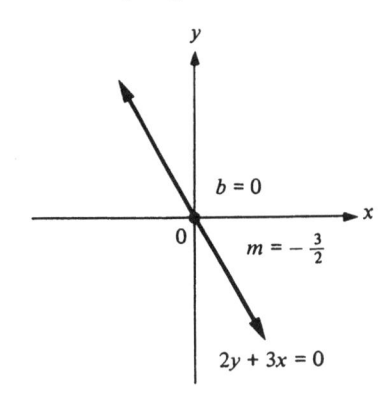

[Graph: line through origin, $b = 0$, $m = -\frac{3}{2}$, equation $2y + 3x = 0$]

37. $y = \frac{-5}{2}x + 5$

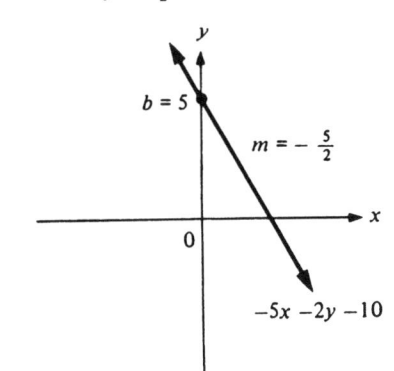

[Graph: $b = 5$, $m = -\frac{5}{2}$, equation $-5x - 2y - 10$]

39. $y = -5$

[Graph: horizontal line, $b = -5$, $m = 0$, equation $y + 5 = 0$]

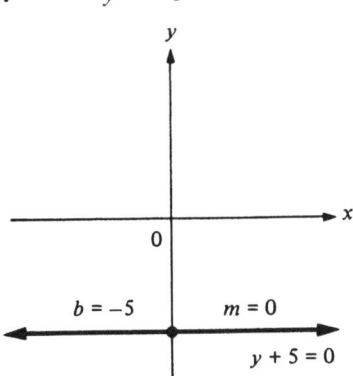

41. $y - 2 = 5(x + 1)$

43. $y - 3 = -3(x - 7)$ **45.** $y + 1 = +\frac{3}{7}(x - 5)$ **47.** $y = -\frac{3}{8}x$ **49.** $y = -5$ **51.** $y = \frac{1}{2}x + \frac{7}{2}$
53. $y = -3x + 5$ **55.** $y = -\frac{3}{7}x$ **57.** $m_1 = m_2 = -\frac{3}{2}$ **59.** $m_1 = 2$; $m_2 = -\frac{1}{2}$; and $2(-\frac{1}{2}) = -1$
61. $F = \frac{9}{5}C + 32$ **63.** (a) $I = 0.07P$; (b) $7,365

PROBLEM SET 5.4 PAGE 243–245

1a. iv   **3a.** $y > \frac{2}{3}x + 2$ **3c.** $y > \frac{1}{2}x + 1$ **5a.** line and below
1b. i   **5b.** line and above
1c. vi   **3b.** $y \geq -x + 3$

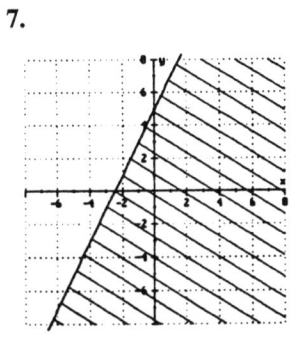

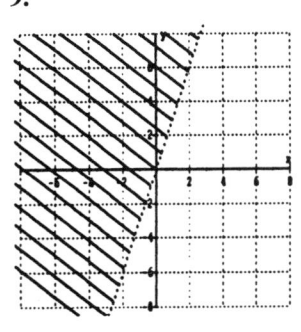

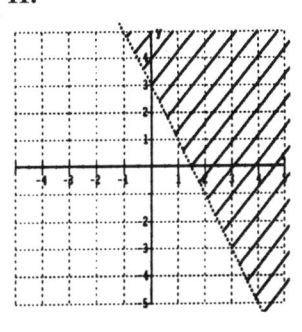

7. [graph] **9.** [graph] **11.** [graph] **13.** [graph]

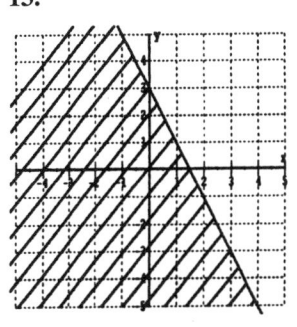

262 SOLUTIONS TO SELECTED PROBLEMS

15.

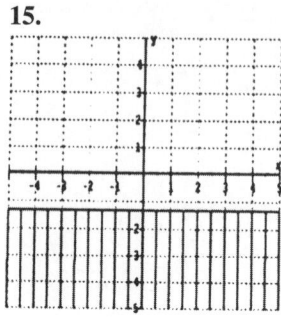

17.

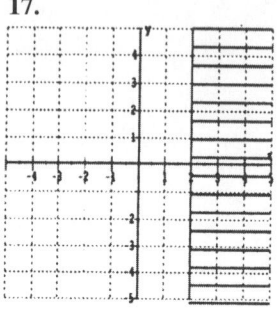

19.

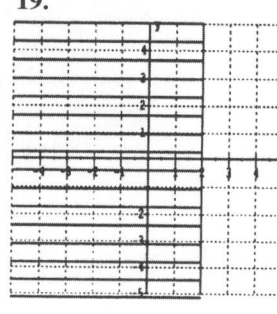

21a. $y \leq 3x + 3$ and $y \geq 3x - 6$
21b. $y \geq 1$ or $y < -2$
21c. $y > -2x - 4$ and $y \leq -2x + 4$

23.

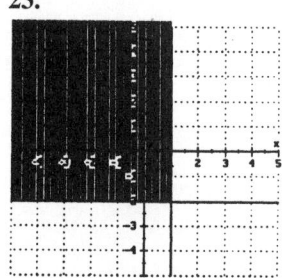

25.

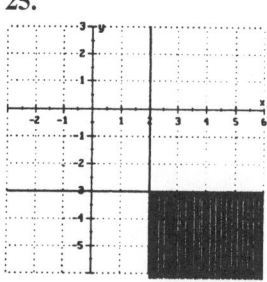

27.

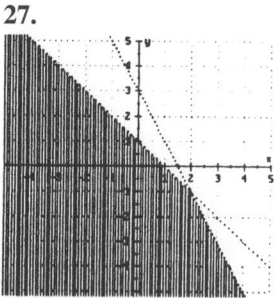

29.

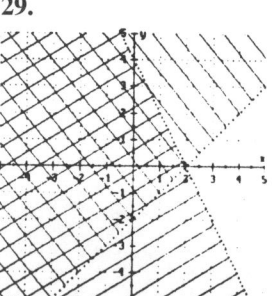

31.

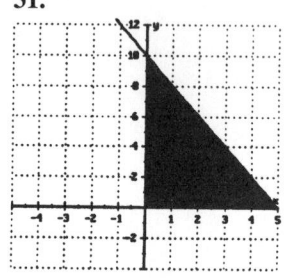

33. $\begin{cases} b + 2A \leq 100 \\ A \geq 0, B \geq 0 \end{cases}$

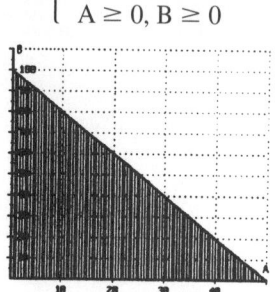

35a. x = # of $8 tickets
y = # of $5 tickets

$\begin{cases} 8x + 5y \geq 3000 \\ x \geq 0, y \geq 0 \end{cases}$

35b.

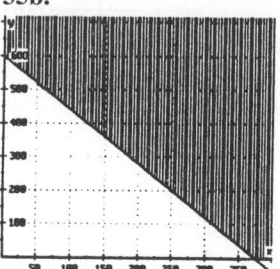

37a. B
37b. A or C
39. Not possible, no graph

41.

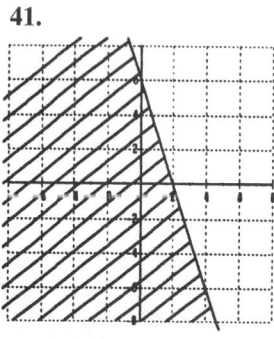

43.

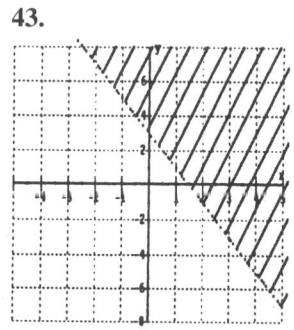

45a. $x < 0$ and $y > 0$
45b. $x > 0$ and $y \neq 0$

47. x = # level I and
y = # level II employees

$\begin{cases} 0 \leq x \leq 20 \\ 0 \leq y \leq 15 \\ 50x + 40y \leq 1400 \end{cases}$

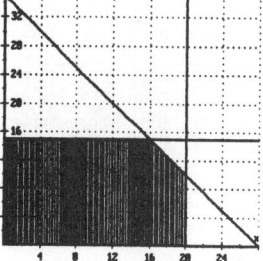

Index

A

Abscissa, 218–219
Absolute value, 18, 19, 20
Addition, 1
    associative property and, 13
    cancellation property and, 15
    commutative property and, 12
    decimals and, 41–43, 45
    fractions, 28–31
    identity property and, 13
    inverse property and, 14–15
    linear equations and, 168
    linear inequalities and, 192
    polynomials and, 124–127
    radicals and, 105–107
    signed numbers and, 18–20
    zero-factor property and, 15
Area
    circles, 68
    parallelograms, 66
    plane figures and, 65–68
    rectangles, 65–66
    squares, 66
    trapezoids, 67
    triangles, 67
Associative property, 13
    multiplication of polynomials, 128–129
    See also Distributive property

B

Binomials, 119
    FOIL method of multiplication and, 132–134
    special products and, 134–135
Braces, 4, 16
Brackets, 4, 16

C

Cancellation property, 15
Cartesian coordinate system, 217–218
    Cartesian coordinate plane and, 218
    coordinates, abscissa/ordinate and, 218–219
    graphing points and, 219–223
    graphing solutions and, 218
    quadrants and, 219–220
    See also Graphs of linear equations
Circles
    area of, 68
    circumference of, 71
    See also Plane figures; Right circular cones; Right circular cylinders
Circumference of a circle, 71
Coefficients of a polynomial, 120
Common factoring procedure, 145–148
Commutative property, 12–13
    multiplication of polynomials, 128–129
    See also Associative property
Complex numbers, 212–213
Composite numbers, 142–143
Compound inequalities, 194–195, 240–241
Conditional equations, 171
Cones. *See* Right circular cones
Conjugates, 212–213
Conversions
    fractions to decimals, 60–64
    fractions to percent, 10–11
Coordinate axis, 7, 217, 218
Coordinates of a point, 7, 217
Counting numbers, 8
Cube root, 96
Cylinders. *See* Right circular cylinders

D

Decimals, 9–11
    addition of, 41–43, 45
    comparison procedure for, 39–40
    conversion to fractions, 60–64
    decimal point and, 35
    division of, 53–58
    fractional place values and, 35–36
    inverse property and, 51, 57–58
    multiplication of, 48–52
    powers of 10 and, 48
    rounding off, 37–38
    subtraction of, 43–44, 45
    zero, insertion of, 38–39
Degree of a term, 120–121
Denominator, 4

Difference of two squares, 150–152
Distance formula, 114–115
Distributive property, 13, 105, 106
    addition/subtraction of polynomials, 124–125
    addition/subtraction of radicals, 105–106
    factoring polynomials and, 146
    multiplication of polynomials, 129–130
    multiplication of radicals, 108–109
Division, 1
    decimals and, 53–58
    exponents and, 88–92
    fractions and, 31–33
    inverse property and, 57–58
    linear equations and, 168
    linear inequalities and, 192
    long division procedure, 138–139
    polynomials and, 136–141
    radicals and, 110–112
    rationalizing denominators and, 110–112
    signed numbers and, 24–25

E

Elements of sets, 6
Equality, 2
    sets of numbers and, 6
    See also Inequalities
Equations. *See* Linear equations in one variable; Quadratic equations
Equivalent equations, 167
Equivalent fractions, 26–27
Equivalent linear inequalities, 192
Expanded forms, 84
Exponents, 83
    applications of, 93, 113–115
    division with, 88–92
    expanded form and, 84
    exponential form and, 83
    like bases property and, 85, 90–91
    multiplication with, 83–86
    negative integer exponents, 89–90
    power of a power property, 85–86
    power of a product property, 86
    power of a quotient property, 91–92
    scientific notation and, 87, 92–93
    zero exponent, 89
    See also Radicals; Roots

F

Factorability test, 156–157
Factoring, 2, 142
    common factoring, polynomials and, 145–148
    composite numbers and, 142–143
    difference of two squares and, 150–152
    greatest common factor and, 143–145
    grouping procedure and, 148
    perfect-square trinomials and, 152–153
    prime numbers and, 142
    quadratic equations and, 199–201
    solving equations, common factoring and, 149
    solving equations, factoring a trinomial and, 157, 162
    test for factorability, 156–157
    trinomial $ax^2 + bx + c$ and, 158–162
    trinomial $x^2 + bx + c$ and, 154–156
Fifth root, 96
Finite sets, 6
First degree equations, 168
FOIL method-multiplying binomials, 132–134
Fourth root, 96
Fractional place values, 35–36
Fraction bar, 4, 16
Fractions, 4, 9
    addition/subtraction of, 28–31
    conversions to decimals, 60–64
    conversion to percent, 10–11
    equivalency of, 26–27
    fractional place values, 35–36
    least common denominator and, 29–31
    like fractions, 28
    multiplication/division of, 31–33
    signed elements of, 27
    simplified fractions, 27–28
    unlike fractions, 28–29
    zero, insertion of, 38–39
    See also Decimals

G

Graph of an inequality, 191
Graphing calculators
    linear equations, 172, 177, 187
    linear inequalities, 196–197, 242
Graph of a number, 7
Graphs of linear equations, 224–230
    applied problems, 242
    calculator use and, 242
    compound inequalities, 240–241
    equation of a line and, 235–237
    intercepts in, 231–233
    linear inequalities in two variables, 238–240
    point-slope form and, 236
    slope-intercept form and, 236–237
    slope of a line and, 233–235
Greatest common factor (GCF), 143–145
Grouping property, 13
Grouping symbols, 4–5, 16

H

Hypotenuse, 113, 114

I

Identity property, 13, 171
Index, 96
Inequalities, 2
    compound inequalities, 194–195, 240–241
    graph of an inequality, 191, 238–242
    interval notation and, 191–192
    linear inequalities, 191–194, 238–240
    properties of inequalities, 192
    satisfying the inequality, 191
    strict/non-strict inequalities, 191
Infinite sets, 6
Integers, 8
    composite number and, 142
    factoring integers, 142–144
    prime number and, 142
Intercepts, 231–233, 236–237
Interval notation, 191–192
Inverse property, 14–15, 51, 57–58
Irrational numbers, 98–99

L

Least common denominator (LCD), 29–31, 170
Like bases property, 85, 90–91
Like fractions, 28
Linear equations in one variable, 167–168
    addition/subtraction properties and, 168, 192
    applied problems, 176–177, 182–187, 195–196
    calculator use and, 172, 177, 187, 196–197
    conditional equations, 171
    equivalent equations, 167
    formulas and, 175–177
    identity classification and, 171
    inconsistent equations and, 171
    inequalities and, 191–195
    interval notation and, 191–192
    literal equations, formulas and, 174–177
    multiplication/division properties, 168, 192
    problem-solving format and, 180–183
    solution/root of, 167
    solving compound inequalities, 194–195
    solving formulas, 175–176
    solving linear equations, 168–170
    solving linear equations/fractions, 170–171
    solving linear inequalities, 192–194
    solving literal equations, 174–175
    symmetric property and, 168
    See also Cartesian coordinate system; Graphs of linear equations; Quadratic equations
Linear inequalities, 191–194, 238–240
Literal equations, 174–177
Long division procedure, 138–139

M

Mathematical modeling, 176–177
Members of sets, 6
Modeling, 176–177
Monomials, 119
Multiplication, 1
    associative property and, 13
    binomials, FOIL method and, 132–134
    cancellation property and, 15
    commutative property and, 12–13
    decimals and, 48–52
    exponents and, 83–86
    fractions and, 31–33
    identity property and, 13
    inverse property and, 14–15, 51
    linear equations and, 168
    linear inequalities and, 192
    polynomials and, 128–131
    radicals and, 108–109
    signed numbers and, 22–24
    zero-factor property, 15
    See also Factoring

N

nth root, 96
Natural numbers, 8
Negative integer exponent, 89–90
Negative numbers, 7, 17–18
    square roots and, 96
    See also Signed numbers
Negative square root, 95
Non-negative integers, 8
Non-strict inequalities, 191
Non-terminating form of rational number, 9–10
Number lines, 7
    integers, 8
    irrational numbers, 9
    natural numbers, 8
    negative side/direction, 7
    origin point, 7
    positive side/direction, 7
    rational numbers, 8, 9–11
    real numbers, 8
    unit length, 7
    whole numbers, 8
Numerals, 2
Numerator, 4
Numerical coefficients of polynomials, 120
Numerical expressions, 3

O

Order of operations, 3, 16
    binomial multiplication, FOIL method and, 132–134

Order of operations *(continued)*
    See also Symbols
Ordinate, 218-219

P

Parallelograms
    area of, 66
    perimeter of, 69–70
    See also Plane figures
Percent, 10–11
Perfect-square trinomials, 152–153
Perimeter
    circumference of circles, 71
    parallelograms, 69–70
    rectangles, 69
    squares, 69
    trapezoids, 70–71
    triangles, 70
Place values, 35–36
Plane figures
    areas of, 65–68
    circles, 68, 71
    parallelograms, 66, 69–70
    perimeters of, 69–71
    pyramids, 73–74
    rectangles, 65–66, 69
    rectangular solids, 72
    right circular cones, 75
    right circular cylinders, 73
    spheres, 75–76
    squares, 66, 69
    trapezoids, 67, 70–71
    triangles, 67, 70
    volumes/surface areas of, 72–76
Point-slope form, 236
Polynomials, 119–120
    addition/subtraction of, 124–127
    associative property and, 128–129
    business applications, 123–124
    commutative property and, 128–129
    degree of, 120–121
    distributive property and, 129–130
    division of, 136–141
    evaluation of, 121, 123
    factoring of, 145–148
    like/similar terms in, 125, 130–131
    long division procedure and, 138–139
    multiple variables in, 131
    multiplication of, 128–131
    multiplying binomials, FOIL method and, 132–134
    notation for, 122–123
    numerical coefficients of, 120
    types of, 119
Positive integers, 8

Positive numbers, 7, 8, 17–18
    See also Signed numbers
Positive square root, 95
Power of the exponent, 83, 85–86, 91–92
Power of a product property, 86
Power of a quotient property, 91–92
Powers of 10, 48
Power to a power property, 85–86
Prime numbers, 142
Principal nth root, 96
Principal square root, 95
Problem-solving format, 180–183
Products, 1, 2, 83
    roots of products, 100–102
    special products, multiplication of binomials, 134–135
    See also Multiplication
Pyramids, 73–74
Pythagorean theorem, 113–114

Q

Quadratic equations
    applied problems, 208–211
    complex numbers/conjugates and, 212–213
    complex solutions to, 212–215
    factoring and, 199–201
    quadratic formula and, 204–207
    roots extraction and, 201–203
    zero-product property and, 199
    See also Cartesian coordinate system; Linear equations with one variable
Quotient, 1, 2

R

Radicals
    addition/subtraction of, 105–107
    applications of, 113–115
    division of, 110–112
    multiplication of, 108–109
    radical symbol, 96
    rationalizing denominators and, 110–112
    roots of products and, 100–102
    roots of quotients, 102–104
    simplified expression of, 101–102
    See also Exponents; Roots
Radical symbol, 96
Radicand, 96–97
Rationalizing denominators, 110–112
Rational numbers, 8
    fractions/decimals, 9
    percent, 10–11
    repeating decimals, 9
    terminating/non-terminating form of, 9–10
    See also Fractions

Real numbers, 8, 12
    associative property and, 13
    cancellation property and, 15
    commutative property and, 12–13
    distributive property and, 13
    identity property and, 13
    inverse property and, 14–15
    number line and, 7–9
    zero-factor property and, 15
Rectangles
    area of, 65–66
    perimeter of, 69
    See also Plane figures
Rectangular solids
    volume/surface area of, 72
    See also Plane figures; Rectangles; Volume
Reduced fractions, 27–28
Right circular cones
    volume/surface area of, 75
    See also Circles; Plane figures; Volume
Right circular cylinders
    volume/surface area of, 73
    See also Circles; Plane figures; Volume
Roots, 95
    cube root, 96
    extraction method, 201-203
    fourth/fifth/sixth roots, 96
    index and, 96, 97
    irrational numbers and, 98–99
    *n*th root, 96
    negative numbers and, 96
    positive/negative roots and, 96
    principal roots, 95, 96, 97
    quadratic equations, root extraction and, 201–203
    radical symbol, 96
    radicand, 96–97
    square root, 95–96
    See also Exponents; Radicals
Roots extraction method, 201-203

S

Satisfying the inequality, 191
Scientific notation, 87, 92–93
Sets of numbers, 6
    elements/members of sets, 6
    equality and, 6
    finite/infinite sets, 6
    number line, real numbers and, 7–9
    rational numbers, 9–11
Signed numbers, 17–18
    absolute value and, 18
    addition of, 18–20
    division of, 24–25

    fractions and, 27
    multiplication of, 22–24
    square roots and, 96
    subtraction of, 20–21
Simplified fractions, 27, 37–38
Simplified radicals, 101–102
Simplifying expressions, 16
Sixth root, 96
Slope-intercept form, 236–237
Slope of a line, 233–235
Special products, 134–135
Spheres
    volume/surface area of, 75–76
    See also Plane figures; Volume
Square root, 95–96
Squares
    area of, 66
    perimeter of, 69
    See also Plane figures
Strict inequalities, 191
Subtraction, 1
    decimals and, 43–44, 45
    fractions and, 28–31
    linear equations and, 168
    linear inequalities and, 192
    polynomials and, 124–127
    radicals and, 105–107
    signed numbers and, 20–21
Surface area
    pyramids, 73–74
    rectangular solids, 72
    right circular cones, 75
    right circular cylinders, 73
    spheres, 75–76
    See also Plane figures; Volume
Symbols, 1
    absolute value, 18
    arithmetic operations, 1
    equality/inequality, 2
    grouping symbols, 4–5, 16
    number lines and, 7–9
    numerals, 2
    numerical expressions, 3
    order of operations, 3, 16
    radical symbol, 96
    simplifying expressions, 16
Symmetric property, 168

T

Terminating form of rational number, 9–10
Trapezoids
    area of, 67
    perimeter of, 70–71

See also Plane figures
Triangles
    area of, 67
    perimeter of, 70
    Pythagorean theorem, 113–114
    See also Plane figures
Trinomials, 119
    $ax^2 + bx + c$ form of, 158–162
    factoring of, 152–162
    perfect-square trinomials, 152–153
    solving equations and, 157, 162
    test for factorability and, 156–157
    $x^2 + bx + c$ form of, 154–156

U

Unit length of a number line, 7
Unlike fractions, 28–29

V

Value of the expression, 3
Volume
    pyramids, 73–74
    rectangular solids, 72
    right circular cones, 75
    right circular cylinders, 73
    spheres, 75–76
    See also Plane figures; Surface area

W

Whole numbers, 8

Z

Zero exponent, 89
Zero-factor property, 15, 22, 25